M & E HANDBOOKS

M & E Handbooks are recommended reading for examination syllabuses all over the world. Because each Handbook covers its subject clearly and concisely books in the series form a vital part of many college, university, school and home study courses.

Handbooks contain detailed information stripped of unnecessary padding, making each title a comprehensive self-tuition course. These are amplified with numerous self-testing questions in the form of Progress Tests at the end of each chapter, each text-referenced for easy checking. Every Handbook closes with an appendix which advises on examination technique. For all these reasons, Handbooks are ideal for pre-examination revision.

The handy pocket-book size and competitive price make Handbooks the perfect choice for anyone who wants to grasp the essentials of a subject quickly and easily.

THE M. & E. HANDBOOK SERIES

PRINCIPLES OF PHYSICAL CHEMISTRY

J. M. GROSS, B.Sc., Ph.D. (Leicester)

*Formerly Senior Lecturer in Chemistry
at the City of London Polytechnic*

and

B. WISEALL, B.Sc., Ph.D. (Leeds), A.R.I.C.

*Principal Lecturer in Chemistry
at the City of London Polytechnic*

MACDONALD AND EVANS

MACDONALD AND EVANS LTD.
Estover, Plymouth PL6 7PZ

First published 1972
Reprinted 1977
Reprinted in this format 1979

©

MACDONALD & EVANS LIMITED
1972

ISBN: 7121 1647 8

Printed in Great Britain by Richard Clay (The Chaucer Press) Ltd.
Bungay, Suffolk

AUTHORS' PREFACE

THIS HANDBOOK presents topics which are basic to a first course in physical chemistry. Although primarily intended for first-year undergraduates, the book should also be of value to students taking National Certificate courses and Part I Grad.R.I.C. It will be useful to sixth-form pupils studying the "A" and "S" Level syllabuses of the various G.C.E. examining boards.

The book tries to emphasise fundamental principles and to discuss applications of these principles. A sound grounding in physical chemistry can be achieved only through the use of mathematics, especially calculus. Many students find difficulty in applying simple mathematics to physical situations and an early chapter on mathematics attempts to remedy this. The elements of thermodynamics are also introduced at an early stage and are extensively used in subsequent chapters.

The book has adopted the use of SI units and the terminology conforms to the recommendations of the Royal Society given in their recent publication *Symbols, Signs and Abbreviations for British Scientific Publications*.

The scheme of the book. Part One deals with the SI system of units and also introduces the mathematics to be used in the remainder of the book. The concept that matter is composed of atoms and molecules is developed in Part Two. Part Three considers the three laws of thermodynamics and applies them primarily to the behaviour of gases. Physical and chemical equilibria, including the properties of ions in solution are dealt with in Parts Four and Five. An account of the structures of atoms and molecules in Part Six precedes an account of the structure of solids and liquids in Part Seven. Chemical kinetics are covered in Part Eight and the final Part is concerned with the chemical properties of surfaces.

The authors are convinced of the value of problems as a test of understanding and as an incentive to apply principles to new situations. Numerous worked examples have been included in the text and numerical and essay-type Progress

viii AUTHORS' PREFACE

Tests appear at the end of each chapter. Fully worked answers
to the numerical Progress Test questions are given in Appendix
IV. Examination questions, taken from Part I B.Sc. (London
University) and Part I Grad.R.I.C., are reproduced in Appendix
III.

Acknowledgments. We are grateful to our colleagues at the
City of London Polytechnic who suggested improvements,
and in particular to Dr C. W. P. C. Crowne, Dr J. D. Barnes,
Dr B. E. Stacy, Dr R. F. W. White and Mr J. V. Westwood.
Our thanks are also due to Mrs Ann Wiseall for typing a
difficult manuscript.

June 1972   J. M. G.
                                                   B.W.

CONTENTS

LIST OF ILLUSTRATIONS

LIST OF TABLES

PART ONE

UNITS AND MATHEMATICS

The first chapter introduces the system of units to be used throughout this book, namely the SI system. The information is presented largely in tabular form, with relevant comments accompanying each table. Following a list of values of physical constants which will be used in the book, the chapter closes with a brief account of dimensional analysis.

Chapter II is an attempt to bring together all the essential mathematical material which will be used in the remainder of the book, and is meant to act as a guide rather than a comprehensive treatment. Mathematical knowledge is important in physical chemistry, and many of the examples mentioned in this chapter are met with in subsequent chapters.

CHAPTER I

THE SI SYSTEM OF UNITS

SI is an abbreviation for *Système International d'Unités*, which is an extension and refinement of the metric system.

NOTE: Part of this chapter has been reproduced from The Royal Society Conference of Editors pamphlet *Metrication in Scientific Journals*.

THE BASIC UNITS

1. Main features of the SI system.

(*a*) There are six basic units, with the probability of a seventh being added in the near future (*see* 2, 3 below). The metre and the kilogramme take the place of the centimetre and the gramme in the metric system.

(*b*) *The unit of force*, the newton (kilogramme × metre ÷ second²) is independent of the Earth's gravitation and the often confusing introduction of the gravitational constant G into equations is no longer necessary.

(*c*) *The unit of energy* in all forms is the joule (newton × metre) and the unit of power the joule per second (watt); thus the erg, calorie, kilowatt hour and horse power are all superseded.

(*d*) *Electrostatic and electromagnetic units* are replaced by SI electrical units.

(*e*) *Multiples of units* are normally restricted to steps of a thousand, and, similarly, fractions to steps of a thousandth.

2. Basic physical quantities and units.

A physical quantity is defined by the product of a numerical value and a unit, *i.e.* physical quantity = numerical value × unit. The six basic physical quantities together with their units are listed in Table 1. A seventh basic physical quantity, *amount of substance*, together with its unit the *mole* is also included in the table. This physical quantity has not yet been officially adopted in SI but it is virtually certain that it will be in the future.

TABLE 1   BASIC SI UNITS

Physical quantity	Symbol for quantity	Name of SI unit	Symbol for SI unit
length	l	metre	m
mass	m	kilogramme	kg
time	t	second	s
electric current	I	ampere	A
thermodynamic temperature	T	kelvin	K
luminous intensity	I_v	candela	cd
amount of substance	n	mole	mol

3. Definitions of the basic SI units.

(a) *Metre*. The metre is the length equal to 1 650 763·73 wavelengths in vacuum of the radiation corresponding to the transitions between the levels $2p_{10}$ and $5d_5$ of the krypton-86 atom.

NOTE: To facilitate the reading of long numbers, the digits are grouped in threes around the decimal point.

(b) *Kilogramme*. The kilogramme is the unit of mass; it is equal to the mass of the international prototype of the kilogramme.

(c) *Second*. The second is the duration of 9 192 631 770 periods of the radiation corresponding to the transition between the two hyperfine levels of the ground state of the caesium-133 atom.

(d) *Ampere*. The ampere is that constant current which, if maintained in two straight parallel conductors of infinite length, of negligible circular cross-section, and placed one metre apart in a vacuum, would produce between these conductors a force equal to 2×10^{-7} newtons per metre of length.

(e) *Kelvin*. The kelvin, the unit of thermodynamic temperature, is the fraction 1/273·16 of the thermodynamic temperature of the triple point of water.

NOTE: The kelvin is a unit both of thermodynamic temperature and of thermodynamic temperature interval. The degree Celsius, symbol °C, is a unit of thermodynamic temperature interval identical with the kelvin. Also

$$\text{Temperature (K)} = 273 \cdot 15 + \text{Temperature (°C)}$$

(f) *Candela*. The candela is the luminous intensity in the perpendicular direction of a surface of 1/600 000 square metre of a black body at the temperature of freezing platinum under a pressure of 101 325 newtons per square metre.

(g) *Mole*. The mole is the amount of substance which contains as many elementary units as there are carbon atoms in 0·012 kilogrammes of carbon-12. *The elementary unit must be specified* and may be an atom, a molecule, an ion, an electron, etc., or a specified group of such entities. For example:

1 mol of NO_2 has a mass equal to 0·046 005 kg
1 mol of N_2O_4 has a mass equal to 0·092 010 kg
1 mol of Na^+ has a mass equal to 0·022 990 kg
1 mol of electrons have a mass equal to $5·486 \times 10^{-7}$ kg

NOTE: It is incorrect to say "1 mol of oxygen" without indicating whether the oxygen is present as atoms (mass 0·015 999 kg) or as molecules (mass 0·031 998 kg).

It is important to realise that amount of substance is not the same as mass. Although the amount of a *particular* substance is proportional to its mass, the amounts of *several* substances are not proportional to their masses. From the definition of the mole, the amount of substance is proportional to the number of specified particles. The proportionality constant is the same for all substances and is called the *Avogadro constant*, L. This term is preferred to *Avogadro's number*, since it is a number divided by an amount of substance and hence has the units of mol^{-1}.

DERIVED UNITS

4. Derived units. All other physical quantities are derived from the seven independent basic physical quantities. The units of these derived quantities follow directly from the units of the basic quantities, and this is one of the major advantages of the SI system. Some of the derived units have special names; for example, the unit of energy is the joule, which is simply (newton × metre). Others, such as the units of density and velocity, do not have special names. Examples of the derived units are shown in Table 2 overleaf..

5. Comments on Table 2.

(a) *Concentration*. A solution with a concentration of 1 mol dm^{-3} is often called a one-molar solution (abbreviated to 1M). However, this notation should not be confused with the use of the adjective "molar," meaning divided by amount of substance (*see* (c) below).

(b) *Specific quantities*. The adjective "specific" before any extensive physical property means divided by mass. If the

TABLE 2 EXAMPLES OF DERIVED SI UNITS

Physical quantity	Name of unit	Symbol for unit	Definition of unit
energy	joule	J	kg m^2 s^{-2}
force	newton	N	kg m s^{-2} = J m^{-1}
power	watt	W	kg m^2 s^{-3} = J s^{-1}
electric charge	coulomb	C	A s
electric potential difference	volt	V	kg m^2 s^{-3} A^{-1} = J A^{-1} s^{-1}
electric resistance	ohm	Ω	kg m^2 s^{-3} A^{-2} = V A^{-1}
electric capacitance	farad	F	A^2 s^4 kg^{-1} m^{-2} = A s V^{-1}
frequency	hertz	Hz	s^{-1}
customary temperature, t	degree Celsius	°C	t °C = T K − 273·15
area	square metre	m^2	
volume	cubic metre	m^3	
density	kilogramme per cubic metre	kg m^{-3}	
velocity	metre per second	m s^{-1}	
acceleration	metre per second squared	m s^{-2}	
pressure	newton per square metre	N m^{-2}	
kinematic viscosity, diffusion coefficient	square metre per second	m^2 s^{-1}	
dipole moment	coulomb metre	C m	
surface tension	newton per metre	N m^{-1}	
heat capacity entropy	} joule per kelvin	J K^{-1}	
specific heat capacity specific entropy	} joule per kilogramme kelvin	J K^{-1} kg^{-1}	
molar heat capacity molar entropy	} joule per mole kelvin	J K^{-1} mol^{-1}	
concentration	mole per cubic metre	mol m^{-3}	
molality	mole per kilogramme	mol kg^{-1}	

extensive physical property is denoted by a capital letter, then the specific quantity is represented by the lower-case letter, e.g.:

volume: V
specific volume: $v = V/m$
heat capacity at constant pressure: C_p
specific heat capacity at constant pressure: $c_p = C_p/m$

NOTE: An extensive physical property is one whose magnitude depends on the size of the system (see VI, 2).

(c) *Molar quantities*. The adjective "molar" before any exten-sive physical property means divided by amount of substance, not by number of moles. However, the unit of amount of substance is *always* the mole. The subscript m is attached to the symbol for the extensive quantity to denote the molar quantity, *e.g.*:

molar volume: $V_m = V/n$
molar heat capacity at constant pressure: $C_{p,m} = C_p/n$

NOTE: In the case of molar heat capacities, the subscript m is often omitted (*see* IV, **16**).

6. Prefixes for SI units. The recommended prefixes which are used to indicate fractions or multiples of the basic or derived units are in general limited to $10^{\pm 3n}$, although prefixes such as 10^{-1} and 10^{-2} are also included (Table 3).

TABLE 3   FRACTIONS AND MULTIPLES

Fraction	Prefix	Symbol	Multiple	Prefix	Symbol
10^{-1}	deci-	d	10	deka-	da
10^{-2}	centi-	c	10^2	hecto-	h
10^{-3}	milli-	m	10^3	kilo-	k
10^{-6}	micro-	μ	10^6	mega-	M
10^{-9}	nano-	n	10^9	giga-	G
10^{-12}	pico-	p	10^{12}	tera-	T

Compound prefixes should not be used, *e.g.* 10^{-9} metre is represented by 1 nm, *not* 1 mμm. The attachment of a prefix to a unit in effect constitutes a new unit. For example:

$$1 \text{ km}^2 = 1 \text{ (km)}^2 = 10^6 \text{m}^2 \; not \; 1 \text{ km}^2 = 1\text{k(m}^2) = 10^3\text{m}^2$$

Although the kilogramme is the basic unit of mass, the gramme will often be used both as an elementary unit (to avoid the absurdity of mkg) and in association with numerical prefixes.

7. Units which are contrary to SI units. Several very familiar units are contrary to the SI system, the most notable for the chemist being the litre, the ångström, the erg, the calorie, the dyne, the atmosphere and the conventional millimetre of mercury. The last unit is the practical unit of pressure and must be converted to N m^{-2} before any calculations can be made. Examples of units which are contrary to SI units are shown in Table 4 together with their SI equivalents.

TABLE 4   EXAMPLES OF UNITS CONTRARY TO SI UNITS

Physical quantity	Name of unit	Symbol for unit	Definition of unit
length	ångström	Å	10^{-10} m $= 10^{-1}$ nm
length	micron	μm	10^{-6} m
volume	litre	l	10^{-3} m^3 $=$ dm^3
force	dyne	dyn	10^{-5} N
pressure	atmosphere	atm	101 325 N m^{-2}
pressure	torr	Torr	(101 325/760) N m^{-2}
pressure	conventional millimetre of mercury	mmHg	$13 \cdot 5951 \times 980 \cdot 665 \times 10^{-2}$ N m^{-2}
pressure	pascal	Pa	N m^{-2}
energy	thermochemical calorie	cal (thermo-chem.)	$4 \cdot 184$ J
energy	erg	erg	10^{-7} J
kinematic viscosity, diffusion coefficient	stokes	St	10^{-4} m^2 s^{-1}
conductance	siemens	S	Ω^{-1}
mass	unified atomic mass unit	u	$1 \cdot 6604 \times 10^{-27}$ kg (approx.)
energy	electron volt	eV	$1 \cdot 6021 \times 10^{-19}$ J (approx.)

8. Comments on Table 4.

(a) *Litre.* The word litre may be used as a special name for the cubic decimetre.

(b) *Pressure.* For most calculations 1 mm Hg can be put equal to (101 325/760) N m^{-2} with very little error.

(c) *Units which cannot be exactly defined in terms of SI units.* The last two units in Table 4 cannot be exactly defined in terms of SI units, since the factors for conversion are experimentally determined quantities and hence are subject to change. The *unified atomic mass unit* is defined as one-twelfth of the mass of an atom of carbon-12. The *electron volt* is defined as the energy acquired by an electron when accelerated through a potential difference of one volt. These are useful units for problems involving electrons and the atomic nucleus.

PHYSICAL CONSTANTS

The recommended values of some physical constants are shown in Table 5.

TABLE 5   RECOMMENDED VALUES OF SOME PHYSICAL CONSTANTS

Quantity	Symbol	Magnitude	Unit
speed of light in a vacuum	c_0	$2 \cdot 997\,925 \times 10^8$	m s^{-1}
permeability of a vacuum	μ_0	$4\pi \times 10^{-7}$	J s^2 C^{-2} m^{-1}
permittivity of a vacuum	$\epsilon_0 = \mu_0^{-1} c_0^{-2}$	$8 \cdot 854\,185 \times 10^{-12}$	J^{-1} C^2 m^{-1}
mass of hydrogen atom	m_H	$1 \cdot 673\,43 \times 10^{-27}$	kg
mass of proton	m_p	$1 \cdot 672\,52 \times 10^{-27}$	kg
mass of neutron	m_n	$1 \cdot 674\,82 \times 10^{-27}$	kg
mass of electron	m_e	$9 \cdot 109\,1 \times 10^{-31}$	kg
charge of proton	e	$1 \cdot 602\,10 \times 10^{-19}$	C
electron radius	$r_e = \mu_0 e^2/4\pi m_e$	$2 \cdot 817\,77 \times 10^{-15}$	m
Boltzmann constant	k	$1 \cdot 380\,54 \times 10^{-23}$	J K^{-1}
Planck constant	h	$6 \cdot 625\,6 \times 10^{-34}$	J s
Bohr radius	$a_0 = h^2/\pi \mu_0 c_0^2 m_e e^2$	$5 \cdot 291\,67 \times 10^{-11}$	m
Rydberg constant	$R_\infty = \mu_0^2 m_e e^4 c_0^3/8h^3$	$1 \cdot 097\,373\,1 \times 10^7$	m^{-1}
	$R_H = R_\infty/(1 + m_e/m_p)$	$1 \cdot 096\,775\,8 \times 10^7$	m^{-1}
Avogadro constant	L	$6 \cdot 022\,52 \times 10^{23}$	mol^{-1}
gas constant	$R = Lk$	$8 \cdot 314\,3$	J K^{-1} mol^{-1}
"ice point" temperature	T_{ice}	$273 \cdot 15$	K
Faraday constant	$F = Le$	$9 \cdot 648\,70 \times 10^4$	C mol^{-1}
gravitational constant	G	$6 \cdot 670 \times 10^{-11}$	N m^2 kg^{-2}

DIMENSIONAL ANALYSIS

9. The fundamental dimensions. The dimensions of all physical quantities may be expressed in terms of the *fundamental* dimensions of mass $[m]$, length $[l]$, time $[t]$, temperature $[T]$, current $[I]$, amount of substance $[n]$ and luminous intensity $[I_v]$. Thus volume has the dimension $[l^3]$, velocity $[lt^{-1}]$, acceleration $[lt^{-2}]$, force $[mlt^{-2}]$, etc.

For an equation containing physical quantities to be valid, the dimensions on the two sides of the equation must be consistent. The examination of an equation in terms of its dimensions is known as dimensional analysis.

10. Rules of dimensional analysis.

(*a*) When terms in an equation are added or subtracted, their dimensions must be consistent so that they also can be added or subtracted.

(*b*) When terms in an equation are multiplied, divided or raised to a power (*see* II, 4), then their dimensions must be treated in an identical manner.

(*c*) Logarithms and exponentials are dimensionless.

(*d*) The differential dx has the same dimensions as x.

11. Examples of dimensional analysis.

EXAMPLE 1: Find the dimensions of R in the expression $R = pV/nT$.

Pressure, p, is force per unit area; dimensions $[mlt^{-2}] \times [l^2]^{-1}$.

Volume, V, is (length)^3; dimensions $[l^3]$.

Amount of substance, n; dimensions $[n]$.

Temperature, T; dimensions $[T]$.

Thus the dimensions of R are $[mlt^{-2}][l^2]^{-1}[l^3]/[n][T]$, *i.e.* $[ml^2t^{-2}n^{-1}T^{-1}]$.

Now ml^2t^{-2} are the dimensions of energy.

Hence R has the dimensions $[(\text{energy})\,n^{-1}\,T^{-1}]$.

In the SI system the units are

$$[\text{N m}^{-2}][\text{m}^3][\text{mol}^{-1}][\text{K}^{-1}] = [\text{N m mol}^{-1}\text{K}^{-1}] = [\text{J mol}^{-1}\text{K}^{-1}]$$

EXAMPLE 2: If R is expressed in J mol^{-1} K^{-1}, and T in K, what are the units of $\Delta H^{\ominus}$ in the following expression?

$$\ln(K_2/K_1) = \frac{\Delta H^{\ominus}}{R}\left(\frac{1}{T_1} - \frac{1}{T_2}\right)$$

$$\Delta H^{\ominus} = \frac{R\ln(K_2/K_1)}{\left(\dfrac{1}{T_1} - \dfrac{1}{T_2}\right)}$$

Units of $\Delta H^{\ominus}$; $[\text{J mol}^{-1}\text{ K}^{-1}]/[\text{K}^{-1}] = [\text{J mol}^{-1}]$

THE HEADING OF TABLES AND THE LABELLING OF GRAPHS

12. Table headings. Since the figures arranged in the columns of a table are pure numbers (*i.e.* they are dimensionless), the headings of the columns must also be pure numbers. This means that the symbol for any physical quantity used as a table heading must be divided by the units used. For example, if temperatures are tabulated in K, the table heading should be T/K; if reciprocal temperatures are tabulated in K^{-1}, the table heading should be K/T; if pressures are tabulated in kN m^{-2}, the table heading should be $p/\text{kN m}^{-2}$.

13. Labelling of graph axes. The same principles are used in the labelling of graph axes, since the points plotted on graphs represent pure numbers.

PROGRESS TEST 1

1. Find the dimensions of each of the following quantities:

(a) a and b in the van der Waals equation

$$\left(p + \frac{n^2a}{V^2}\right)(V - nb) = nRT$$

where the symbols have their usual significance.

(b) A in the Clausius–Clapeyron equation

$$\ln p = -\frac{\Delta H_t}{RT} + A$$

where ΔH_t is the latent heat of transition and has the dimensions of energy $\times$ (amount of substance)$^{-1}$.

2. Find the units of each of the following quantities:
(a) E_p in the equation

$$E_p = e^2/4\pi\epsilon_0 d$$

where e is the charge on the electron, ϵ_0 is the permittivity of a vacuum, and d is the distance.
(b) E in the equation

$$E = mv^2/2$$

where m is the mass and v is the velocity.

3. Is the following equation for the root mean square speed of gas molecules dimensionally correct:

$$(\overline{c^2})^{\frac{1}{2}} = (3RT/M)^{\frac{1}{2}}$$

where M is the molar mass, dimensions $[m\,n^{-1}]$?

CHAPTER II

ESSENTIAL MATHEMATICS

SUMMATIONS AND AVERAGES

1. Summations—the $\sum$ notation. The instruction "find the sum of the first ten values of a quantity" (x, say) can be written algebraically as

$$x_1 + x_2 + x_3 + \ldots + x_9 + x_{10}$$

It is obvious that this notation will be inconvenient if the sum of a large number of terms is required, and the shorthand notation normally used is

$$\sum_{r=1}^{r=10} x^r$$

This is an instruction to add together all terms of the form x_r, starting with the term for which $r = 1$ and increasing r by 1 for each successive term until $r = 10$.

2. The average or arithmetic mean. The average of a set of values is simply obtained by adding all the values together and dividing by the number of values.

(a) In the above example the average value of x is given by

$$\bar{x} = \left(\sum_{r=1}^{r=10} x_r \right) \Big/ 10$$

(b) If there are n_1 items with value x_1, n_2 with value x_2, etc., the average is given by

$$\bar{x} = (n_1 x_1 + n_2 x_2 + \ldots + n_{10} x_{10})/(n_1 + n_2 + \ldots + n_{10})$$

$$= \sum_{r=1}^{r=10} n_r x_r \Big/ \sum_{r=1}^{r=10} n_r$$

3. The root mean square value. This is obtained by adding the squares of the values, dividing by the number of values and taking the square root.

(*a*) For the example given in 2 (*a*) above, the root mean square value is given by:

$$(\overline{x^2})^{\frac{1}{2}} = \left\{ \left(\sum_{r=1}^{r=10} x_r{}^2 \right) \middle/ 10 \right\}^{\frac{1}{2}}$$

(*b*) For the example given in 2 (*b*) above:

$$(\overline{x^2})^{\frac{1}{2}} = \left(\sum_{r=1}^{r=10} n_r x_r{}^2 \middle/ \sum_{r=1}^{r=10} n_r \right)^{\frac{1}{2}}$$

NOTE: The position of the bar is important in distinguishing between the mean square value, $\overline{x^2}$, and the square of the mean or average value, $\bar{x}^2$. $(\overline{x^2})^{\frac{1}{2}}$ is sometimes given the special symbol x_{rms}.

This particular form of average is most useful in representing the average speed of a number of gas molecules (*see* IV, **13–14**). Thus if the root mean square speed of all the molecules in a gas is $(\overline{c^2})^{\frac{1}{2}}$ and the total mass of the molecules is M, the total kinetic energy of the gas can be expressed as $M\overline{c^2}/2$. This has the same value as the sum of the kinetic energies of the individual molecules, each moving with its own particular speed. If the total kinetic energy had been expressed as $M\bar{c}^2/2$ this would not have been the case.

EXAMPLE 1: If four molecules travel at 1 m s^{-1}, three at 2 m s^{-1}, six at 3 m s^{-1} and two at 4 m s^{-1}, calculate (*a*) the root mean square speed, (*b*) the average speed of the molecules.

$$(a) \ (\overline{c^2})^{\frac{1}{2}} = \left(\sum_{r=1}^{r=4} n_r c_r{}^2 \middle/ \sum_{r=1}^{r=4} n_r \right)^{\frac{1}{2}}$$
$$= \{(4 \times 1^2 + 3 \times 2^2 + 6 \times 3^2 + 2 \times 4^2)/$$
$$(4 + 3 + 6 + 2)\}^{\frac{1}{2}}$$
$$= 2\cdot73 \text{ m s}^{-1}$$

$$(b) \quad \bar{c} = \sum_{r=1}^{r=4} n_r c_r \middle/ \sum_{r=1}^{r=4} n_r$$
$$= (4 \times 1 + 3 \times 2 + 6 \times 3 + 2 \times 4)/$$
$$(4 + 3 + 6 + 2)$$
$$= 2\cdot40 \text{ m s}^{-1}$$

INDICES

Expressions such as $x \times x \times x$ and $x \times x \times x \times x$ may be written in the shorthand form as x^3 and x^4 respectively, where the numbers [3] and [4] are called *indices* or *exponents*.

4. The laws of indices.

(a) *The multiplication law*: to multiply, add indices.
$$x^a \times x^b = x^{a+b} \quad e.g. \ 3x^2 \times 2x^3 = 6x^5$$

(b) *The division law*: to divide, subtract indices.
$$x^a \div x^b = x^{a-b} \quad e.g. \ 36x^6 \div 9x^2 = 4x^4$$

(c) *The power law*: to raise to a power, multiply indices.
$$(x^a)^b = x^{ab} \quad e.g. \ (2x^2)^3 = 8x^6$$

5. Negative and fractional indices.

(a) *Negative indices*. The expression x^{-a} is simply the reciprocal of x^a, i.e. $1/x^a$, e.g. $3x^{-2} = 3/x^2$; $(3x)^{-2} = 1/(3x)^2 = 1/9x^2$.

(b) *Fractional indices*. The expression $x^{1/a}$ is simply the ath root of x, i.e. $\sqrt[a]{x}$, e.g. $2x^{\frac{1}{2}} = 2(\sqrt[4]{x})$; $3x^{-\frac{1}{2}} = 3/\sqrt[3]{x}$.

NOTE: $x^0 = 1$ for all values of x except 0, e.g. $2 \cdot 37^\circ = 1$; $129 \cdot 1^\circ = 1$.

NUMBERS AND LOGARITHMS

6. Large and small numbers. Very large and very small numbers occur frequently in physical chemistry, *e.g.* Avogadro's constant has a value of 602 252 000 000 000 000 000 000 mol^{-1}, while the equilibrium internuclear separation in the hydrogen molecule is 0·000 000 000 074 m.

Such numbers are customarily expressed as the product of a number between 1 and 10, and 10 raised to the relevant power (the exponent). On this basis Avogadro's constant becomes $6 \cdot 022 \ 52 \times 10^{23} \ mol^{-1}$, and the internuclear separation in the hydrogen molecule $7 \cdot 4 \times 10^{-11}$ m. Among the advantages of this method are:

(a) The exponent of a number is identical with the characteristic of the logarithm of that number (*see* 8 below).

(b) Mathematical operations are facilitated by the laws of indices, *e.g.* the product of the above two numbers is

$$7 \cdot 4 \times 10^{-11} \times 6 \cdot 022 \ 52 \times 10^{23} = 7 \cdot 4 \times 6 \cdot 022 \ 52 \times 10^{23-11}$$
$$= 44 \cdot 55 \times 10^{12} = 4 \cdot 455 \times 10^{13}$$

7. Number systems; the definition of e. The familiar number system is based on 10. However, other number systems may be used, *e.g.* almost all computers work with a number system based on 2, and a number system based on the universal constant e is important in connection with logarithms (*see* **12** below).

e is defined as the limit of the series $(1 + 1/n)^n$, where n is a very large number. The concept of a series limit can be seen from Table 6.

TABLE 6   VALUES OF $(1 + 1/n)^n$ FOR INCREASING VALUES OF n

n	$(1 + 1/n)^n$
1	2·000
2	2·250
3	2·369
5	2·487
10	2·594
100	2·705
1000	2·717
∞ (infinity)	2·718 28 . . .

As n increases, the value of $(1 + 1/n)^n$ approaches the limit 2·718 28 more and more closely but never quite reaches it. For most purposes e may be taken as having an approximate value of 2·718. Table 6 illustrates the *law of exponential growth*, where the increase in a property (in this case $(1 + 1/n)^n$) is proportional to the amount of the property that is increasing.

8. Common logarithms. As we have seen in **4** above, multiplication or division can be performed by the addition or subtraction of exponents, *e.g.*

(a) $16 \times 32 = 2^4 \times 2^5 = 2^{4+5} = 2^9 = 512$
(b) $100 \div 1000 = 10^2 \div 10^3 = 10^2 \times 10^{-3} = 10^{2-3}$
$= 10^{-1} = 0 \cdot 1$

Multiplication and division using logarithms extend this method to include non-integral exponents. Thus common logarithm tables are simply lists of exponents of 10, *e.g.* the statements $\log_{10} 5 = 0 \cdot 6990$; $\log_{10} 2000 = \log_{10} (2 \times 10^3) = 3 \cdot 3010$; $\log_{10} 100 = \log_{10} 10^2 = 2 \cdot 0000$ are equivalent to $10^{0 \cdot 6990} = 5$; $10^{3 \cdot 3010} = 2000$; $10^{2 \cdot 0000} = 100$.

The shorthand form of $\log_{10}$, lg, will be used in the remainder of the book. The part of the logarithm preceding the decimal point is called the *characteristic*, while that after the decimal point is called the *mantissa*. Only the mantissa is obtained from logarithm tables, the characteristic being obtained independently. Thus for lg 2000 the mantissa is obtained from logarithm tables (lg 2 = 0·3010), while the characteristic is 3, since the number is in the thousands, *i.e.* lg 2000 = 3·3010.

In general, if $10^x = y$, then $x = $ lg y and x may be found by looking up y direct in the logarithm tables.

9. Antilogarithms. If x is given, y may be found using the logarithm tables in the reverse order, *i.e.* by looking up the number whose logarithm is x. This operation is called *finding the antilogarithm*. From the above examples, antilg 0·6990 = 5; antilg 3·3010 = 2000; antilg 2·0000 = 100. Only the mantissa is searched for in the logarithm tables, and the characteristic gives the number of digits before the decimal point; *e.g.* evaluate antilg 3·3010. The number corresponding to lg 0·3010 in the logarithm tables is 2000, and since the characteristic is 3 the number must be in the thousands and is therefore 2000.

10. The use of logarithms.

(*a*) *Multiplication and division.* When multiplying two numbers, *add* their logarithms; when dividing two numbers, *subtract* their logarithms.

EXAMPLE 2: Evaluate 230 × 18 ÷ 3·7.
Taking logarithms,

lg (230 × 18 ÷ 3·7) = lg 230 + lg 18 − lg 3·7
                     = 2·3617 + 1·2553 − 0·5682 = 3·0488

and antilg 3·0488 = 1119.

i.e.   230 × 18 ÷ 3·7 = 1119

(*b*) *Powers.* When raising a number to a power, the logarithm of the number is *multiplied* by the power.

EXAMPLE 3: Evaluate:

(*a*) $18·3^4$ = 4 lg 18·3 = 4 × 1·2625 = 5·0500;
    antilg 5·0500 = 11 220
(*b*) $18·3^{\frac{2}{3}}$ = ($\frac{2}{3}$) lg 18·3 = ($\frac{2}{3}$) × 1·2625 = 0·8417;
    antilg 0·8417 = 9·252

NOTE: lg 1 = 0.

11. Negative logarithms. The logarithms of numbers less than 1 may be expressed in two ways. Consider the number 0·05.

(a) $\lg 0{\cdot}05 = \lg(5 \times 10^{-2}) = \lg 5 + \lg 10^{-2}$
$= 0{\cdot}6990 - 2 = \bar{2}{\cdot}6990$
(b) $\lg 0{\cdot}05 = 0{\cdot}6990 - 2 = -1{\cdot}3010$

Method (a) uses the bar notation, where the bar over the 2 indicates that it alone is considered negative; the mantissa is positive. This form is used when logarithms are added and subtracted.

EXAMPLE 4: Evaluate $2{\cdot}17 \times 0{\cdot}007$.

$$\lg(2{\cdot}17 + 0{\cdot}007) = \lg 2{\cdot}17 + \lg 0{\cdot}007$$
$$= 0{\cdot}3365 + \bar{3}{\cdot}8451 = \bar{2}{\cdot}1816$$

and antilg $\bar{2}{\cdot}1816 = 0{\cdot}01519$

When logarithms are multiplied or divided, the bar notation is not suitable, since part of the number is positive and part negative. Instead, method (b) is used, where the logarithm is converted to a completely negative number.

EXAMPLE 5: Evaluate $3{\cdot}612 \times \lg \bar{3}{\cdot}617$.

$$3{\cdot}612 \times \lg \bar{3}{\cdot}617 = 3{\cdot}612 \times (0{\cdot}617 - 3) = 3{\cdot}612 \times (-2{\cdot}383)$$
$$= -(3{\cdot}612 \times 2{\cdot}383)$$

The product inside the brackets is found using logarithm tables in the normal way;

$$\lg(3{\cdot}612 \times 2{\cdot}383) = 0{\cdot}5577 + 0{\cdot}3772 = 0{\cdot}9349$$
$$\text{antilg } 0{\cdot}9349 \quad = 8{\cdot}604$$

Hence $3{\cdot}612 \times \lg \bar{3}{\cdot}617 = -8{\cdot}604$.

EXAMPLE 6: Calculate the pH of a solution containing 5×10^{-4} M (*i.e.* mol dm^{-3}) of H_3O^+ ions.

pH is defined as $\quad \text{pH} = -\lg[H_3O^+]$

where the square brackets indicate concentration in mol dm^{-3}

$$\therefore \text{pH} = -\lg(5 \times 10^{-4}) = -(0{\cdot}6990 - 4) = -(-3{\cdot}301)$$
$$= 3{\cdot}301$$

12. Natural logarithms. In physical chemistry we are often concerned with natural logarithms, that is, logarithms to the base e. This arises because of the importance of e in differentiation and integration (*see* 17–23 below). In general, if $e^x = y$, then $x = \log_e y$. Natural logarithms will be represented by ln and e^x as exp x in the remainder of this book.

If tables of natural logarithms are not available, we may

look up the number in common logarithm tables and use the conversion factor

$$\ln N = \ln 10 \times \lg N = 2 \cdot 303 \lg N$$

EXAMPLE 7: The variation of the rate constant with temperature for a reaction is given by the expression

$$k = A \exp(-E/RT),$$

where A, E and R are constants. Derive an expression for E.

$$k = A \exp(-E/RT)$$

Taking natural logarithms, $\ln k = \ln A - E/RT$
Converting to common logarithms,

$$2 \cdot 303 \lg k = 2 \cdot 303 \lg A - E/RT$$

Hence
$$E = 2 \cdot 303\, RT(\lg A - \lg k)$$
$$= 2 \cdot 303\, RT \lg(A/k)$$

EQUATIONS

13. Functions of variables. In writing the expression $y = f(x)$ we indicate that there is some procedure by which one number, x (the independent variable), may be converted to another number, y (the dependent variable). This might, for example, be by a mathematical equation:

$$y = f(x) = 2x^2 + 5x + 6$$

where $f(x)$ contains only terms in x and constants.

Thus the pressure of one mole of an ideal gas is a function of temperature and volume (*i.e.* a function of two independent variables):

$$p = f(T, V)$$

In fact $p = RT/V$, where R is the gas constant.

14. Solution of equations. We shall be concerned mainly with two types of equation:

(*a*) Linear equations $y = ax + b$
(*b*) Quadratic equations $y = ax^2 + bx + c$

If these equations can be reduced to the form

$$ax + b = 0 \quad \text{then } x = -b/a$$
$$\text{and } ax^2 + bx + c = 0 \quad \text{then } x = [-b \pm (b^2 - 4ac)^{\frac{1}{2}}]/2a$$

For the quadratic equation there are two solutions for x, but when the equation describes a chemical or physical process one of these may usually be discarded by inspection.

EXAMPLE 8: For the equilibrium $H_2 + I_2 \rightleftharpoons 2HI$, the equilibrium constant K is given by $K = 4x^2/(a - x)(b - x)$, where a and b are the initial amounts of H_2 and I_2 respectively and x is the amount of H_2 and I_2 that has reacted. Find x in terms of K, a and b.

The equation for the equilibrium constant may be rearranged to give

$$Kab - Kax - Kbx + Kx^2 = 4x^2$$
$$x^2(4 - K) + x(Ka + Kb) - Kab = 0$$

Hence

$$x = [-(Ka + Kb) \pm \{(Ka + Kb)^2 - 4(4 - K)(-Kab)\}^{\frac{1}{2}}]/2$$
$$(4 - K)$$

Since x cannot be a negative number, the solution with the minus sign before the square root can be deleted.

15. Graphical methods. Functions such as $y = f(x)$ may be represented simply and visually by means of a graph. Thus the simultaneous variation of two quantities may be represented by a line on a two-dimensional diagram.

The graph of y against x for the expression $y = ax + b$ is shown in Fig. 1 and is a straight line. The line is made up

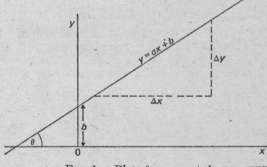

FIG. 1.—*Plot of $y = ax + b$*

of a number of points and each point is obtained by calculating the value of y for a given value of x. The slope of the line, a, is given by $\Delta y/\Delta x$, where Δy is the vertical increase corresponding to Δx, the horizontal increase.

NOTE: Δy and Δx represent changes in y and x respectively.

a is also equal to tan θ, where θ is the angle the line makes with the positive direction of the x axis. b represents the intercept with the y axis, *i.e.* the distance between the origin and the point of intersection of the line and y axis.

16. Examples from the ideal gas equation. For one mole of an ideal gas $pV = RT$, where p, V and T are all variables. It is customary to keep one variable constant and to plot the other two.

At constant volume, $p = RT/V = k'T$. A plot of p against T yields a straight line of slope k' passing through the origin.

At constant temperature, $p = RT/V = k''/V$. A plot of p against V yields a rectangular hyperbola. It is usually convenient to convert such a curve into a linear form, and this may be done by plotting p against $1/V$. The slope of the line which passes through the origin is k''.

Linear graphs are important, since quantities may be evaluated from their slopes and intercepts, and experimental observations can readily be compared with theoretical calculations. This can also be done with non-linear graphs, *e.g.* the hyperbola above, but the procedure is much more difficult, and such graphs are usually converted to a linear form.

DIFFERENTIATION

17. The graphical significance of differentiation. If the function $y = f(x)$ gives a linear graph, then the slope is given by $\Delta y/\Delta x$ (*see* **15** above). On the other hand, if the graph is curved the slope is continuously changing. If Δy and Δx are

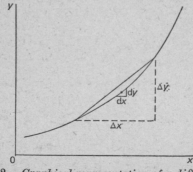

Fig. 2.—*Graphical representation of a differential*

made infinitesimally small, they are represented by dy and dx and are called *differentials*. The ratio dy/dx will then represent the slope of the curve at any point, *i.e.* it will represent the slope of the tangent to the curve at that point (*see* Fig. 2). dy/dx is called the *derivative* and may be found by plotting the graph and finding the tangent to the curve at the desired point.

NOTE: The above treatment suggests that dy/dx is a fraction, and indeed it can be subjected to normal algebraic operations. However, if $\Delta y = \Delta x = 0$, dy/dx does not have the meaningless form $0/0$.

18. The rules of differentiation. The above graphical method is a clumsy and inconvenient way of finding dy/dx, and a better way is to apply some general rules once the relationship between y and x is known. Some of these rules, which may be proved, are listed in Table 7.

TABLE 7 RULES OF DIFFERENTIATION

	Function	Derivative (dy/dx)
1	$y = x^n$	nx^{n-1}
2	$y = ax^n$	anx^{n-1}
3	$y = u + v$	$du/dx + dv/dx$
4	$y = a$	0
5	$y = uv$	$udv/dx + vdu/dx$
6	$y = u/v$	$(vdu/dx - udv/dx)/v^2$
7	$y = \exp x$	$\exp x$
8	$y = \exp ax$	$a \exp ax$
9	$y = \ln x$	$1/x$

where a is a constant and u and v are functions of x.

EXAMPLE 9:

(a) $y = x^4 - x^2$
$$dy/dx = d(x^4)/dx - d(x^2)/dx = 4x^3 - 2x$$

(b) $y = x^4(x^2 - 4)$
$$dy/dx = x^4 d(x^2 - 4)/dx + (x^2 - 4)d(x^4)/dx$$
$$= x^4(2x) + (x^2 - 4)(4x^3) = 6x^5 - 16x^3$$

For many purposes derivatives may be looked upon as normal fractions and may be cancelled out, as seen in the following example.

(c) $y = (2x + 2)^{-\frac{1}{3}}$ Let $s = 2x + 2$, then $y = s^{-\frac{1}{3}}$
$$ds/dx = 2 \quad \text{and} \quad dy/ds = (-\tfrac{1}{3})(s^{-\frac{4}{3}})$$

Hence $dy/dx = (ds/dx) \times (dy/ds) = (-\tfrac{2}{3})(s^{-\frac{4}{3}})$
$$= (-\tfrac{2}{3})(2x + 2)^{-\frac{4}{3}}$$

The derivative may be separated into its component differentials, and, for example, the solution to Example 9(a) may be written

$$dy = (4x^3 - 2x)dx$$

Also rule 5 in Table 7 may be written

$$dy = u\,dv + v\,du$$

19. Partial differentiation. In our discussion so far we have considered equations containing only one independent variable. Many chemical and physical phenomena are described by equations containing two or more independent variables.

If $y = f(x, z)$ and x and z are variables independent of each other, the *partial derivative of y with respect to x* is obtained by treating z in $f(x, z)$ as a constant and taking the normal derivative with respect to x. This partial derivative is written $(\partial y/\partial x)_z$ and is the rate of change of y with respect to x, keeping z constant. Likewise, $(\partial y/\partial z)_x$ represents the partial derivative of y with respect to z, keeping x constant.

The change in y (dy) brought about by simultaneous changes in x and z (dx and dz respectively) is given by

$$dy = (\partial y/\partial x)_z\,dx + (\partial y/\partial z)_x\,dz$$

Thus the *total differential*, dy, is the sum of two partial differentials. This equation may be written in a different form by dividing both sides by dx

$$(dy/dx) = (\partial y/\partial x)_z + (\partial y/\partial z)_x(dz/dx)$$

EXAMPLE 10: The volume of one mole of an ideal gas is a function of both temperature and pressure, *i.e.* $V = f(T, p)$. Find the infinitesimal change in V for simultaneous infinitesimal changes in T and p.

Since T and p can be varied independently of the other

$$dV = (\partial V/\partial T)_p\,dT + (\partial V/\partial p)_T\,dp \cdot \quad . \quad . \quad \text{(II, 1)}$$

Now
$V = RT/p$, hence $(\partial V/\partial T)_p = R/p$ and $(\partial V/\partial p)_T = -RT/p^2$
Substituting these values into (II, 1) gives

$$dV = (R/p)dT - (RT/p^2)dp$$

Now $R/p = V/T$ and $RT/p^2 = V/p$
Hence $dV = (V/T)dT - (V/p)dp = V(dT/T - dp/p)$
$$= V(d \ln T - d \ln p)$$

See Table 7, rule 9, or Table 8, rule 4.

INTEGRATION

20. The basic principles. The process of integration is the reverse of differentiation. If the relationship between dy and dx is known, then integration allows the relationship between y and x to be found.

Consider the function $y = f(x)$. Suppose that y is divided up into small parts each of magnitude Δy. Then $\Sigma \Delta y = y$. If y is made infinitesimally small, *i.e.* equal to dy, the summation must now be performed over an infinite number of parts and is known as *integration*, symbol $\int$. Thus $\int dy = y$.

Suppose $y = 3x^3$, then $dy/dx = 9x^2$, which may be rewritten

$$dy = 9x^2 dx$$

This expression may be integrated:

$$\int 9x^2 dx = \int dy = y$$

i.e.
$$9x^2 dx = 3x^3 \qquad \text{. . . (II, 2)}$$

Thus differentiation followed by integration results in the original function. This is not strictly true, since the equation $y = 3x^3 + c$, where c is a constant, may also be differentiated to give $dy/dx = 9x^2$. Hence a constant must be included in the integration process. Therefore (II, 2) becomes

$$\int 9x^2 dx = 3x^3 + c$$

where c is known as the *integration constant*. The rules of integration follow directly from those of differentiation and are given in Table 8 overleaf.

EXAMPLE 11: Evaluate:

(a) $(9x^3 + 2x^2 + x^{\frac{1}{2}} + 1)dx$

$$= \int 9x^3 dx + \int 2x^2 dx + \int x^{\frac{1}{2}} dx + \int dx$$

$$= (\tfrac{9}{4})x^4 + (\tfrac{2}{3})x^3 + (\tfrac{2}{3})x^{\frac{3}{2}} + x + c$$

(b) $\exp(6x)dx = (\tfrac{1}{6})\exp(6x) + c$

(c) $\int \dfrac{3dx}{x+2} = 3\int \dfrac{dx}{x+2} = 3\ln(x+2) + c$

TABLE 8   RULES OF INTEGRATION

	Integral	*Function*
1	$\int \mathrm{d}x$	$x + c$
2	$\int x^n \mathrm{d}x$	$[x^{n+1}/n + 1] + c$ when $n \neq -1$
3	$\int (ax + b)^n \mathrm{d}x$	$[(ax + b)^{n+1}/a(n + 1)] + c$ when $n \neq -1$
4	$\int \dfrac{\mathrm{d}x}{x}$	$\ln x + c$
5	$\int \dfrac{\mathrm{d}x}{ax + b}$	$(1/a)\ln(ax + b) + c$
6	$\int (u + v)\mathrm{d}x$	$\int u\mathrm{d}x + \int v\mathrm{d}x$
7	$\int \exp(x)\mathrm{d}x$	$\exp x + c$
8	$\int \exp(ax)\mathrm{d}x$	$(1/a)\exp ax + c$

where a and b are constants and u and v are functions of x.

21. Integration between limits. In many problems it is necessary to evaluate the integral between two particular values of x. When this is done, the integration constant conveniently drops out. Such an integral is known as a *definite* integral. Thus $x^2\mathrm{d}x$ may be integrated between the limits $x = 1$ and $x = 3$:

$$\int_1^3 x^2\mathrm{d}x = [x^3/3]_1^3 = \tfrac{27}{3} - \tfrac{1}{3} = \tfrac{26}{3}$$

The two numbers are substituted into the integrated expression, and the value obtained from the lower limit ($\tfrac{1}{3}$ in the above example) is subtracted from the value obtained from the upper limit. (*See* **23** below for the graphical significance of integration between limits.)

22. Integration by partial fractions. This is a method of integration in which the expression is split into the sum of simpler expressions by the method of partial fractions. The method can readily be illustrated by means of an example.

EXAMPLE 12: Evaluate $\int \dfrac{dx}{x^2 - 4}$

This expression can be more easily integrated if it is split up into partial fractions

$$1/(x^2 - 4) \equiv 1/(x + 2)(x - 2) \equiv A/(x + 2) + B/(x - 2)$$
$$\equiv \{A(x - 2) + B(x + 2)\}/\{(x + 2)(x - 2)\}$$

where A and B are to be evaluated.

NOTE: $y \equiv f(x)$ expresses the fact that the two expressions merely represent two different ways of writing the same quantity. In particular, y is equal to $f(x)$ for *all* values of x. The symbol $\equiv$ is called an *identity* symbol.

Since the numerators on opposite sides of the identity sign must be identical

$$1 \equiv A(x - 2) + B(x + 2)$$
i.e. $$1 \equiv x(A + B) + 2B - 2A$$

Equating coefficients of x on opposite sides of the identity sign: $0 \equiv A + B$.
Equating constant terms on opposite sides of the identity sign: $1 \equiv 2B - 2A$.

Hence $A = -\tfrac{1}{4}$ $B = \tfrac{1}{4}$

Thus $\int \dfrac{dx}{x^2 - 4} = \tfrac{1}{4}\int \dfrac{dx}{x - 2} - \tfrac{1}{4}\int \dfrac{dx}{x + 2}$

$$= \tfrac{1}{4}\{\ln(x - 2) - \ln(x + 2)\} + c$$
$$= \tfrac{1}{4}\ln\{(x - 2)/(x + 2)\} + c$$

23. Graphical integration. A plot of p against V for a gas at constant temperature is shown in Fig. 3. Suppose that the area under the curve between the limits V_1 and V_2 is required. This may be obtained by dividing the area into rectangles each dV wide and p high. For any given rectangle the value of p will obviously depend on the particular value of V at which the rectangle is drawn. The area will be

$$\int_{V_1}^{V_2} p\,dV$$

and may be found by graphical integration, *i.e.* by measuring the area under the curve.

If, however, the relationship between p and V is known, the integral may be evaluated in the normal way (*see* **21** above). For one mole of an ideal gas, $p = RT/V$. Therefore

$$\int_{V_1}^{V_2} p\,\mathrm{d}V = \int_{V_1}^{V_2} \frac{RT\,\mathrm{d}V}{V} = RT \int_{V_1}^{V_2} \frac{\mathrm{d}V}{V} \text{ at constant temperature}$$
$$= RT(\ln V_2 - \ln V_1) = RT \ln(V_2/V_1)$$

The area under this curve has a special significance, since it represents the work done by a gas in expanding from V_1 to V_2 (*see* VI, **7**).

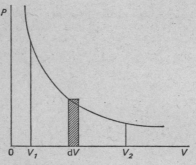

Fig. 3.—*Graphical integration*

DIFFERENTIAL EQUATIONS

Many of the fundamental equations in physical chemistry contain derivatives or differentials and are known as *differential equations*. The solution of such equations involves integration.

24. Examples of differential equations.

EXAMPLE 13: The rate of a first-order chemical reaction is given by an expression of the form $-\mathrm{d}x/\mathrm{d}t = kx$, where k is a constant and x is the concentration of the reactant at time t. Solve this expression for x.

The differential equation $-\mathrm{d}x/\mathrm{d}t = kx$ expresses the fact that the dependent variable, *i.e.* the reactant concentration, is *decreasing* (hence the minus sign) at a rate which is proportional to the magnitude of the independent variable, *i.e.* the time.

Thus the reactant concentration decreases with time. The x's and dx's can be collected together on one side of the equation, with dt on the other side. This operation is known as separating the variables.

$$dx/x = -kdt$$

On integrating this expression, the differentials disappear:

$$\int dx/x = -k\int dt \qquad \qquad \text{. . . (II, 3)}$$

Hence $$\ln x = -kt + c \qquad \qquad \text{. . . (II, 4)}$$

Equation (II, 4) is the solution of the differential equation.

Suppose that at time $t = 0$, $x = x_0$, then $c = \ln x_0$.
Equation (II, 4) becomes $\ln x = -kt + \ln x_0$ . . . (II, 5)
$$\therefore \ln(x/x_0) = -kt$$
Taking antilogarithms, $\quad x/x_0 = \exp(-kt) \qquad$ . . . (II, 6)
or $$x = x_0 \exp(-kt)$$

The problem may be tackled in a slightly different way, which does not involve evaluating the integration constant, c. Equation (II, 3) may be integrated between the limits $x = x_0$ at $t = 0$ and $x = x$ at $t = t$ and becomes

$$\int_{x_0}^{x} dx/x = -k\int_{0}^{t} dt$$

$$[\ln x]_{x_0}^{x} = -k[t]_{0}^{t}$$

$$\therefore \ln x - \ln x_0 = -k[t - 0] = -kt$$

which is identical with (II, 5).

EXAMPLE 14: The rate of a second-order chemical reaction may be represented by $dx/dt = k(a - x)(b - x)$, where a and b are the concentrations of the two reactants at zero time and x is the amount which has reacted. Find the solution of the differential rate equation.

Separating the variables, $dx/(a - x)(b - x) = kdt$. This expression may be integrated by the method of partial fractions.

$$1/(a - x)(b - x) \equiv A/(a - x) + B/(b - x) \quad \text{. . . (II, 7)}$$

From which $A = 1/(b - a) = -1/(a - b)$; $B = 1/(a - b)$

Substituting these values into (II, 7) and integrating between the limits $x = 0$ at $t = 0$ and $x = x$ at $t = t$ gives

$$\int_0^x \frac{dx}{(a-x)(b-x)} = -\frac{1}{a-b}\int_0^x \frac{dx}{a-x} + \frac{1}{a-b}\int_0^x \frac{dx}{b-x} = \int_0^t k\,dt$$

$$-\frac{1}{a-b}[-\ln](a-x)]_0^x + \frac{1}{a-b}[-\ln(b-x)_0^x = k[t]_0^t$$

$$-\frac{1}{a-b}[-\ln(a-x) + \ln a] + \frac{1}{a-b}[-\ln(b-x) + \ln b]$$
$$= k[t-0]$$

which reduces to

$$\frac{1}{a-b}\ln\left[\frac{b(a-x)}{a(b-x)}\right] = kt$$

EXAMPLE 15: Solve the differential equation $d(\ln K)/dT = \Delta H^\ominus/RT^2$, where K is the equilibrium constant and $\Delta H^\ominus$ is the standard heat of reaction.

Separating the variables, $\quad d(\ln K) = \Delta H^\ominus dT/RT^2$
$$\int d(\ln K) = \int \frac{\Delta H^\ominus dT}{RT^2}$$

If $\Delta H^\ominus$ is assumed to be independent of temperature, then

$$\int d(\ln K) = \frac{\Delta H^\ominus}{R}\int \frac{dT}{T^2}$$

If K_1 is the equilibrium constant at T_1 and K_2 that at T_2

$$\int_{K_1}^{K_2} d(\ln K) = \frac{\Delta H^\ominus}{R}\int_{T_1}^{T_2} \frac{dT}{T^2}$$

$$\therefore \ln(K_2/K_1) = \frac{\Delta H^\ominus}{R}\left(-\frac{1}{T_2} + \frac{1}{T_1}\right)$$

$$= \frac{\Delta H^\ominus}{R}\left(\frac{T_2 - T_1}{T_1 T_2}\right)$$

VECTORS

25. Definition of a vector. If an object moves from point A to point B (Fig. 4) the change in position is measured by a physical quantity called the *displacement*. The displacement is specified by:

(a) A *magnitude* as indicated by the length of the line *AB*.
(b) A *direction* as indicated by the angular orientation of the line, θ.

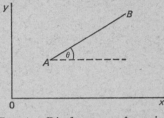

FIG. 4.—*Displacement of a point*

Displacement is an example of a physical quantity called a *vector* and has a magnitude and direction independent of the choice of co-ordinate system.

26. Vectors and scalars. A physical quantity which is completely specified by a magnitude (and a unit) is called a *scalar*. A scalar also represents the magnitude of a vector. Vectors are written in heavy type while scalars are written in italics. Thus velocity **c** is a vector, while its magnitude is c, the speed. An arrow is often placed above a symbol to indicate a vector.

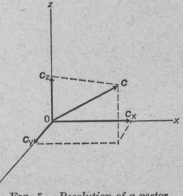

FIG. 5.—*Resolution of a vector*

27. Resolution of a vector. A vector may be resolved into two or more vectors which have the given vector as their sum. Thus the velocity of a molecule c may be resolved into components c_x, c_y and c_z along the x, y and z axes (*see* Fig. 5) and

$$c = c_x + c_y + c_z$$

The magnitude of the vector, *i.e.* its speed, is given by

$$c^2 = c_x{}^2 + c_y{}^2 + c_z{}^2$$

where c_x, c_y and c_z represent the scalar components of the projection of the vector on to the x, y and z axes. The above equation finds use in the kinetic theory of gases (*see* IV).

PROGRESS TEST 2

1. The potential of an Fe^{3+}/Fe^{2+} electrode is given by

$$E = E^{\ominus} + 0 \cdot 059 \lg \frac{[Fe^{3+}]}{[Fe^{2+}]}$$

If $E = 0 \cdot 82$ V, $E^{\ominus} = 0 \cdot 78$ V and $[Fe^{2+}] = 0 \cdot 015$ M, calculate $[Fe^{3+}]$. **(8–11)**

2. Calculate the hydrogen ion concentration in a solution of pH 4·31. **(8–11)**

3. After time t, the concentration c of a reactant in a first-order reaction is given by $c = c_0 \exp(-kt)$ where c_0 is the initial reactant concentration and k is the rate constant. Derive an expression for k. **(12)**

4. The Langmuir adsorption isotherm can be written

$$\theta = \frac{bp}{1 + bp}$$

Rearrange the equation to give a linear relationship. **(15–16)**

5. The quantity of heat, q, necessary to raise the temperature of a fixed amount of gas is given by $q = aT + bT^2 + cT^3$ where a, b and c are constants. Derive an expression for the heat capacity, C, of the gas given $C = \mathrm{d}q/\mathrm{d}T$. **(17–18)**

6. The rate of decay of a radioactive isotope is given by $-\mathrm{d}N/\mathrm{d}t = \lambda N$ where N is the number of atoms at time t and λ is the decay constant. Integrate this expression between the limits $N = N_0$ at $t = 0$ and $N = N$ at $t = t$. **(24)**

7. The work done by the reversible isothermal expansion of an ideal gas is given by $w = RT \int \mathrm{d}V/V$. Given $R = 8 \cdot 314$ J K^{-1} mol^{-1}, $T = 300$ K, and initial and final volumes of 1 dm^3 and 10 dm^3, find w. **(24)**

PART TWO

ATOMS, MOLECULES AND THE PROPERTIES OF GASES

In the next three chapters we review the evidence supporting the fundamental concept that matter is composed of atoms and molecules.

Chapter III contains classical and modern evidence for the atomic concept. The determination of relative atomic and molecular masses (also called atomic and molecular weights) is discussed in the latter half of the chapter, and it should be noted that our very ability to determine and assign relative atomic and molecular masses provides powerful evidence for the real existence of atoms and molecules.

The properties of gases are considered in Chapters IV and V. Gases are the simplest collections of atoms or molecules, and some of the most elegant of the early evidence for the atomic concept was that implied by the success of the kinetic theory in explaining their properties.

CHAPTER III

ATOMS AND MOLECULES

INTRODUCTION

Physical chemistry may loosely be defined as the study of the principles underlying the properties of and the changes undergone by matter. The observations made on matter can be summarised in the form of *experimental laws* (*e.g.* Boyle's law) and *theories* advanced to explain these laws (*e.g.* the kinetic theory of gases). A theory is considered satisfactory if it correlates all the known facts about a particular phenomenon, and in addition predicts new facts which may be experimentally verified.

1. Matter. The many different kinds of matter of which the universe is composed may be classified into (*a*) *homogeneous* matter and (*b*) *heterogeneous* matter.

Homogeneous matter has the same properties throughout, *i.e.* it constitutes a single *phase*, while heterogeneous matter consists of several phases each with distinct properties and separated from other phases by demonstrable boundaries.

2. Substances. A species of matter which is homogeneous and has a definite set of physical and chemical properties under specified conditions is known as a *pure substance, e.g.* sodium chloride, mercury, lead, pure water, etc. Pure substances cannot be broken down into simpler substances by physical processes such as distillation, crystallisation, etc.

A solution of sodium chloride in water is not a pure substance, for, even though it is homogeneous throughout, its composition may vary widely. Here the term *mixture* is appropriate. A mixture may be either homogeneous or heterogeneous, the only requirement being that it contains more than one pure substance.

3. Elements and compounds. Pure substances may be subdivided into two groups, *elements* and *compounds*. An element

33

cannot be broken down into simpler substances by *ordinary chemical methods*. Compounds, on the other hand, are composed of two or more elements and can be broken down chemically into their constituent elements. Chemical methods are those requiring energies of the order of 10^3 kJ or less. Elements may be broken down into smaller particles (subatomic particles, *see* **9** below) by the use of energies far in excess of 10^3 kJ.

All natural matter is ultimately composed of ninety-two elements, although in recent years about ten new elements have been produced synthetically, using very high energies.

Elements are themselves composed of atoms, and it is instructive at this stage to review the atomic theory of matter.

THE ATOMIC THEORY

4. Dalton's atomic theory. The name usually associated with the atomic concept is that of Dalton, who in 1808 published the following postulates:

(*a*) *An atom is the smallest particle of an element capable of independent existence.*

NOTE: This does not necessarily imply that atoms are not themselves made up of still more fundamental particles.

(*b*) *Atoms of the same element have identical properties, e.g.* mass, volume, chemical reactivity, etc.

(*c*) *Atoms of different elements have different properties.*

(*d*) *Chemical combination occurs by atoms joining together in simple integral numbers to give molecules.*

NOTE: For the present purpose it is sufficient to define a molecule as a group of atoms bonded to one another.

5. Deductions from the atomic theory. Since all the atoms present before a chemical reaction occurred would still be present afterwards (though arranged differently), two important experimental observations could be explained:

(*a*) *The law of conservation of mass* (Lavoisier, 1774). Mass can never be created or destroyed in a chemical reaction.

(*b*) *The law of definite proportions* (Proust, 1799). A particular compound always contains the same elements in the same proportions by mass irrespective of its method of preparation.

Two further laws, whose proposal was prompted by the atomic theory, were:

(c) *The law of multiple proportions* (Dalton, 1810). If two elements combine to form more than one compound, the different masses of one element which combine with a fixed mass of the other are in the ratio of small integers.

(d) *The law of reciprocal proportions* (Berzelius, 1812). The ratio of the masses of two elements A and B which combine with a third element is the same as the ratio of the masses in which A and B themselves combine.

EVIDENCE FOR THE ATOMIC THEORY

6. Evidence from the law of combining volumes (Gay-Lussac, 1808).

(a) *The law of combining volumes.* This important piece of evidence, which at the time was rejected by Dalton as inconsistent with his atomic theory, states that when gases combine together at constant temperature and pressure they do so in volumes which bear a simple ratio to each other and to the volume of the gaseous product.

Since all gaseous elements were thought to be monatomic, this law, coupled with Berzelius' assumption that equal volumes of gases contained equal numbers of *atoms*, led to apparent contradictions of the atomic theory, *e.g.*

1 volume hydrogen + 1 volume chlorine = 2 volumes hydrogen chloride

∴ n atoms hydrogen + n atoms chlorine = $2n$ molecules hydrogen chloride

∴ 1 atom hydrogen + 1 atom chlorine = 2 molecules hydrogen chloride.

Since each molecule of hydrogen chloride must contain some hydrogen and some chlorine, the last step apparently involves splitting both the hydrogen and the chlorine atoms.

(b) *Avogadro's hypothesis* (first put forward in 1811). The anomaly was finally resolved in 1858 by Cannizaro's assumption that the common gaseous elements were diatomic and the consequent restatement of Avogadro's hypothesis, *i.e.* equal volumes of all gases at the same temperature and pressure contain equal numbers of *molecules*.

If the correct distinction is made between atoms and molecules both Gay-Lussac's law and Avogadro's hypothesis support the atomic concept.

7. Indirect evidence. Further indirect evidence was available in the form of:

(a) *Faraday's laws of electrolysis* (c. 1833):

(i) The mass of any substance liberated during electrolysis is proportional to the quantity of electricity passed.

(ii) The masses of different substances liberated by a given quantity of electricity are proportional to their molar masses divided by a small integer.

The consequences of these laws are discussed in detail in XII.

(b) *The period table* (first Mendeleef form, 1869). If the elements are arranged according to their atomic numbers (relative atomic masses in the earlier forms of the periodic table), systematic and regular variations in their chemical and physical properties become apparent.

8. Physico-chemical evidence. With the advent of more sophisticated physical techniques (about 1870 onwards) an increasing amount of more direct evidence became available from:

(a) *Diffraction techniques* (*see* also XIX, **24–30**). X-rays are electromagnetic radiation with wavelengths of the order of 10^{-1} nm and should therefore be capable of diffraction in much the same way as visible light. The observation of diffraction patterns on passage of X-rays through crystals can be explained only by assuming that the crystals are made up of regular arrays of particles with spacings of about 10^{-1} nm.

NOTE: The spacings perform the same function as the slits in the more familiar diffraction experiment with visible light.

Electron beams have similar wavelengths and, owing to their negative charge, can be focused by electrostatic and magnetic "lenses." They are utilised in electron microscopes, which have provided excellent photographs of some of the larger naturally occurring molecules such as viruses and fibres. Very recently electron microscope photographs have been taken of individual thorium atoms.

(b) *Mass spectrometry.* The principles and operation of the mass spectrograph, an instrument which separates a beam of particles into components of different mass, are discussed in **21** below. The results obtained can be unambiguously interpreted only if the real existence of atoms and molecules is assumed.

MODIFICATIONS OF THE ATOMIC THEORY

9. Sub-atomic physics (*see* also XV, 2–5). Early this century it became apparent that the atom itself could be divided into smaller sub-atomic particles. The atom is now known to consist essentially of a dense positively charged nucleus occupying about 10^{-15} of the total volume of the atom but accounting for almost all its mass, surrounded by sufficient negatively charged electrons to maintain electrical neutrality. The nucleus itself is composed of positively charged protons and neutral neutrons of approximately equal mass. The atomic number Z of an element represents both the number of protons and the number of extranuclear electrons, while the mass number A represents the total number of protons and neutrons.

In spite of the bewildering array of sub-atomic particles recently discovered (including the positron, neutrino and several types of meson and boson), these three fundamental particles are sufficient to describe most phenomena of chemical interest.

10. Isotopes (*see* also XV, 4–5). The mass of an atom of a given element need not be constant. Atoms of the same element must necessarily have the same number of protons in the nucleus, since this characterises the element, but the number of neutrons and hence the mass of the atom can vary. *Nuclides, i.e.* nuclear species—or, more loosely, atoms—with the same number of protons but different numbers of neutrons, are called *isotopes*. Naturally occuring oxygen consists of three isotopes $^{16}_{8}O$ (99·76%), $^{17}_{8}O$ (0·04%), $^{18}_{8}O$ (0·20%), where the superscript indicates the mass number, the subscript the atomic number and the figures in parentheses the relative abundances. The subscript may be omitted, since the symbol uniquely determines the element.

NOTE: ^{16}O may be written: oxygen-16 (*see* I, 3).

11. The equivalence of mass and energy. When atoms combine together to form a molecule the mass of the molecule is very slightly less than the sum of the masses of the constituent atoms. This is because the energy (E) required for the formation of the molecule is produced by the disappearance of an equivalent amount of mass (m). This equivalence of mass and energy is expressed by the relationship

$$E = mc_0{}^2$$

where c_0 is the velocity of light. For most chemical reactions E is less than 10^3 kJ mol^{-1} and hence m is less than 10^{-11} kg, an amount much too small to be detected. Thus the law of conservation of mass can be considered to hold for all chemical reactions.

12. Non-stoichiometric compounds. Compounds which contain discrete molecules invariably have a fixed composition and are known as *stoichiometric* compounds. Thus carbon monoxide always has a carbon to oxygen ratio of 1:1. This ratio can be changed only in integral units, *e.g.* 1:2, and a new compound is produced (CO_2 in this case). Many solid compounds do not contain discrete molecules and do not necessarily obey the law of definite proportions. Thus FeS never has exactly the composition indicated—sulphur is always present in excess. Formulae of the type Fe_6S_7 and $Fe_{11}S_{12}$ have been deduced by chemical analysis. This is an example of a *non-stoichiometric* compound and arises because some Fe atoms are absent from the crystal lattice (*see* XIX, **4**). Furthermore, the exact composition depends on the method of preparation.

The variations in atomic composition for non-stoichiometric compounds do not affect the chemical properties of the compound, although the physical properties may be modified.

RELATIVE ATOMIC AND MOLECULAR MASSES

The atomic weight of an element is based on a relative rather than an absolute scale, and the term "relative atomic mass," A_r, is preferred. Likewise, the term "relative molecular mass," M_r, is preferred to "molecular weight." For elements with more than one naturally occurring isotope, the relative atomic mass is a weighted average of the masses of the component isotopes.

13. Standards for relative atomic and molecular masses.

(a) *The hydrogen standard.* The earliest definition of the relative atomic mass of an element was the ratio of the mass of one atom of that element to the mass of the hydrogen atom, which was taken as exactly unity.

(b) *The oxygen standard.* Oxygen was soon accepted as a better standard, since more elements combine directly with oxygen than with hydrogen, and the oxides so formed are easily analysed. Also the oxygen standard depends on the measurement of mass, whereas the hydrogen standard normally involves the measurement of a volume of gas—a less convenient procedure.

Until recently there were two oxygen scales in use. The chemical standard was based on the mass of the naturally occurring oxygen atom, which was taken as exactly 16·000 00 units. On the other hand, the physical standard was based on the mass of the oxygen-16 isotope, which was taken as exactly 16·000 00 units. Since naturally occurring oxygen is a mixture of three isotopes, the two scales do not quite agree.

(c) *The carbon standard.* Both oxygen scales have now been abandoned and replaced by one based on the carbon-12 isotope. The accepted standard is now one-twelfth of the mass of the carbon-12 isotope, which is taken as exactly 12·000 00 units. This has necessitated only a small change in relative atomic masses based on the oxygen standard. Relative atomic and molecular masses are frequently determined mass spectroscopically, and carbon is a much more suitable standard for use with this technique (*see* **21** below).

14. The mole. The mole has been defined in I, 3 and is the *only* unit of amount of substance. The amount of substance in moles (often called the number of moles) is found by dividing the actual mass of the substance by the mass of one mole of the same substance (the molar mass, symbol M). In the SI system the mass units are kilogrammes. Because of the manner in which the mole is defined, the amount of substance may also be obtained by dividing the mass of the substance in grammes by the relative molecular mass.

EXAMPLE 1: Calculate the amount of (a) 100 kg of NaCl, (b) 10^{-2} g of H_2O, (c) 10^{-6} g of aluminium trichloride.

(a) Molar mass of NaCl = 0·058 44 kg mol^{-1}
amount of NaCl = 100/0·058 44 = 1·71 × 10^3 mol
$$= 1·71 \text{ k mol}$$
(this is the same as (mass in grammes)/(relative molecular mass) = 100 × 10^{-3}/58·44).

(b) Molar mass of H_2O = 0·018 kg mol^{-1}
amount of H_2O = $10^{-2} \times 10^{-3}/0\cdot018$ = $5\cdot56 \times 10^{-6}$ mol
$$= 5\cdot56 \ \mu \text{ mol}$$

(c) Aluminium trichloride has the formula $AlCl_3$ in the solid but Al_2Cl_6 in the vapour.

If $AlCl_3$, molar mass = 0·133 kg mol^{-1}
amount of $AlCl_3$ = $10^{-6} \times 10^{-3}/0\cdot133$ = $7\cdot52 \times 10^{-9}$ mol
$$= 7\cdot52 \text{ n mol}$$

If Al_2Cl_6, molar mass = 0·266 kg mol^{-1}
amount of Al_2Cl_6 = $10^{-6} \times 10^{-3}/0\cdot266$ = $3\cdot76 \times 10^{-9}$ mol
$$= 3\cdot76 \text{ n mol}$$

The last example stresses the importance of stating the formula of the compound if this is in doubt.

15. The determination of relative atomic and molecular masses. It is convenient to consider the determination of relative atomic and molecular masses as loosely divided into approximate and accurate methods, though the distinction is somewhat arbitrary and there is considerable overlap between the two descriptions.

Under the heading of approximate determinations we consider chemical (gravimetric) methods, methods based on simple vapour density determinations and colligative methods.

Accurate determinations include vapour density methods coupled with the method of limiting densities or limiting pressures, X-ray methods, mass spectroscopy and microwave spectroscopy.

Although mass spectroscopy is now of paramount importance in the determination of relative atomic and molecular masses, it should not be thought that the other methods are of solely historical interest. In particular, the osmotic pressure method is one of the most important ways of measuring the relative molecular masses of polymers. Other colligative properties provide information on relative molecular masses in solution, which may be different from those measured by the mass spectrometer, while gravimetric and vapour density methods form the basis of most common analytical techniques.

APPROXIMATE METHODS FOR RELATIVE ATOMIC AND MOLECULAR MASSES

16. Chemical methods for relative atomic masses (based on oxygen standard). Such methods are implicitly based on the gravimetric laws (*see* **5** above) and normally involve use of the relationship:

relative atomic mass = equivalent mass × valency

. . . (III, 1)

(*a*) *Determination of equivalent mass.* The equivalent mass (the number of parts by mass of an element which combine with eight parts by mass of oxygen) may be found quantitatively by preparation of the oxide (either directly from the element or via the nitrate), by reduction of the oxide, by displacement of hydrogen from a dilute acid or alkali, by synthesis of the chloride, or by electrolysis using Faraday's laws (*see* **7** (*a*) above).

(*b*) *Determination of valency.* For the present purpose, the definition of valency implied by (III, 1) is sufficient (*i.e.* the ratio of the relative atomic mass to the equivalent mass) with the rider that the valency must be an integer (*see* XVI, **3** for a more rigorous definition).

(*i*) *Dulong and Petit's method.* Knowing the equivalent mass, the valency can be determined by finding an approximate value of the relative atomic mass from the *law of Dulong and Petit, i.e.* for solid elements,

relative atomic mass × molar heat capacity

$$\approx 27 \text{ J K}^{-1} \text{mol}^{-1}$$

A more accurate value of the relative atomic mass can then be obtained from (III, 1).

(*ii*) *Misterlich's method.* There are many exceptions to the law of Dulong and Petit, and an alternative method for determining valency utilises *Misterlich's law of isomorphism, i.e.* isomorphous compounds have the same type of chemical formula. Thus chromic oxide, ferric oxide and aluminium oxide are all isomorphous and the valency of the metal atom is two in each case.

NOTE: Isomorphous compounds have similar crystal shapes and form overgrowths and mixed crystals with each other.

17. Vapour density and eudiometric methods for relative molecular masses (usually based on hydrogen standard). The relationship between vapour density and relative molecular mass is derived as follows:

$$\text{vapour density of A} = \frac{\text{mass of a certain volume of A}}{\text{mass of the same volume of hydrogen}}$$

Introducing Avogadro's hypothesis that equal volumes of all gases at the same temperature and pressure contain the same number of molecules, this expression simplifies to

$$\text{vapour density of A} = \frac{\text{mass of one molecule of A}}{\text{mass of one molecule of hydrogen}}$$

$$= \frac{\text{mass of one mole of A}}{\text{mass of one mole of hydrogen}}$$

$$= \frac{\text{molar mass of A (kg mol}^{-1})}{0.002 \text{ (kg mol}^{-1})}$$

$$= \frac{\text{molar mass of A (g mol}^{-1})}{2 \text{ (g mol}^{-1})}$$

$$\ldots \text{(III, 2)}$$

The molar mass in g mol^{-1} is numerically equal to the relative molecular mass. Of the numerous methods for determining approximate vapour densities of easily vaporisable liquids the best known are those due to *Victor Meyer* and to *Hofmann*, in which a known mass of liquid is converted into vapour, the volume of which is measured, and that due to *Dumas*, in which a known volume of vapour is weighed. Methods for relative atomic and molecular masses based on these determinations are:

(a) *Cannizaro's method*. The relative molecular masses of a large number of compounds of the element in question are found, and each is analysed to find the mass of the element contained in one mole of the compound. The assumption is then made that, if sufficient compounds have been considered, at least one should contain only one atom of the element. In such a case the measured mass occurring in one mole of the compound is the molar mass of the element.

(b) *Eudiometry*. Vapour density methods are often useful in the analysis of organic mixtures by eudiometry and in predicting degrees of association and dissociation of vapours, as illustrated in the Progress Tests.

18. Colligative methods for relative molar masses in solution (*see* also X, 4–11). Such methods are useful for substances which vaporise only at very high temperatures or which decompose before reaching their boiling points. When a non-volatile substance is dissolved in a solvent the solution exhibits the following properties: the vapour pressure and freezing point are lowered relative to the pure solvent, the boiling point is raised, and the solution exerts an osmotic pressure when separated from the pure solvent by a semi-permeable membrane. These four colligative properties are very closely associated, and depend on the number of solute particles in a given amount of solvent rather than on the nature of the solute. The theory of colligative properties is discussed in X, 4–11, and the results are quoted here merely to show their application to the determination of relative molecular masses.

(a) *Vapour pressure lowering*

$$\Delta p = p_1^{\bullet} x_2 \qquad \qquad . \; . \; . \; \text{(III, 3)}$$

where Δp and $p_1^{\bullet}$, the decrease in vapour pressure and the vapour pressure of the pure solvent, are measured, and x_2, the mole fraction of solute, is calculated.

$$x_2 = n_2/(n_1 + n_2) \approx n_2/n_1 \text{ (if the solution is dilute)}$$

where n_1 and n_2 are the amounts of solvent and solute respectively. Since $n = m/M$, the molar mass and hence the relative molecular mass of the solute may be found if the masses of the solvent and solute are known together with the molar mass of the solvent.

(b) *Freezing point depression and boiling point elevation.* For dilute solutions

$$\Delta T = K' x_2 \qquad \qquad . \; . \; . \; \text{(III, 4)}$$

where ΔT is the freezing point depression or boiling point elevation and K' is a constant specific to the solvent used. x_2 may again be calculated and hence M_r determined.

(c) *Osmotic pressure.* For dilute solutions

$$\pi V = RT x_2 \qquad \qquad . \; . \; . \; \text{(III, 5)}$$

where π is the osmotic pressure, V the volume of one mole of the solvent, T the temperature and R is a constant whose value is very close to that of the universal gas constant.

ACCURATE METHODS FOR RELATIVE ATOMIC AND MOLECULAR MASSES

19. Vapour density methods.

(*a*) *Regnault's method*. A dry evacuated quartz globe of known volume is weighed, filled with pure dry test gas at a known pressure and reweighed. Buoyancy corrections, necessitated by changes in external conditions during the experiment, are minimised by counterpoising with a similar globe filled with air. Corrections for departures from ideality are made, since the ideal gas equation (*see* IV, **6, 8**) in the form

$$M = \rho RT/p$$

provides only an approximate description of the behaviour of real gases around atmospheric pressure.

The density, ρ, is measured at a series of pressures below atmospheric pressure, the ratio (ρ/p) is plotted against the pressure, and the graph is linearly extrapolated to give the ratio $(\rho/p)_0$ at zero pressure. The accurate molar mass M, and hence the relative molecular mass, can then be calculated from the equation

$$M = (\rho/p)_0 RT \qquad \text{. . . (III, 6)}$$

(*b*) *The buoyancy balance*. In this method the pressures at which two gases have the same density are compared. The test gas is admitted into a vessel containing a light evacuated quartz globe attached to a pointer which can swing about a fulcrum. The pressure is adjusted to bring the pointer to some arbitrary zero position and the procedure is then repeated using a gas of accurately known relative molecular mass. The ideal gas equation is used in the form

$$p = \rho RT/M$$

so that, for two gases of relative molecular masses $(M_r)_1$ and $(M_r)_2$ which have equal density at pressures p_1 and p_2,

$$p_1/p_2 = M_2/M_1 = (M_r)_2/(M_r)_1 \quad \text{. . . (III, 7)}$$

A correction for departure from ideality is achieved by determining the ratio $(p_1/p_2)_0$ by linear extrapolation to zero pressure of a plot of (p_1/p_2) against p_1.

20. X-ray methods.
X-ray crystallography is considered in detail in XIX, **24–30**. For the present it may be noted that

this technique can be used to measure with extremely high precision the volume per molecule in a crystal. If this volume is multiplied by Avogadro's constant, the molar volume, V_m, is obtained. The density of the crystal is determined by standard methods and the molar mass M (and hence the relative molar mass, M_r) obtained from the relationship

$$\rho = M/V_m$$

21. The mass spectrometer. In the mass spectrometer, shown schematically in Fig. 6, ionised particles are accelerated through a magnetic field in order to give them a circular trajectory. The heavier the ion, the wider is the arc traversed, so that ions of different mass can be separated.

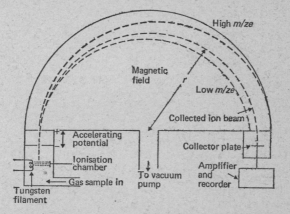

Fig. 6.—*Schematic diagram of mass spectrometer. The poles of the magnet are above and below the plane of the paper*

Gas is admitted continuously into the chamber, where it is ionised by a stream of accelerated electrons from a hot tungsten wire. The pressure is maintained at about 10^{-5} mmHg (1 mN m^{-2}) so that the ions formed are unlikely to collide with each other before reaching the collecting plate. After acceleration through a potential applied across a pair of parallel slits, the ions pass through a magnetic field at right angles to their path and are thus made to take a circular trajectory. As a result of the accelerating potential V, a positive ion of charge ze acquires kinetic energy given by

$$zeV = mv^2/2 \qquad \ldots \text{(III, 8)}$$

In the magnetic field of strength B, the magnetic force on the ion, $Bzev$, exactly balances the centrifugal force, mv^2/r, i.e.

$$mv^2/r = Bzev \qquad \ldots \text{(III, 9)}$$

Eliminating v between (III, 8) and (III, 9) gives

$$r = (2Vm/B^2ze)^{\frac{1}{2}} \qquad \ldots \text{(III, 10)}$$

Thus, for a given accelerating potential and magnetic field strength, the radius of the circular path is determined by the mass to charge ratio of the ions. Rearranging (III, 10) gives

$$m/ze = B^2r^2/2V$$

By adjusting either the accelerating potential or the magnetic field strength, ions with any desired m/ze can be made to fall on the collector plate. Each ion reaching the plate sets up a minute electric current, which is amplified and fed to a recorder which plots ion current against accelerating potential or magnetic field strength.

Most ions leaving the ionisation chamber will be singly charged, and under these conditions ions of increasing mass can be focused on the collector plate and recorded. The relative atomic or molecular mass may then be found by comparison with the carbon-12 ion. Complex molecules break down under the ionising electron beam and yield a characteristic *cracking pattern* from which the original molecule may be identified.

Since the mass spectrometer detects individual ions, the masses and percentage abundances of the various isotopes of an element can be measured directly. These can then be compared with the relative atomic masses as determined some other way. The mass spectrometer can measure to a high degree of precision, accuracies of 1 part in 10^6 being common.

22. Microwave spectroscopy. Although a consideration of microwave spectroscopy is beyond the scope of this book, it should be mentioned that this technique is capable of providing relative atomic and molecular masses to better than 1 part in 10^5.

CHEMICAL FORMULAE AND EQUATIONS

23. Chemical formulae. A system of symbols to represent elements has been developed over the years. Such symbols, either separately or combined with others, may (*a*) represent one atom or molecule of a substance, (*b*) represent one mole of a substance, or (*c*) merely be an abbreviation for the name of a substance.

Substances such as sodium chloride and ferrous sulphide do not contain discrete molecules, and the formulae NaCl and FeS express only the *relative* numbers of atoms of each element present in the substance. Even this may not be strictly true; it will be remembered that ferrous sulphide was used as an example of a non-stoichiometric compound and that FeS represents its approximate formula.

Other substances such as chlorine, Cl_2, methane, CH_4, etc., consist of discrete molecules, and the *molecular* formula gives not only the relative numbers of atoms present but also the total number of atoms in the molecule.

Structural formulae which give a two-dimensional representation of the arrangement of atoms in molecules are discussed in XVI, 1.

24. The balanced chemical equation. A chemical equation is the shorthand representation of a chemical reaction. Thus the reaction between solutions of silver nitrate and calcium chloride may be represented by

$$AgNO_3 + CaCl_2 = AgCl(s) + Ca(NO_3)_2 \quad . \quad . \quad . \quad (III, 11)$$

The substances on the left-hand side of the equation are the *reactants* and those on the right-hand side the *products*. As it stands, this equation is not *balanced*, *i.e.* the numbers of atoms of a given kind are not the same on both sides of the equation. The balanced equation is

$$2AgNO_3 + CaCl_2 = 2AgCl(s) + Ca(NO_3)_2 \quad . \quad . \quad . \quad (III, 12)$$

The numbers in front of the formulae are the *stoichiometric coefficients* (the number 1 is omitted) and represent the relative number of molecules or moles which participate in the chemical reaction. The stoichiometric coefficients need not be integers when they represent relative numbers of moles, and the synthesis of ammonia may be represented as

$$\tfrac{1}{2}N_2(g) + \tfrac{3}{2}H_2(g) = NH_3(g) \quad . \; . \; . \; (III, 13)$$

or as
$$N_2(g) + 3H_2(g) = 2NH_3(g) \quad . \; . \; . \; (III, 14)$$

Equations (III, 13) and (III, 14) both indicate that three times as many hydrogen molecules as nitrogen molecules are used up in the reaction. A balanced chemical equation is necessitated by the law of conservation of mass. In addition, charge must also be conserved. Thus equation (III, 12) may be written in terms of ions, *i.e.*

$$2Ag^+ + 2NO_3^- + Ca^{2+} + 2Cl^-$$
$$= 2AgCl(s) + Ca^{2+} + 2NO_3^-$$

Since the calcium and nitrate ions are common to both sides of the equation, an abbreviated form is

$$Ag^+ + Cl^- = AgCl(s)$$

where the stoichiometric coefficient 2 has been cancelled out.

25. Mass relationships in chemical equations. Equation (III, 12) indicates that two molecules of silver nitrate react with one molecule of calcium chloride to produce two molecules of silver chloride and one molecule of calcium nitrate. The term mole may be exchanged for molecule and moles may then be converted into masses. Thus 2 mol ($2 \times 0.169\,87$ kg) of silver nitrate react with 1 mol ($0.110\,98$ kg) of calcium chloride to produce 2 mol ($2 \times 0.143\,32$ kg) of silver chloride and 1 mol ($0.164\,90$ kg) of calcium nitrate. Knowing the mass relationship, it is possible to calculate that, for example, 10 g of silver chloride could be produced from ($2 \times 0.169\,87/2 \times 0.143\,32$) $\times 10 = 11.83$ g of silver nitrate.

26. Other symbols in chemical equations. The letters s, l, and g in parentheses after a formula specify the solid, liquid and gaseous states. If they are absent, then the liquid state is understood. For reactions in aqueous solution, the symbol aq indicates that the substance concerned is associated with a very large amount of water and the solution is effectively infinitely dilute, *e.g.* $CuSO_4(aq)$ means an infinitely dilute solution of copper sulphate.

The reactants and products in a chemical reaction may be separated by signs other than $=$ when some special feature of the reaction is to be emphasised. Thus a single arrow $\longrightarrow$ indicates that the reaction proceeds almost entirely from left to right, while a double arrow $\rightleftharpoons$ indicates that the reaction is in equilibrium. These signs will be encountered frequently in later chapters.

PROGRESS TEST 3

1. Show that the following results are in agreement with the law of definite proportions:

(a) On heating in oxygen, 0·560 g of silver gave 0·602 g of silver oxide.

(b) 1·06 g of silver were converted to the nitrate which on ignition gave 1·14 g of the same oxide. (5)

2. Two compounds of element A with element B contain 32·2% A and 49·0% A respectively. Show that these data are in agreement with the law of multiple proportions. (5)

3. The equivalent mass of lead is 103·5 and its specific heat is 0·130 $Jk^{-1} g^{-1}$. Calculate the relative atomic mass. (16)

4. A mixture of methane, ethylene and acetylene has a vapour density of 11·3 kg m^{-3}. When 10 cm^3 of this mixture and 30 cm^3 of oxygen were sparked together over aqueous caustic potash, the volume contracted to 5·5 cm^3 and then disappeared when pyrogallol was introduced. All volumes were measured under the same conditions of temperature, pressure and humidity. Calculate the composition of the original mixture. (17)

5. The vapour pressure of water is 30 mmHg at 25°C. How much sodium chloride must be dissolved in 1 kg of water to lower the vapour pressure to 29 mmHg? What is the osmotic pressure of the resulting solution? (18)

6. Use the following data to find the relative molecular mass of carbon dioxide: $p = 1$ atm, $\rho = 1·981$ kg m^{-3}; $p = 0·7$ atm, $\rho = 1·382$ kg m^{-3}; $p = 0·4$ atm, $\rho = 0·788$ kg m^{-3}; $p = 0·2$ atm, $\rho = 0·393$ kg m^{-3}. The relative molecular mass and $(\rho/p)_0$ for oxygen at the same temperature are 31·998 and 1·428 kg m^{-3} atm^{-1} respectively. (19)

7. Define what is meant by: (a) an element (3), (b) a compound (3), (c) isotopes (10), (d) non-stoichiometric compounds (12), (e) a balanced chemical equation (24).

8. List the postulates in Dalton's atomic theory. (4)

9. State the law of (a) definite proportions (5), (b) multiple proportions (5), (c) reciprocal proportions (5), (d) combining volumes (6).

10. Discuss the recent modifications to Dalton's atomic theory. (9–12)

11. What is the present standard of relative atomic and molecular masses? (14)

12. List the methods which may be used for the determination of approximate relative atomic and molecular masses. (16–18)

13. Describe the mass spectrometer. (21)

CHAPTER IV

THE PROPERTIES OF IDEAL GASES

1. The gaseous state. The gaseous state is the simplest state of matter. Since intermolecular separations are large (relative to the size of the molecules), the molecules are essentially independent of each other and it is possible to describe the behaviour of the gas without referring to the nature of the individual molecules. Gases which conform perfectly to this pattern are called *ideal gases* and can all be described by a single *equation of state* showing the mathematical relationship between the values of the properties (pressure, temperature, volume, etc.) of the gas. The laws governing the behaviour of such gases (Boyle's, Charles', Avogadro's, Graham's and Dalton's), their heat capacities and the phenomenon of the Brownian motion can be elegantly explained using Maxwell's kinetic theory. Since this theory is essentially molecular in nature, powerful evidence for the existence of atoms and molecules is thereby provided.

In real gases neither molecular size nor intermolecular forces can be ignored, and insight into these factors is gained by consideration of the modifications to the kinetic theory necessary to explain such properties as liquefaction, viscosity and thermal conductivity. It is obvious that real gases should tend to become ideal as intermolecular effects are minimised, *i.e.* at high temperatures and low pressures.

IDEAL GAS LAWS

2. Boyle's law (1662). The volume of a given mass of gas at constant temperature is inversely proportional to its pressure. Algebraically

$$V \propto 1/p \text{ or } pV = K \qquad . \quad . \quad . \quad \text{(IV, 1)}$$

where p is pressure of the gas, V the volume and K a constant. The numerical value of K depends on the value of the constant temperature, the mass of gas taken and the units of p and V.

Boyle's law may also be expressed in the symmetrical form

$$p_1 V_1 = p_2 V_2$$

3. Law of Charles (1787) and Gay-Lussac (1802).

Following an earlier observation by Charles, Gay-Lussac found that at constant pressure the volume of a given mass of gas (V) increased by about $1/273$ of its volume at $0°C$ (V_0) for each degree Celsius rise in temperature. A more recent value of this fraction is $1/273 \cdot 15$. Thus

$$V = V_0 + \frac{V_0}{273 \cdot 15}t = V_0\left(\frac{273 \cdot 15 + t}{273 \cdot 15}\right)$$

or

$$\frac{V}{V_0} = \frac{273 \cdot 15 + t}{273 \cdot 15} \qquad \cdots \text{(IV, 2)}$$

This equation suggests the basis for a scale of temperature, the absolute scale of temperature (symbol T, unit K), which is defined such that

$$T(\text{K}) = 273 \cdot 15 + t \ (°\text{C})$$

This absolute scale of temperature is identical with the theoretical thermodynamic scale of temperature, which may be formulated from the second law of thermodynamics (*see* VIII). Since $273 \cdot 15 \text{ K} = 0°C$, (IV, 2) may be written

$$V/V_0 = T/T_0$$

and in general $V_1/V_2 = T_1/T_2$; $V_2/V_3 = T_2/T_3$, etc.

Hence $\qquad V/T = \text{constant}$

The law of Charles and Gay-Lussac may be stated as follows: the volume of a given mass of gas at constant pressure is directly proportional to the absolute temperature, *i.e.* $V = K'T$.

4. The absolute zero of temperature.

The above law indicates that the volume of any gas will be zero at 0 K, and that at still lower temperatures the volume would become negative. Clearly, this is impossible and 0 K is therefore the lowest temperature attainable. The law of Charles and Gay-Lussac cannot be tested experimentally at low temperatures, since all gases condense as the temperature approaches 0 K;

however, theoretical considerations indicate that 0 K is the lowest conceivable temperature. Temperatures of the order of 10^{-4} K have been reached in specially designed apparatus.

5. Avogadro's hypothesis.

Avogadro's hypothesis, which was given in III, 6, may be stated in a different way. The volume of a gas (V) is directly proportional to the amount of gas (n) present at constant temperature and pressure, i.e. $V \propto n$. The volume occupied by one mole of an ideal gas at standard temperature and pressure (273·15 K and 101·325 kN m^{-2} (one atmosphere)) is 22·414 dm^3.

6. The ideal gas law.

Boyle's law, Charles and Gay-Lussac's law and Avogadro's hypothesis may be combined to give the mathematical relationship between p, V, T and n for an ideal gas.

Boyle's law: $V \propto 1/p$ at constant T and n
Charles and Gay-Lussac's law: $V \propto T$ at constant p and n
Avogadro's hypothesis: $V \propto n$ at constant T and p

Hence
$$V \propto (1/p)(T)(n)$$
i.e.
$$V \propto nT/p \quad \text{or} \quad V = RnT/p \quad . . . \text{(IV, 3)}$$

where R is the constant of proportionality and is known as the *universal gas constant*.

Equation (IV, 3) is an equation of state known as the *ideal gas law* and is usually written in the form

$$pV = nRT \qquad . . . \text{(IV, 4)}$$

NOTE: n, the amount of substance, is often called the number of moles, since the only unit of amount of substance is the mole.

R may readily be evaluated from the information that at 273·15 K and 101·325 × 10^3 N m^{-2} one mole of an ideal gas occupies 0·022 414 m^3.

$$R = pV/nT = (101{\cdot}325 \times 10^3 \times 0{\cdot}022\ 414)/(1 \times 273{\cdot}15)$$
$$= 8{\cdot}314 \text{ J K}^{-1} \text{ mol}^{-1}$$

The ideal gas law may also be expressed in the symmetrical form

$$p_1 V_1/T_1 = p_2 V_2/T_2 \qquad . . . \text{(IV, 5)}$$

where p_1, p_2 and V_1, V_2 are the pressures and volumes of a given mass of gas at temperatures T_1 and T_2.

For some purposes it is convenient to work in terms of numbers of molecules rather than moles. This is achieved by dividing R by the Avogadro constant, L (the number of molecules per mole), to obtain k, the *Boltzmann constant*. The ideal gas law is then expressed as

$$pV = NkT \qquad \cdot \cdot \cdot \text{(IV, 6)}$$

where N refers to the number of molecules present and k is effectively the gas constant per molecule.

EXAMPLE 1: Calculate the mass of hydrogen gas which will exert a pressure of $0 \cdot 15$ atm in a one-litre flask at $27°C$.

Using the ideal gas law in the form

$$n = m/M = pV/RT$$

where m is the mass of gas and M is the molar mass

$$m = MpV/RT$$

Converting to SI units:

$M = 2 \text{ g mol}^{-1} = 2 \times 10^{-3} \text{ kg mol}^{-1}$; $V = 1 \text{ litre} = 10^{-3} \text{ m}^3$;
$p = 0 \cdot 15 \text{ atm} = 20 \cdot 265 \text{ kN m}^{-2}$
$\therefore m = (2 \times 10^{-3} \times 20 \cdot 265 \times 10^3 \times 10^{-3})/(8 \cdot 314 \times 300)$
$\quad = 1 \cdot 6 \times 10^{-5} \text{ kg}$

EXAMPLE 2: If a given mass of gas occupies $1 \cdot 6 \text{ dm}^3$ at $27°C$ and $70 \cdot 928 \text{ kN m}^{-2}$ pressure, find the volume occupied at $87°C$ and $131 \cdot 59 \text{ kN m}^{-2}$ pressure.

From (IV, 5) $(70 \cdot 928 \times 1 \cdot 6)/300 = (131 \cdot 59 \times V_2)/360$
From which $\qquad\qquad V_2 = 1 \cdot 12 \text{ dm}^3$

7. Dalton's law of partial pressures (1801).

The total pressure in a mixture of gases is equal to the sum of the partial pressures of the constituents. The partial pressure of each constituent is defined as the pressure which this constituent would exert if it alone occupied a volume equal to that of the mixture.

Algebraically, for a mixture of two ideal gases exerting partial pressures p_1 and p_2, the total pressure p is given by

$$p = p_1 + p_2 = (n_1 + n_2)RT/V$$

or more generally $p = \sum_i p_i = (RT/V)\sum_i n_i \qquad \cdot \cdot \cdot \text{(IV, 7)}$

where i is the index identifying each constituent in the mixture.

A useful corollary of this law can be obtained by combining the expressions

$$p = (RT/V)\sum_i n_i \text{ and } p_1 = n_1 RT/V$$

$$p_1/p = n_1/\sum_i n_i$$

$$\therefore \; p_1 = p(n_1/\sum_i n_i)$$

The quantity in parentheses is called the *mole fraction*, x_1, of component 1, for the obvious reason that it indicates what fraction component 1 is of the total number of moles. In a similar fashion $p_2 = px_2 = p(n_2/\sum_i n_i)$, $p_3 = px_3 = p(n_3/\sum_i n_i)$, etc.

EXAMPLE 3: A 5 dm³ bulb at 27°C contains 1 g of hydrogen and 1 g of helium. Calculate the pressure of the mixture and the partial pressures of the constituents.

Amount of hydrogen = $n_1 = \frac{1}{2}$ = 0·5 mol
Amount of helium = $n_2 = \frac{1}{4}$ = 0·25 mol
$$\therefore \; \sum_i n_i = n_1 + n_2 = 0\cdot75 \text{ mol}$$

Using Dalton's law in the form

$$p = (RT/V)\sum_i n_i$$

$$p = (8\cdot314 \times 10^{-3} \times 300/5 \times 10^{-3})(0\cdot75) = 374\cdot130 \text{ kN m}^{-2}$$

Using Dalton's law in the form

$$p_1 = p(n_1/\sum_i n_i)$$

$$p_1 = 374\cdot130(0\cdot5/0\cdot75) = 249\cdot420 \text{ kN m}^{-2}$$
$$p_2 = 374\cdot130(0\cdot25/0\cdot75) = 124\cdot710 \text{ kN m}^{-2}$$

8. Graham's law of diffusion (1833).

The rate of diffusion of a gas through a porous barrier is inversely proportional to the square root of its density. Algebraically

$$R \propto \rho^{-\frac{1}{2}}$$

where R is the rate of diffusion of a gas of density ρ.

The ideal gas law may be written in the form

$$pV = mRT/M$$

Since $\rho = m/V$, the above equation becomes

$$M = \rho RT/p$$

Thus the density of an ideal gas is directly proportional to its molar mass and therefore to its relative molecular mass.

Graham's law of diffusion may be written

$$R \propto M_r^{-\frac{1}{2}}$$

and for two gases 1 and 2,

$$R_1/R_2 = \{(M_r)_2/(M_r)_1\}^{\frac{1}{2}} \quad . \quad . \quad . \quad \text{(IV, 8)}$$

Both these expressions indicate that light gases diffuse faster than heavier ones. Thus if an equimolar mixture of 1 and 2 is allowed to diffuse through a porous barrier, the sample initially collected will be enriched in the lighter component by a factor $\{(M_r)_2/(M_r)_1\}^{\frac{1}{2}}$.

KINETIC THEORY OF IDEAL GASES

In order to explain the above behaviour it is necessary to set up a "model" system, *i.e.* a mathematical model which simulates the physical processes occurring in the gas. The properties of the model are then calculated and compared with experimental observation. Any lack of agreement indicates that one or more of the assumptions made in setting up the model must be modified.

9. Assumptions of the kinetic theory model. The assumptions made in setting up the kinetic theory model are that:

(*a*) The gas consists of identical molecules which are considered as non-interacting point masses, so that intermolecular forces and intermolecular collisions can be ignored.

(*b*) Molecules are in constant random motion and the associated translational kinetic energy is determined only by the absolute temperature.

(*c*) Collisions with the walls of the container are perfectly elastic, *i.e.* involve no loss of translational kinetic energy.

(*d*) The pressure on the walls of the container is due solely to their bombardment by the gas molecules.

10. Calculation of the pressure. Using simple classical mechanics we can calculate the pressure exerted by our model gas in terms of the mean square speed of its molecules.

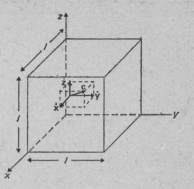

FIG. 7.—*Velocity components in the kinetic theory of gases*

We consider a single gas molecule of mass m in a cube of side l (Fig. 7). The velocity c in any direction can be resolved into components along the x, y and z axes, since

$$c = \dot{x} + \dot{y} + \dot{z}$$

and each component can be considered separately (*see* II, **27**).

NOTE: $\dot{x} = dx/dt$, *i.e.* the velocity component in the x direction.

The magnitude of the velocity, *i.e.* the speed, is given by

$$c^2 = \dot{x}^2 + \dot{y}^2 + \dot{z}^2$$

Since there are no intermolecular collisions, a molecule travelling along the x axis will continue along this path with momentum $m\dot{x}$ until it collides with the wall perpendicular to the x axis. The collision is perfectly elastic so that though the direction of the velocity component is reversed its magnitude is unchanged.

New momentum of molecule $= -m\dot{x}$

$\therefore$ Change of momentum of molecule $=$ final momentum $-$
   initial momentum $= -m\dot{x} - m\dot{x} = -2m\dot{x}$

The change of momentum of the wall on collision will be the reverse of this, *i.e.* change of momentum of wall $= 2m\dot{x}$.

As will be shown in **11** below, it is important to use molecular speeds rather than velocities. This will not affect the magnitude of the momentum change of the wall $= 2m\dot{x}$.

On average the molecule will hit a wall each time it travels a distance l, *i.e.* every $l/\dot{x}$ seconds. Thus the number of impacts is $\dot{x}/l$ per second.

Now, force exerted on wall $F'_x =$ rate of change of momentum of wall

$\qquad\qquad\qquad\qquad = $ (change of momentum of wall on impact) $\times$ (number of impacts per second)

$$\therefore F'_x = (2m\dot{x})(\dot{x}/l) = 2m\dot{x}^2/l$$

The gas as a whole will contain many molecules, all with different velocities (N_1 with velocity component $\dot{x}_1$, N_2 with velocity component $\dot{x}_2$, etc.) and using assumptions **9** (*a*) and **9** (*c*) we can generalise the above expression:

$$F_x = (2m/l)\sum_i N_i \dot{x}_i^2$$

where N_i is the number of molecules with velocity component $\dot{x}_i$. Similarly, the forces exerted on the walls perpendicular to the y and z axes will be

$$F_y = (2m/l)\sum_i N_i \dot{y}_i^2 \text{ and } F_z = (2m/l)\sum_i N_i \dot{z}_i^2$$

$$\therefore \text{ Total force, } F = (2m/l)\sum_i (N_i\dot{x}_i^2 + N_i\dot{y}_i^2 + N_i\dot{z}_i^2)$$

$$= (2m/l)\sum_i N_i c_i^2 \qquad \ldots \text{ (IV, 9)}$$

To simplify this expression we note that here the mean square speed for one mole is given by

$$\overline{c^2} = (\sum_i N_i c_i^2)/L$$

Thus, multiplying top and bottom of (IV, 9) by Avogadro's constant

$$F = \frac{2mL}{l}\frac{\sum_i N_i c_i^2}{L} = \frac{2mL\overline{c^2}}{l} \qquad \ldots \text{ (IV, 10)}$$

Now, pressure = force/area where the area is the *total* area over which the force is applied = $6l^2$

$$\therefore p = \frac{2m L\overline{c^2}}{l} \times \frac{1}{6l^2} = \frac{m L\overline{c^2}}{3l^3} = \frac{m L\overline{c^2}}{3V}$$

$$\therefore pV = \tfrac{1}{3} m L\overline{c^2} = \tfrac{1}{3} M\overline{c^2} = \tfrac{2}{3}(\tfrac{1}{2} M\overline{c^2}) = \tfrac{2}{3} E_t \ . \ . \ . \ (IV, 11)$$

where M is the molar mass (and not the relative molecular mass) and E_t is the translational energy per mole.

11. Speed and velocity. Although the use of component velocities was essential in the derivation of (IV, 11), it is preferable to use speeds in all subsequent discussions. This is because the average velocity of all the gas molecules will be zero unless the gas is moving as a whole, whereas the average speed will certainly not be zero. The use of speeds instead of velocities will not affect the *magnitude* of pV in (IV, 11).

12. Prediction of the behaviour of ideal gases. The behaviour of ideal gases may be predicted from (IV, 11) as follows:

(*a*) At constant temperature E_t and $\overline{c^2}$ are constant — assumption 9 (*b*).

$$\therefore pV = \text{constant} \ . \ . \ . \ \textit{Boyle's law}$$

(*b*) At constant pressure E_t and $\overline{c^2}$ are determined only by the temperature — assumption 9 (*b*)

$$\therefore V \propto T \ . \ . \ . \ \textit{Charles' law}$$

(*c*) Since E_t is determined only by the temperature — assumption 9 (*b*)

$$2E_t/3 \propto T$$
$$\therefore pV \propto T \ . \ . \ . \ \textit{ideal gas law}$$

(*d*) Dalton's law of partial pressures is implicit in assumptions 9 (*a*) and 9 (*b*).

(*e*) For two gases, 1 and 2, at the same pressure, volume and temperature

$$(E_t)_1 = (E_t)_2 \ . \ . \ . \ \text{assumption 9 (}b\text{)}$$
$$M_1\overline{c^2}_1/2 = M_2\overline{c^2}_2/2$$
$$\therefore (\overline{c_1^2})^{\frac{1}{2}}/(\overline{c_2^2})^{\frac{1}{2}} = (M_2/M_1)^{\frac{1}{2}} \ . \ . \ . \ (IV, 13)$$

Since it is reasonable to suppose that the rate of diffusion is proportional to the root mean square speed of the gas molecules, and also the molar mass is proportional to the relative moleular mass, (IV, 12) may be rewritten

$$R_1/R_2 = \{(M_r)_2/(M_r)_1\}^{\frac{1}{2}} \quad . \quad . \quad . \quad Graham's\ law$$

DISTRIBUTION OF MOLECULAR SPEEDS

In deriving the kinetic theory equations it was assumed that different molecules in a gas would have different speeds. This is because, even if at some initial time all the molecules had the same speed, irregularities in the surface of the container and collisional exchange of speed would soon lead to a random distribution.

13. The Maxwell distribution. The statistical analysis of the speed distribution was first carried out by Maxwell (1860), who calculated the fraction of molecules dn/n with speeds in a narrow range between c and $c + dc$ as

$$\frac{dn}{n} = 4\pi \left(\frac{m}{2\pi kT}\right)^{\frac{3}{2}} \exp(-mc^2/2kT)c^2 dc \quad . \quad . \quad . \quad (IV, 14)$$

NOTE: In terms of molar masses this becomes

$$\frac{dn}{n} = 4\pi \left(\frac{M}{2\pi RT}\right)^{\frac{3}{2}} \exp(-Mc^2/2RT)\,c^2 dc$$

Apart from constants, (IV, 14) can be considered as the product of two factors: $\exp(-mc^2/2kT)$ and $4\pi c^2 dc$.

(a) $exp(-mc^2/2kT)$ factor. $mc^2/2$ is the translational kinetic energy, E_t, of a molecule of mass m and speed c, so that the exponential factor can be written as $\exp(-E_t/kT)$. This is an example of the Boltzmann factor, the importance of which cannot be too strongly emphasised, since for all systems of chemical interest the fraction of molecules dn/n with energy E (any type of energy) is proportional to $exp(-E/kT)$.

(b) $4\pi c^2 dc$ factor. If the expression $c^2 = \dot{x}^2 + \dot{y}^2 + \dot{z}^2$ is plotted in three dimensions with $\dot{x}$, $\dot{y}$ and $\dot{z}$ as the Cartesian axes, a sphere is obtained. Thus molecules with speeds between c and $c + dc$ must lie in a spherical shell of thickness dc. The volume of this shell, which will be proportional to the number of molecules with speeds between c and $c + dc$, is $4\pi c^2 dc$.

The two factors act in opposite directions. The c^2 factor favours high speeds and ensures that there are very few molecules with speeds near zero, whereas the Boltzmann factor favours low speeds and limits the number of molecules with high speeds.

14. Graphical representation of the speed distribution. Fig. 8 shows the Maxwell distribution for a gas at two different temperatures. The speed corresponding to the maximum on the curve is the most probable speed, $\hat{c}$; the mean speed, $\bar{c}$, and root mean square speed, $(\overline{c^2})^{\frac{1}{2}}$, are also shown and are related to each other by $\hat{c}:\bar{c}:(\overline{c^2})^{\frac{1}{2}} = 1:1\cdot13:1\cdot22$. If the curve were symmetrical, $\hat{c}$ would be equal to $\bar{c}$, but $(\overline{c^2})^{\frac{1}{2}}$ would always exceed the other two because the squaring process gives increased weight to the higher speeds. Increase in temperature broadens the distribution and shifts the maximum towards higher values of c. The shape of the distribution also depends on the mass of the molecules; at a given temperature a heavy gas has a narrower distribution than a lighter gas.

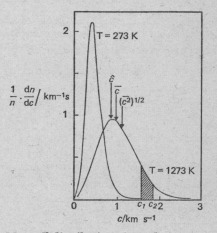

FIG. 8.—*Maxwell distribution of speeds for gas molecules at two temperatures*

The area between the curve and the c axis is proportional to the total number of molecules, while the shaded area between the ordinates c_1 and c_2 (*see* Fig. 8) is proportional to the number of molecules with speeds in the range c_1 to c_2.

THE BROWNIAN MOTION

15. Perrin's experiment. Evidence for the kinetic theory in both gases and liquids was obtained following the work of the botanist Brown (1827), who discovered that pollen grains in water were in constant, random motion. This is caused by the continual bombardment of the suspended particles by the molecules of the medium, and molecular motions can therefore be studied by observing their effects on the relatively large suspended particles.

Perrin obtained a value of 6.5×10^{23} mol^{-1} for Avogadro's constant on the assumption that the kinetic theory could be applied to suspensions of gamboge particles. This is in good agreement with values obtained from determinations not involving use of the kinetic theory.

KINETIC THEORY OF HEAT CAPACITIES

The molar heat capacity of a substance is defined as the quantity of heat required to raise the temperature of unit amount (1 mol) of the substance by 1 K. For gases it is necessary to distinguish between the molar heat capacities at constant volume C_V and at constant pressure C_p since, according to the kinetic theory, heat supplied to a gas at constant volume is used solely in increasing the kinetic energy of the molecules (i.e. in increasing the temperature), whereas heat supplied at constant pressure also enables the gas to do work by expansion.

16. Monatomic ideal gases. The kinetic theory successfully predicts the heat capacities of monatomic ideal gases as follows. From (IV, 11)

$$pV = 2E_t/3 = RT \quad \text{for one mole}$$
$$\therefore E_t = 3RT/2 \; (= 3kT/2 \text{ for one molecule}) \; \ldots \; (IV, 14)$$

Any increase in temperature, dT, at constant volume can only appear as an increase in the translational kinetic energy of the molecule, dE_t. Differentiating (IV, 14) gives

$$dE_t = 3RdT/2$$

This equation may be rearranged and written in terms of partial derivatives

$$(\partial E_t / \partial T)_V = 3R/2$$

But $(\partial E_t / \partial T)_V$ is the molar heat capacity at constant volume, C_V

$$\therefore C_V = 3R/2 \approx 12\cdot5 \text{ J K}^{-1} \text{ mol}^{-1}.$$

When the temperature is raised at constant pressure the increase in the translational kinetic energy $(dE_t)_p$ is the sum of that at constant volume $(dE_t)_V$ plus the energy used up when the gas does work by expansion against the atmospheric pressure dw, *i.e.*

$$\therefore (dE_t)_p = (dE_t)_V + dw \quad \dots \text{(IV, 15)}$$

For one mole of an ideal gas, $pV = RT$.
Differentiating, $pdV = RdT$, since p and R are constant.
Now pdV is the work done by expansion (*see* VI, 7), hence (IV, 15) becomes

$$(dE_t)_p = (dE_t)_V + pdV = (dE_t)_V + RdT$$

or $$(\partial E_t / \partial T)_p = (\partial E_t / \partial T)_V + R$$

i.e. $$C_p = C_V + R = 3R/2 + R = 5R/2$$

i.e. about $20\cdot8$ J K^{-1} mol^{-1}.

Thus the kinetic theory predicts a value of γ (the ratio of the molar heat capacities at constant pressure and constant volume) for a monatomic ideal gas as,

$$\gamma = \frac{C_p}{C_V} = \frac{5R/2}{3R/2} = 1\cdot67$$

Some values of γ are shown in Table 9.

TABLE 9   HEAT CAPACITY RATIOS (γ) FOR GASES AT ROOM TEMPERATURE

Gas	γ	Gas	γ	Gas	γ
He	1·66	H_2	1·41	H_2O	1·31
Ne	1·64	CO	1·40	CO_2	1·30
Ar	1·67	Cl_2	1·36	N_2O	1·28

17. Deviations from $\gamma = 1\cdot67$. For di- and triatomic gases, γ is consistently less than $1\cdot67$ and decreases with increasing complexity of the gas molecule. The reason for this discrepancy is that heat supplied to such gases may also cause increases in rotational and vibrational energy.

NOTE: This implies that assumption **9** (c) is not obeyed; on collision, translational energy may be converted into rotational or vibrational energy.

The "extra" internal absorption of energy E' will take place at both constant pressure and constant volume. Thus

$$\gamma = \frac{C_p}{C_V} = \frac{(5R/2) + E'}{(3R/2) + E'}$$

and the larger the value of E' the greater will be the departure from the value of $\gamma = 1.67$ calculated above.

A more quantitative insight may be obtained using the *principle of equipartition of energy*.

18. Principle of equipartition of energy.

This states that *the total energy of a molecule is divided equally among the different degrees of freedom*. Here "degrees of freedom" refers to the number of independent square terms (terms involving the square of a co-ordinate) required to express the total energy of the molecule. For example, the translational kinetic energy per mole is $3RT/2$, and this can be expressed as the sum of three independent square terms, $m\dot{x}^2/2$, $m\dot{y}^2/2$, $m\dot{z}^2/2$. Thus each square term or degree of freedom contributes $R/2$ to the molar heat capacity.

Rotational kinetic energy is the sum of terms of the form $I\omega^2/2$, where I is the moment of inertia about a particular axis of rotation and ω is the corresponding angular velocity. Thus each independent type of rotation contributes $R/2$ to the molar heat capacity.

Vibrational energy is partly kinetic ($mv^2/2$, where v is the vibrational velocity) and partly potential ($kx^2/2$ for a displacement x against a restoring force kx), so that each independent mode of vibration should contribute R to the molar heat capacity.

19. Diatomic ideal gases.

Translational motion contributes $3R/2$ as for a monatomic gas. Rotational motion occurs in two mutually perpendicular and independent ways (*see*

Fig. 9) and contributes R to the molar heat capacity. Thus if we ignore the vibrational contribution for the moment:

$$C_V = 3R/2 + R = 5R/2$$
$$C_p = C_V + R = 7R/2$$
$$\therefore \gamma = \frac{7R/2}{5R/2} = 1 \cdot 40$$

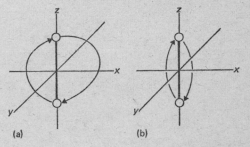

(a)   (b)

FIG. 9.—*Rotation of a diatomic molecule.* (a) *Rotation about the y axis, cutting the x axis.* (b) *Rotation about the x axis, cutting the y axis*

Vibrational motion is possible in only one direction and should therefore contribute R to the molar heat capacity, giving $C_V = 7R/2$ and $C_p = 9R/2$.

Hence $\gamma = \frac{9R/2}{7R/2} = 1 \cdot 29$

From Table 9 it is evident that vibrational degrees of freedom contribute less than R to the molar heat capacity of chlorine, and make no contribution to the molar heat capacities of hydrogen and carbon monoxide. The reason for this anomaly is discussed below.

20. Polyatomic ideal gases. For a polyatomic molecule there are three types of translational motion, three types of rotational motion (two for linear molecules—*see* Fig. 9) and $3N$–6 modes of vibration ($3N$–5 for a linear molecule), where N is the number of atoms in the molecule. Figs. 10 and 11 show the vibrational modes of linear and non-linear triatomic molecules. It is obvious from Table 9 that these vibrations cannot each contribute R to the molar heat capacity, and comparison of the results for di- and triatomic molecules indicates that a vibrational degree of freedom may contribute anything from zero to R, depending on the molecule.

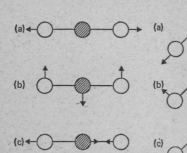

FIG. 10.—*Vibrational modes of a linear triatomic molecule (e.g. CO_2).* (a) *Symmetric stretching mode.* (b) *Bonding mode (two equivalent vibrations).* (c) *Asymmetric stretching mode*

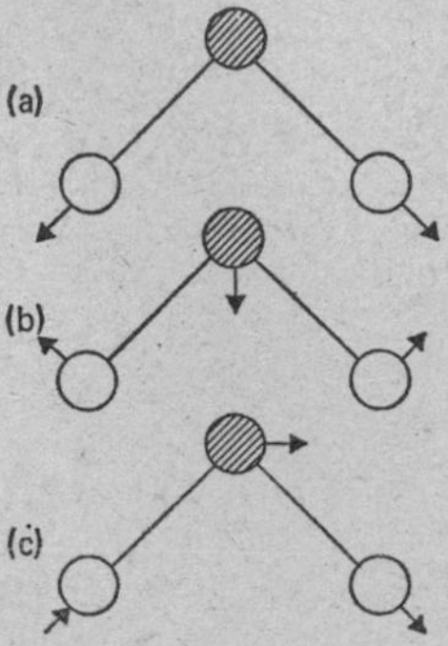

FIG. 11.—*Vibrational modes of a non-linear triatomic molecule (e.g. H_2O).* (a) *Symmetric stretching mode.* (b) *Breathing mode.* (c) *Asymmetric stretching (rocking) mode*

The reason for this anomaly lies in the equipartition principle, which, since it was derived before the advent of quantum mechanics, assumed that rotational and vibrational energies were continuous.

21. Thermal energies of gas molecules. For monatomic gases the average *molar* kinetic energy is $3RT/2$, while the average *molecular* kinetic energy is $3kT/2$. The average molecular thermal energy of di- and polyatomic molecules will be greater than $3kT/2$ because they possess rotational and vibrational energies in addition to translational energy and may be of the order of $3kT$ or $4kT$. It is useful to remember that *the average molecular thermal energy of a gas is of the order of kT*.

QUANTUM THEORY OF HEAT CAPACITIES

22. Quantisation of energy (*see* XV for a more detailed discussion of quantisation). The translational, rotational, vibrational and electronic energies of a molecule are, in fact, quantised, *i.e.* the molecule can exist only in certain allowed energy levels, each separated from the other by a discrete unit of energy called a quantum. The translational energy levels are so closely spaced that for all practical purposes they may be taken as continuous (*see* Table 10). The allowed rotational and vibrational energy levels of a diatomic molecule are given by:

$$E_r = \frac{h^2}{8\pi^2 I} \times J(J + 1)$$

$$E_v = h\nu_0(v + 1/2)$$

where h is Planck's constant, I the molecular moment of inertia, ν_0 the characteristic frequency (a complicated function of the nuclear mass and the electronic charge distribution) and J and v are respectively the rotational and vibrational *quantum numbers* and may take any integral value.

23. Rotational contribution to the molar heat capacity. The equipartition principle will only be a good approximation if the energy levels are closely spaced relative to the average energy available at the temperature of the measurement (about kT). Rotational quanta for most molecules are relatively small and at room temperature many energy levels are available for the redistribution of translational energy occurring on collision, so that the rotational contribution to the molar heat capacity is reasonably well predicted by the equipartition principle (but contrast hydrogen and nitrogen—*see* Progress Test 4, Question 7).

24. Vibrational contribution to the molar heat capacity. Vibrational quanta are generally very much larger. The value of ν_0, and hence the vibrational spacing, decreases with increasing mass of the nuclei and with decreasing strength of the internuclear bond. Thus for light, tightly bound molecules such as CO at room temperature only a very small number of molecules can acquire sufficient energy by collision to populate any but the $v = 0$ level. These molecules have only the *zero point energy* $h\nu_0/2$ (*see also* XVII, 1) independent of temperature, so that there is no vibrational contribution to the molar heat capacity. At temperatures in excess of about 1000 K the vibrational spacings begin to be comparable with kT, and as the temperature is raised further the vibrational contribution to the molar heat capacity approaches the value of R predicted by the equipartition principle. For heavier, less tightly bound molecules such as chlorine the vibrational spacings are sufficiently small to contribute to the molar heat capacity at room temperature, and at 1000 K a vibrational contribution of R is approximately realised.

25. Electronic contribution to the molar heat capacity. A fourth type of molecular energy has not so far been considered —*electronic energy* (*see* XV for an account of electronic energies). Molecules usually have widely spaced electronic energy levels and only the lowest level is occupied at normal temperatures. For example, kT at 1000 K $\approx 1\cdot38 \times 10^{-20}$ J, while typical electronic energy spacings are about 10^{-18} J. Hence the electronic contribution to the thermal energy of the molecules and therefore to the heat capacity is almost always zero. Table 10 summarises the various molecular energy level spacings.

TABLE 10   TYPICAL MOLECULAR ENERGY LEVEL SPACINGS

	E/J molecule^{-1}*	E/J mol^{-1}†
Electronic	10^{-18}	6×10^5
Vibrational	10^{-20}	6×10^3
Rotational	10^{-23}	6
Translational	10^{-41}	6×10^{-18}

* kT at 300 K $= 4 \times 10^{-21}$ J
† RT at 300 K $= 2\cdot4 \times 10^3$ J

PROGRESS TEST 4

1. The following set of pV data refer to a gas at 27°C. Show that the clearest graphical demonstration that the gas obeys Boyle's law involves a plot of pV against p or pV against $1/V$. (2)

p/kN m^{-2}	506·725	405·300	303·975	202·650	101·325	50·673
V/dm^3	4·92	6·15	8·20	12·30	24·60	49·2

2. When m g of a gas A is introduced into a bulb it is found to exert a certain pressure. If $2m$ g of gas B produces the same pressure in an identical bulb at the same temperature, what is the ratio of the relative molecular masses of A and B? (6)

3. Two flasks of equal volume are connected by a tube of negligible volume. Initially both flasks are at 27°C and contain 0·7 mol of hydrogen gas, the pressure being 0·5 atm. One of the flasks is then immersed in a hot oil bath at 127°C while the other is kept at 27°C. Calculate the final pressure of, and the amount of hydrogen in, each flask. (6)

4. 2·08 g of phosphorus pentachloride are vaporised at 227°C in a 1 dm³ flask. Some dissociation occurs according to the equation

$$PCl_5(g) = PCl_3(g) + Cl_2(g)$$

The partial pressure of chlorine in the resulting mixture is 20·265 kN m⁻². Find the partial pressures of the other constituents and the total pressure of the mixture. (6–7)

5. Starting from the basic kinetic theory equation $pV = mL\overline{c^2}/3$, show that the root mean square speed can be expressed as $(\overline{c^2})^{\frac{1}{2}} = (3RT/M)^{\frac{1}{2}}$. Calculate the root mean square speed, at 25°C, of a nitrogen molecule and of a free electron, taking the mass of the electron to be 5×10^{-4} times that of a proton. (10–11)

6. Suppose that at some initial time all the molecules in a container have the same translational energy, 2×10^{-21} J. As time passes, the motion becomes chaotic and finally the energies are distributed in a Maxwellian way. (a) Compute the final temperature of the system. (b) Using the substitution $E_t = mc^2/2$, convert the speed distribution into an energy distribution and find the fraction of molecules having energies in the narrow range between $1·98 \times 10^{-21}$ and $2·02 \times 10^{-21}$ J. (10–11, 13)

7. Calculate the number of rotational energy levels with energies less than kT at 100 K and 300 K for the hydrogen and nitrogen molecules given: $h = 6·5 \times 10^{-34}$ J s, $k = 1·38 \times 10^{-23}$ J K⁻¹, $L = 6 \times 10^{23}$ mol⁻¹ and the internuclear separations in the hydrogen and nitrogen molecules are 0·075 nm and 0·109 nm respectively. (22)

8. State in words and symbols (a) Boyle's law (2), (b) the law of Charles and Gay-Lussac (3), (c) Dalton's law of partial pressures (7), (d) Graham's law of diffusion (8).

9. Write down the ideal gas law for (a) an amount of gas n, (b) a mass of gas m, (c) a number of molecules N. (6)

10. List the assumptions of the kinetic theory of gases. (9)

11. Using the kinetic theory of gases, derive an expression for the pressure in terms of the translational kinetic energy. (10)

12. Show graphically the distribution of molecular speeds at two temperatures. (14)

13. Calculate C_p for a monatomic ideal gas. (16)

14. State the principle of equipartition of energy and illustrate its use by reference to a diatomic ideal gas. (18–19)

15. Outline the quantum theory of heat capacities. (22–25)

CHAPTER V

THE PROPERTIES OF REAL GASES

The success of the simple kinetic theory in predicting the equation $pV = nRT$ provides justification of our model of the ideal gas as a collection of non-interacting point masses. In order to modify this model so that it describes real gases we must first investigate the extent to which real gases depart from ideal behaviour.

1. Deviations from ideality. Deviations from ideality may be shown on a plot of pV against p (*see also* Progress Test 4, Question 1) or, more clearly, by a plot of the compressibility factor $Z = pV/RT$ against p. The shapes of these two plots for a given gas will be identical, since, at a fixed temperature, RT is a constant. Notice that the compressibility factor is the ratio of the observed volume V to the volume V_{id} calculated for an ideal gas at the same temperature and pressure, since

$$V_{id} = RT/p$$
$$\therefore V/V_{id} = pV/RT = Z$$

Obviously, for an ideal gas $Z = 1$ independent of temperature and pressure.

Fig. 12 shows a plot of Z against p for several gases. At pressures around one atmosphere ($101 \cdot 325$ kN m^{-2}) most gases behave ideally. However, at higher pressures, as the kinetic theory assumptions of point mass and no intermolecular forces become less and less realistic, Z becomes greater than unity. It is also noticeable that in the low pressure region Z is less than unity (except for hydrogen), particularly for the more easily liquefiable gases.

THE VAN DER WAALS EQUATION

Effects due to finite size and intermolecular forces are taken into account as follows.

2. Effects due to finite size. Since molecules are of finite size, no two molecules can simultaneously occupy the same point in space. Thus one can consider half the molecules as "excluding" a certain volume of the container from the other

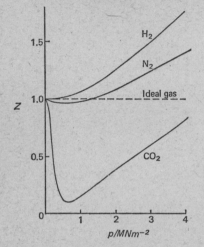

Fig. 12.—*Compressibilities of gases at 273 K*

half. If b is the "excluded volume," then the actual volume available to the molecules is $(V - b)$, and incorporating this into the ideal gas law gives

$$p(V - b) = RT$$

This removes one obvious defect, since the equation no longer predicts zero volume for the gas at 0 K.

If there are N molecules of a gas, and these are regarded as hard spheres of radius $d/2$, then the presence of one molecule will exclude a volume $4\pi d^3/3$ from the centre of any other molecule. This can readily be seen from Fig. 13(a). Since half the molecules exclude this volume from the other half, the total excluded volume is

$$(N/2)(4\pi d^3/3) = 2\pi d^3 N/3 = b$$

Since b may be determined experimentally, the above equation allows d, i.e. the size of a molecule, to be calculated.

3. Effects due to intermolecular interactions. Attractive forces exist between all types of molecule. There is in fact more than one type of force, depending on whether the molecules are polar or non-polar, and the general name *van der Waals forces* is used to describe them (*see* XVIII, **8** for a detailed account of these forces).

Intermolecular interactions are taken into account by considering their effect on the speeds with which molecules strike the walls. A molecule in the centre of the gas is attracted symmetrically in all directions and so experiences no net force. On approaching the wall, however, a net attraction is experienced back towards the centre of the vessel, since there are more molecules on the side remote from the wall. The resulting decrease in speed means that a real gas will exert less pressure than an ideal gas under the same conditions. This effect can be allowed for by a term of the form a/V^2, since both the number of molecules exerting the retarding force and the number upon which the force is exerted depend upon concentration (*i.e.* both are proportional to $1/V$). a is a constant for a particular gas, independent of temperature and pressure. Thus the van der Waals equation may be expressed as

$$\left(p + \frac{a}{V^2}\right)(V - b) = RT \quad \ldots \text{ (V, 1)}$$

For an amount of gas n

$$\left(p + \frac{n^2 a}{V^2}\right)(V - nb) = nRT$$

4. Validity of the van der Waals equation. The equation is first rearranged to a form more appropriate for comparison with Fig. 12.

$$p = \frac{RT}{V - b} - \frac{a}{V^2}$$

$$\therefore Z = \frac{pV}{RT} = \frac{V}{V - b} - \frac{a}{RTV}$$

(on multiplying by V/RT). The comparison is then carried out as follows:

(*a*) At low gas densities

$$\frac{V}{V - b} \approx 1, \quad \frac{a}{RTV} \approx 0 \quad \therefore Z \approx 1$$

i.e. the equation successfully predicts the approach of real gases towards ideal behaviour as the density decreases.

(*b*) At high temperatures and pressures

$$\frac{a}{RTV} \ll \frac{V}{V-b}$$

$$\therefore Z \approx \frac{V}{V-b} > 1 \quad \text{(since } b \text{ is positive)}$$

i.e. the equation successfully predicts the positive deviations from ideality.

(*c*) At low temperatures and moderate pressures

$$\frac{V}{V-b} \approx 1$$

$$\therefore Z \approx 1 - \frac{a}{RTV} < 1 \quad \text{(since } a \text{ is positive)}$$

i.e. the equation successfully predicts the negative deviations from ideality.

(*d*) Other things being equal, an increase in temperature should result in an increase in $V/(V-b)$ and a decrease in a/RTV. Thus the equation predicts that there will be some temperature at which the two effects will exactly balance (*i.e.* the effects due to molecular size exactly balance those due to intermolecular attraction) so that Z is unity over a wide range of pressures. This temperature is called the *Boyle temperature*, T_B.

INTERMOLECULAR COLLISIONS

Since our model of real gases now involves molecules of finite size, it is of interest to consider phenomena based on intermolecular collisions.

5. The collision number. The number of collisions per second, or *collision number*, Z, is calculated as follows. Suppose that there are N molecules per unit volume, each of diameter d, and that one molecule A has a speed c, while the others are stationary. A collision will occur each time A approaches so close to another molecule that the distance between their *centres* is d. Thus we can think of A as sweeping out a "collision cylinder" of radius d and length c every second. Whenever the *centre* of another molecule comes within this cylinder, collision occurs (*see* Fig. 13).

Volume of collision cylinder $= \pi d^2 c$

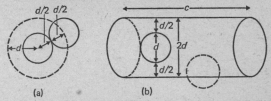

(a)   (b)

FIG. 13. *Intermolecular collisions in a gas.* (a) *Collision sphere.* (b) *Collision cylinder*

Number of molecules per unit volume in the cylinder $= N$

∴ Number of collisions of A with other molecules (all except A stationary) $= N\pi d^2 c$.

If all the molecules are in motion, with an average speed $\bar{c}$, the collision number is determined by the velocity of the molecules relative to each other. On average the molecules will collide at right angles, and under these conditions $c_{rel} = 2^{\frac{1}{2}}\bar{c}$. Z', the number of collisions of A with other molecules (all in motion), is given by

$$Z' = 2^{\frac{1}{2}}N\pi d^2 \bar{c} \qquad \ldots \text{(V, 2)}$$

But the true collision number, Z, is the number of collisions per second of *all* molecules with all other molecules.

$$\therefore Z = NZ'/2$$

where the factor N is introduced because there are N molecules per unit volume and the factor $\frac{1}{2}$ so that each collision is not counted twice (once as A colliding with B and once as B colliding with A).

Thus $\qquad Z = \frac{2^{\frac{1}{2}}}{2}N^2\pi d^2 \bar{c}$

$\bar{c}$ is obtained by summing (by means of an integration) the product of the number of molecules with speed c by the speed, and dividing by the total number of molecules.

$$\bar{c} = (8RT/\pi M)^{\frac{1}{2}} \qquad \ldots \text{(V, 3)}$$

$$\therefore Z = \frac{2^{\frac{1}{2}}}{2}N^2\pi d^2\left(\frac{8RT}{\pi M}\right)^{\frac{1}{2}}$$

$$= 2N^2 d^2(\pi RT/M)^{\frac{1}{2}} \qquad \ldots \text{(V, 4)}$$

NOTE: The dimensions of Z are (volume)$^{-1}$ (time)$^{-1}$.

Similarly, for a mixture of two gases, 1 and 2, the collision number is given by

$$Z_{12} = N_1 N_2 d_{12}^2 (8\pi RT/\mu)^{\frac{1}{2}} \quad . \quad . \quad . \quad (V, 5)$$

where N_1 and N_2 are the numbers of 1 and 2 per unit volume, $d_{12} = (d_1 + d_2)/2$, the reduced mass μ is defined by $\mu = M_1 M_2/(M_1 + M_2)$ and the factor $\frac{1}{2}$ has been dropped.

Such estimates of collision number are important in the calculation of reaction rates (*see* XXII, 4–5), since a chemical reaction between two molecules can occur only when they collide.

6. The mean free path, λ. The mean free path is the average distance travelled by a molecule between successive collisions. For a molecule of speed c which makes Z' collisions per second,

$$\lambda = c/Z' = 1/2^{\frac{1}{2}} N\pi d^2 \quad . \quad . \quad . \quad (V, 6)$$

It can be shown on purely theoretical grounds that λ is related to the viscosity of the gas, and further that the viscosity should be independent of gas density but proportional to the square root of the temperature. The experimental confirmation of these two predictions, neither of which is obvious at first sight, provided striking evidence for the kinetic theory model of real gases.

THE CRITICAL STATE

All gases, including the so-called permanent gases H_2, He, O_2, etc., may be liquefied by the simultaneous use of high pressure and low temperature. The essential conditions for the liquefaction of gases may be found by considering the variation of volume with pressure at a number of temperatures. The family of curves (isotherms) obtained is shown in Fig. 14. The T_3 and T_4 isotherms are more or less as expected, being approximately hyperboloid, but the low-temperature isotherms are of a rather different shape.

7. T_1 isotherm. At low pressures (*e.g.* along AB) the system is entirely gaseous and the volume decreases with increasing pressure in the normal fashion (*i.e.* AB is roughly hyperboloid). At B liquefaction commences; vapour is converted into liquid of very much greater density, so that BC is approximately parallel to the volume axis. Liquefaction is complete at C and CD shows the small effect of pressure on a liquid.

To summarise, along AB there is only vapour, along CD only liquid and along BC liquid and vapour co-exist. Since BC is parallel to the volume axis, there can be only one pressure at which liquid and vapour co-exist. This pressure is called the *vapour pressure* of the gas.

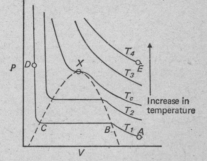

FIG. 14.—*Isotherms of a real gas near the critical temperature*

8. T_2 isotherm. This is qualitatively similar except that the horizontal portion is shorter (*i.e.* the range of volumes over which gas and liquid can co-exist is smaller) and the vapour pressure higher.

9. T_c isotherm. Here the horizontal portion has shrunk to a point of inflexion, *i.e.* there is only one volume, V_c, at which gas and liquid can co-exist. This point, marked X on Fig. 14, is called the *critical point*. Since the isotherms for temperatures greater than T_c show no horizontal portion, gas and liquid cannot co-exist above this temperature, so that T_c may be defined as the maximum temperature at which liquefaction can take place. T_c is referred to as the *critical temperature*; the pressure, p_c, necessary to liquefy the gas at this temperature is called the *critical pressure*, and the volume of one mole at the critical temperature and pressure is called the *critical volume*. At the critical point the gas exists in its *critical state*, where the temperature, pressure and volume all have their critical values.

The reason for the lack of any sharp transition from vapour to liquid above the critical temperature is that in this region the average thermal energy of the molecules is greater than the intermolecular attractive forces.   Below the critical temperature the reverse is true, and if two molecules are brought sufficiently close together by compression the attractive forces will act to bring them into still closer proximity.   Since the attractive forces become stronger the smaller the separation, the effect is cumulative and a sharp transition occurs.

NOTE: The van der Waals forces, acting between neutral, non-polar molecules, fall off in proportion to the inverse sixth power of the intermolecular separation (see also XVIII, 8).

10. Principle of continuity of states. The dome-shaped curve shown by the dashed line in Fig. 14 connects the end points of the horizontal portions of the isotherms.   Since a horizontal portion represents two phases in equilibrium (see IX, 12), all points inside the dome correspond to an equilibrium between liquid and vapour.   For such points it is always possible to draw a sharp distinction between liquid and vapour, since there is a definite boundary surface between the two phases.

For points lying outside the dome only one phase (liquid or vapour) is present, and here it is not possible to draw any sharp distinction between liquid and vapour.   This fact is known as the *principle of continuity of states* and may be demonstrated by reference to Fig. 14.

Points A and D both lie on the T_1 isotherm, and we have already described A as representing only vapour and D as representing only liquid.   If the transformation from A to D is carried out isothermally, an observable change of state (liquefaction) occurs along BC.   However, we may also carry out the transformation via the path AED, *i.e.* by heating the gas from T_1 to T_4 at constant volume (causing an increase in pressure from A to E) and then decreasing the temperature from T_4 to T_1 at constant pressure (causing the volume to decrease from E to D).   The transformation from A (gas only) to D (liquid only) has now been accomplished, but the system has at no time passed through the region inside the dome in which gas and liquid co-exist.   That is, liquefaction as we normally think of it has not occurred.   The transformation from one state to the other has been carried out in a perfectly continuous fashion.

An important implication of the principle of continuity of states is that one may think of a liquid near the critical point as just a highly compressed gas. If an equation of state can be found which accurately describes the gas close to its critical temperature and pressure, that equation should also describe the liquid (however, *see* XX, 2).

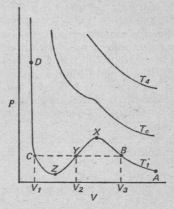

Fig. 15.—*Isotherms of a van der Waals gas near the critical temperature*

THE VAN DER WAALS ISOTHERMS

If values are assigned to the van der Waals constants a and b, theoretical isotherms can be calculated as shown in Fig. 15. We can compare these with the isotherms of a real gas shown in Fig. 14.

11. T_4 isotherm. At temperatures well above T_c, both the van der Waals and real gas isotherms are approximately hyperboloid, since both gases behave more or less ideally under these conditions.

12. T_c isotherm. Again the van der Waals isotherm qualitatively reproduces all the features shown by real gases.

13. T_1 isotherm. At temperatures below T_c the van der Waals isotherm exhibits a maximum and minimum in place of the horizontal portion shown by real gases.

NOTE: No simple equation of state can be expected to reproduce the horizontal discontinuity BYC corresponding to liquefaction.

Regions AB and CD accurately represent the p–V conditions in a real gas, whereas BX, XZ and ZC represent hypothetical metastable conditions (for a definition of metastable see IX, 16). BX and ZC are experimentally realisable, BX representing superheated vapour (vapour at a higher pressure than that which would normally cause liquefaction), while ZC represents superheated liquid (liquid at a lower pressure than that which would normally cause vaporisation). XZ cannot be experimentally realised, since it represents a system in which a decrease in pressure causes a decrease in volume.

14. Evaluation of the critical constants. The critical constants may be evaluated from the van der Waals equation as follows:

$$\left(p + \frac{a}{V^2}\right)(V - b) = RT \quad \ldots \text{ (V, 7)}$$

Multiplying out the parentheses and multiplying by V^2/p

$$V^3 - \left(b + \frac{RT}{p}\right)V^2 + \left(\frac{a}{p}\right)V - \frac{ab}{p} = 0$$

or, at the critical temperature,

$$V^3 - \left(b + \frac{RT_c}{p_c}\right)V^2 + \left(\frac{a}{p_c}\right)V - \frac{ab}{p_c} = 0 \quad \ldots \text{ (V, 8)}$$

Thus, at any temperature below T_c, a curve of the form of the T_1 isotherm is represented by a cubic equation and must therefore have three roots. These are shown in Fig. 15 as V_1, V_2, V_3 and the equation describing the curve may be written in terms of its roots as

$$(V - V_1)(V - V_2)(V - V_3) = 0$$

NOTE: V_2 has no physical significance, i.e. V_2 is not a real root (see II, 14).

As the temperature is increased the roots V_1, V_2 and V_3 approach each other until at the critical temperature

$$V_1 = V_2 = V_3 = V_c$$
$$\therefore (V - V_c)(V - V_c)(V - V_c) = 0$$

or, on expanding this expression,

$$V^3 - 3V_cV^2 + 3V_c^2V - V_c^3 = 0 \quad \cdots \quad (V, 9)$$

Equations (V, 8) and (V, 9) are simply alternative descriptions of the critical isotherm and must therefore be identical, *i.e.*

$$V^3 - \left(b + \frac{RT_c}{p_c}\right)V^2 + \left(\frac{a}{p_c}\right)V - \frac{ab}{p_c}$$
$$\equiv V^3 - 3V_cV^2 + 3V_c^2V - V_c^3$$

As is usual with identities (*see* II, 22), we can equate co-efficients on opposite sides of the identity sign as follows:

(*a*) Coefficients of V^2: $3V_c = b + RT_c/p_c$
(*b*) Coefficients of V^1: $3V_c^2 = a/p_c$
(*c*) Coefficients of V^0 (*i.e.* constant terms): $V_c^3 = ab/p_c$

From these three equations it is a simple matter to express the critical constants in terms of the van der Waals constants as:

$$V_c = 3b; \; p_c = a/27b^2; \; T_c = 8a/27Rb \quad \cdots \quad (V, 10)$$

and to derive the equation

$$p_cV_c = 3RT_c/8 \qquad \cdots \quad (V, 11)$$

Equation (V, 11) predicts a compressibility factor for the van der Waals gas of

$$Z = p_cV_c/RT_c = 3/8 = 0{\cdot}375$$

Since values of Z for most gases lie in the range $0{\cdot}27$–$0{\cdot}30$ it is apparent that, while the van der Waals equation provides a useful insight into molecular interactions, its description of the behaviour of real gases is only qualitative.

GENERALISATION OF THE VAN DER WAALS EQUATION

The isotherms of most gases near the critical temperature are very similar in form, though they lie in different regions of the graph. By choosing the correct scale it should therefore be possible to superimpose the isotherms for all gases and so obtain at each temperature a single isotherm which is approximately true for all gases.

15. The reduced equation of state. The above situation can be realised if the pressure, volume and temperature of each gas are expressed as a fraction of its critical pressure, volume and temperature, *i.e.* as the reduced variables:

(*a*) Reduced pressure $\pi = p/p_c$ or $p = \pi p_c$
(*b*) Reduced volume $\phi = V/V_c$ or $V = \phi V_c$
(*c*) Reduced temperature $\theta = T/T_c$ or $T = \theta T_c$

Substituting the reduced variables in the van der Waals equation

$$\left(\pi p_c + \frac{a}{V_c^2}\right)(\phi V_c - b) = R\theta T_c$$

and substituting for p_c, V_c and T_c from (V, 10)

$$\left(\frac{\pi a}{27b^2} + \frac{a}{\phi^2 9b^2}\right)(3b\phi - b) = \frac{8aR\theta}{27b}$$

Simplifying,

$$\left(\pi + \frac{3}{\phi^2}\right)(3\phi - 1) = 8\theta \quad \text{. . . (V, 12)}$$

This equation is known as the *reduced equation of state* and is effectively an expression of van der Waals' equation, not involving any constants specific to the gas under consideration. This completely general equation of state should thus apply equally well to all gases, and in fact the agreement for non-associated substances is quite good.

16. The law of corresponding states. This is a verbal re-statement of (V, 12) and states that all gases at the same reduced temperature and pressure have the same reduced volume; the gases are then said to be in *corresponding states*.

An important implication of this law is that any comparison of properties depending on pressure or volume should be carried out with the substances at the same *reduced temperature*. For example, it is more correct to compare substances at their boiling points than at the same temperature, since the boiling points of many substances are approximately two-thirds of their critical temperatures, *i.e.* the substances are in corresponding states.

OTHER EQUATIONS OF STATE

The van der Waals equation is only one of the many equations of state which have been proposed over the years, though it should be pointed out that all subsequent equations have made use of the two fundamental postulates of van der Waals—that molecules have finite size and interact with each other. Of the other equations of state the best known are:

17. The Berthelot equation.

$$p(V - b) = RT - \frac{a}{TV^2}$$

This is somewhat better than the van der Waals equation at pressures around a few atmospheres.

18. The Dieterici equation.

$$p(V - b) = RT \exp(-a/RTV)$$

This provides a good description of gases around their critical points but is less convenient to use because of the exponential factor.

NOTE: Both the above expressions predict values of $p_c V_c/RT_c$ of around 0·27.

19. The Beattie–Bridgman equation.

$$pV = RT + (\beta/V) + (\gamma/V^2) + (\delta/V^3)$$

This equation involves five empirical constants, apart from R, which determine the values of β, γ and δ, and is the most accurate of the empirical equations.

20. The Virial equation.

$$pV = RT\{1 + (B/V) + (C/V^2) + (D/V^3) + \ldots\}$$

This equation is completely general and can be expanded to as many terms as necessary. The virial coefficients B, C, etc., are functions of temperature. B represents contributions from interactions between pairs of molecules, C simultaneous interactions between three molecules, and so on.

PROGRESS TEST 5

1. The normal density of carbon dioxide is 1.981 kg m^{-3} at 25°C. Calculate the average distance between the centres of the molecules and compare this distance with the molecular diameter calculated from the van der Waals constant $b = 4.27 \times 10^{-5}$ m^3 mol^{-1}. Calculate also the collision number and mean free path under these conditions. (2, 5, 6)

2. Gas A has $T_c = 200$ K, $p_c = 5.066$ MN m^{-2}. Gas B has $T_c = 400$ K, $p_c = 4.043$ MN m^{-2}. Which gas has the smaller value of (a) the van der Waals constant a, (b) the van der Waals constant b, (c) the critical volume; (d) which gas most closely approximates to ideality at room temperature and 2.026 MN m^{-2} pressure? (14)

3. What are the two chief causes for the deviation of real gases from ideal behaviour? How are they allowed for in the van der Waals equation? (2, 3)

4. Show how the van der Waals equation predicts the deviations from ideality of real gases. (4)

5. Derive expressions for (a) the collision number (5), (b) the mean free path (6).

6. Draw the isotherms of a real gas near the critical temperature and explain the main features of the graph. (7–10)

7. Derive the relationships between the critical constants and the van der Waals constants. (14)

8. Derive the reduced equation of state and explain its significance. (15)

PART THREE

THERMODYNAMICS

Thermodynamics is built on a framework of three laws based on experimental observations. The first law is considered in Chapters VI and VII. Chapter VI contains a discussion of the relationship between heat and work, while the heat changes involved in chemical and physical processes are discussed in Chapter VII. This chapter closes with an account of the relationship between heats of reaction and bond energies.

Chapter VIII deals with the second and third laws and introduces the important thermodynamic functions, the entropy and the Gibbs free energy. The application of thermodynamics to an industrial process is considered at the end of this chapter.

CHAPTER VI

THE FIRST LAW OF THERMODYNAMICS

Thermodynamics is concerned with the laws governing the interconversion of various types of energy. The laws of thermodynamics enable far-reaching conclusions to be made regarding the limitations of energy interconversion and the practical consequences of such limitations. Because the laws of thermodynamics relate to the properties of bulk matter they are in no way influenced by theories of atomic and molecular structure.

A discussion of the laws of thermodynamics and their consequences is postponed until some important thermodynamic terms have been defined.

DEFINITIONS

1. System. A *system* is any part of the universe chosen for thermodynamic study and is separated from the *surroundings* by a *boundary*. Systems may be classified as follows:

(a) *Isolated*, when there is no transfer of either energy or matter across the boundary, *i.e.* an isolated system is the system *plus* the surroundings.

(b) *Closed*, when there is transfer of energy but not of matter across the boundary.

(c) *Open*, when both matter and energy are transferred across the boundary.

The term *body* refers to either an isolated or a closed system these are the only types of system we shall consider). A *phase* is a region having uniform properties and is separated from other phases by a boundary. A system may be either *homogeneous* or *heterogeneous*, depending on whether it is composed of one or several phases.

2. Properties of a system. The macroscopic or bulk properties of a system (volume, pressure, mass, etc.) may be divided into two classes:

(a) *Extensive* properties, such as mass and volume, depend on the quantity of matter present in the system.

(b) *Intensive* properties such as pressure and density, do not depend on the quantity of matter present in the system.

A system is said to be in a state of *equilibrium* if its properties do not change with time.

3. State of a system. A system is said to be in a certain *state* when all its properties have definite values. It is not necessary to specify every possible property of a system in order to characterise its state. Thus, for a given mass of a pure substance, it suffices to specify two independent variables, usually temperature and pressure; the other properties are automatically fixed.

4. Change in state. A change in state occurs when one or more properties of the system changes, specification of the initial and final states being sufficient to define the change. The change from the initial to the final state may take place by a variety of *paths*, and a *process* is the mechanism by which the change in state is effected. A *cyclic* process is one by which a system undergoes a series of changes in state and in doing so returns to its initial state.

5. Isothermal and adiabatic processes. Processes which take place at constant temperature are called *isothermal* processes. *Adiabatic* processes are those in which there is no exchange of heat between the system and its surroundings, although the temperature does change.

SOME IMPORTANT CONCEPTS

6. Mechanical work. The mechanical work dw performed by a system on its surroundings is the product of the force exerted by the system f, and the displacement of the force ds, i.e.

$$dw = fds$$

By convention, the sign of dw is *positive* if work is done *by* the system *on* the surroundings, and *negative* if work is done *on* the system *by* the surroundings.

7. Expansion or "pV" work. The expansion of a gas enclosed in a cylinder by a frictionless piston (Fig. 16) produces a type of mechanical work known as *expansion* or *"pV"*

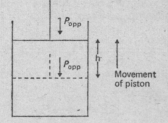

FIG. 16.—*Expansion work*

work. Suppose the constant opposing pressure (force per unit area) on the piston of cross-sectional area A is p_{opp}, and that expansion of the gas causes the piston to move a distance h. The work done is given by

$$w = p_{opp} \times A \times h = p_{opp}\Delta V$$

where ΔV is the change in volume.

For an infinitesimal change in volume, dV

$$dw = p_{opp}dV$$

while for a change between the limits V_1 and V_2

$$w = \int_{V_1}^{V_2} p_{opp}dV$$

The graphical representation of expansion work was considered in II, 23.

8. Energy. Energy is the capacity for doing work. Many forms of energy are known. Thus mechanical energy is the sum of two forms—*kinetic* energy, which depends upon the motion of a body, and *potential* energy, which depends upon the position of the body in a force field. Other common forms of energy include *electrical*, *chemical* and *thermal* energy or *heat*, and one type of energy may be transformed into another using a suitable device. Mechanical, electrical and chemical energy may be completely converted into heat, but the converse is not true.

The relationship between the various forms of energy was systematically investigated by Joule (1840–78), who found that the same amount of heat was produced by the expenditure of the same amount of any other form of energy. Thus there is an exact relationship between heat and the other forms of energy.

NOTE: Mass (m) and energy (E) are related by the equation $E = mc_0^2$ where c_0 is the velocity of light. When mass is destroyed an equivalent amount of energy takes its place.

A quantity of heat, q, may be either absorbed or evolved by a body; by convention q is taken as *positive* if heat is *absorbed* and vice versa.

9. Temperature.

The concept of hotness can be put on a quantitative scale by assigning numbers to represent various degrees of hotness. Such a number may be called a *temperature*, with large numbers corresponding to greater degrees of hotness. The definition of a scale of temperature requires the choice of some property of a substance which changes reproducibly with hotness. Thus the volume occupied by a thread of mercury forms the basis of the common thermometer calibrated at the freezing and boiling points of water and graduated in 100 divisions on the Celsius scale. Such an empirical scale is unsatisfactory because it depends on the substance used, whereas the concept of temperature does not. The *absolute* scale of temperature is defined with reference to the expansion of an ideal gas and is independent of the working substance, since all ideal gases are defined by a single equation of state (*see* IV, 6). This scale of temperature has been considered in IV, 3–4.

A theoretical scale of temperature, the thermodynamic scale, may be defined in terms of the second law of thermodynamics and is identical with the absolute scale.

10. Exact and inexact differentials.

To distinguish between these two types of differential, we shall consider the case of a

boulder at the top of a hill. The potential
boulder may be expressed as a function of
sea level:

$$E_p = mgh + \text{constant}$$

where m is the mass and g the acceleration due to gravity. If E_p is changed by dE_p because of a change in h of dh, then

$$dE_p = mgdh$$

mg is a constant and hence dE_p depends only on dh, *i.e.* on the initial and final values of h and not on the path taken. dE_p is called an *exact differential*, since dh gives the change in E_p directly.

To move the boulder up the hill, work must be done against (a) gravity, (b) frictional forces, and the amount of work required will obviously vary depending on the path taken. Thus the expenditure of an amount of work dw results in a change in height dh which depends on how the work is carried out. dw is known as an *inexact differential* and this may be stressed by writing Dw.

Any property of a thermodynamic system which gives rise to an exact differential is known as a *function of state*, since it depends only on the initial and final states and not on the path taken between the states.

11. Net change over a cyclic path. An important difference between exact and inexact differentials is the net change over a cyclic or closed path. If the boulder is raised to the top of the hill and then pushed down, then, so long as the initial and final heights are the same, the net change in potential energy over the closed path will be zero. This may be written

$$\oint dE_p = 0$$

where $\oint$ represents the integral around a closed path.

On the other hand, the amount of work done over a closed path is not fixed and depends on the path, *i.e.*

$$\oint Dw \neq 0$$

Likewise, the amount of heat taken in or given out over a closed path depends on the path, and

$$\oint Dq \neq 0$$

cesses. Suppose that the external pressure acting ▬▬▬▬▬ in Fig. 16 is only infinitesimally smaller than that exerted by the enclosed gas. Under these conditions the expansion of the gas will take place infinitesimally slowly, and the system will always be in equilibrium with itself and with its surroundings. The direction of the change may be reversed by an infinitesimal change in the conditions. Such a process is called a *reversible* process, since all changes that occur in any part of the process are exactly reversed when the process is carried out in the reverse manner. Reversible processes are important because they can be looked on as a succession of equilibrium states. Under these conditions an equation of state (such as the ideal gas law) may be used to describe the process. Ordinary processes are *irreversible*, although reversibility may be approached closely in some instances (*see* XIV, **9**).

THE FIRST LAW OF THERMODYNAMICS

13. Verbal statement of the first law. This law, based on the experiments of Joule and others, may be stated as follows:

(*a*) *Energy may be neither created nor destroyed, although it may be changed from one form to another.* This statement is often called *the law of conservation of energy.*

An alternative statement is:

(*b*) The algebraic sum of the energy changes in any isolated system is zero.

14. Mathematical statement of the first law. If a system absorbs an amount of heat Dq from the surroundings, and performs work Dw on the surroundings, the energy difference Dq — Dw must remain in the system as a distinct form of energy which is called the *internal energy*, symbol U, *i.e.*

$$dU = Dq - Dw \qquad . \ . \ . \ \text{(VI, 1)}$$

The sign conventions regarding heat and work have been given in **6** and **8** above. Although Dq and Dw are inexact differentials, their difference, dU, is an exact differential and depends only on the state of the system (*see* **15** below).

For finite changes (VI, 1) may be written

$$\Delta U = U_2 - U_1 = q - w \ . \ . \ . \ \text{(VI, 2)}$$

where the subscripts 1 and 2 refer to the initial and final states respectively. U_2 and U_1 cannot be measured separately, only their difference ΔU.

NOTE: Equation (VI, 2) does not involve Δq or Δw (i.e. $q_2 - q_1$ or $w_2 - w_1$), since heat and work are not properties of the initial and final states but appear *during* the change in state.

15. Proof that dU is an exact differential. The situation that would arise if dU were not an exact differential (i.e. if U were not a function of state) is readily seen from the following. Suppose a system moves from state 1 to state 2 by path a and back to state 1 by path b (Fig. 17). If path a *requires*

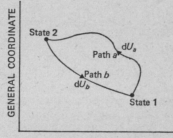

FIG. 17.—*Illustration that dU is an exact differential*

an amount of internal energy dU_a, while path b *gives out* an amount dU_b, then if $dU_b > dU_a$ a net amount of internal energy will be obtained from the system without the disappearance of any other forms of energy. This creation of energy is contrary to the first law of thermodynamics, and the conclusion is that $dU_b = dU_a$, i.e. that dU is an exact differential.

16. Changes at constant volume. If a system can perform only expansion work, then at constant volume $dV = 0$, hence $Dw = 0$ (*see* 7 above), and (VI, 1) becomes

$$dU = Dq_v = dq_v$$

where dq_v is an exact differential under these limited conditions.

For a finite change

$$\Delta U = q_v$$

Thus the heat absorbed at constant volume is equal to the increase in internal energy.

17. Changes at constant pressure, the enthalpy. If only expansion work is performed, then for a finite change at constant pressure

$$\Delta U = q_p - p\Delta V \qquad \cdots \text{(VI, 3)}$$
$$U_2 - U_1 = q_p - p(V_2 - V_1)$$
$$\therefore q_p = (U_2 + pV_2) - (U_1 + pV_1)$$

U, p and V depend only on the state of the system; hence $U + pV$ is a function of state and is represented by the symbol H, the *enthalpy* or *heat content*.

$$\therefore q_p = H_2 - H_1 = \Delta H \qquad \cdots \text{(VI, 4)}$$

For an infinitesimal change

$$dH = Dq_p = dq_p$$

The heat absorbed at constant pressure is equal to the increase in enthalpy.

18. The relationship between ΔU and ΔH. Since $H = U + pV$, for a finite change at constant pressure we may write

$$\Delta H = \Delta U + \Delta(pV) = \Delta U + p\Delta V \cdots \text{(VI, 5)}$$

The heat absorbed at constant pressure exceeds that absorbed at constant volume by the amount $p\Delta V$.

(a) *Condensed phases.* For reactions involving liquids and solids only, $\Delta V = \{(\text{volume of products}) - (\text{volume of reactants})\}$ will be very small, and hence $\Delta H \approx \Delta U$.

(b) *Gases.* For gaseous reactions, ΔV will be significant. Assuming the gases to behave ideally, then at constant temperature and pressure

$$pV_r = n_r RT \quad \text{for the reactants}$$
and
$$pV_p = n_p RT \quad \text{for the products}$$
Hence
$$p(V_p - V_r) = (n_p - n_r)RT$$
or
$$p\Delta V = \Delta n RT$$

Equation (VI, 5) may be rewritten

$$\Delta H = \Delta U + \Delta n RT \qquad \cdots \text{(VI, 6)}$$

NOTE: For reactions involving both gases and condensed phases Δn refers only to the change in the amount of gas.

19. Changes under adiabatic conditions. Under these conditions $Dq = 0$ and (VI, 1) becomes

$$dU = -Dw = -dw$$

and for finite changes

$$\Delta U = -w$$

The internal energy change under adiabatic conditions is equal to the work done on the system.

APPLICATION OF THE FIRST LAW TO IDEAL GASES

20. Definition of an ideal gas (*see* also IV). A gas is said to be ideal if it has the following properties:

(*a*) It obeys the ideal gas law, $pV = nRT$.

(*b*) Its internal energy is a function of temperature alone and is independent of volume and of pressure at constant temperature, *i.e.* $(\partial U/\partial V)_T = 0$ and $(\partial U/\partial p)_T = 0$.

(*b*) may be predicted from the kinetic theory of ideal gases, where interactions between gaseous molecules are absent, and hence the energy of the molecules does not depend on their mutual separations at constant temperature.

21. The heat capacity. The molar heat capacity has been defined in IV, 16. For monatomic ideal gases C_p and C_v were defined in terms of the translational kinetic energy. In general, for one mole of any gas

$$C = Dq/dT \qquad \text{. . . (VI, 7)}$$

Hence
$$C_V = Dq_v/dT = (\partial U/\partial T)_V \qquad \text{. . . (VI, 8)}$$

and
$$C_p = Dq_p/dT = (\partial H/\partial T)_p \qquad \text{. . . (VI, 9)}$$

22. The relationship between C_p and C_V (*see* also IV, 16). If only expansion work is performed, then at constant pressure for an infinitesimal change

$$dH = dU + pdV \qquad \text{. . . (VI, 10)}$$

For an ideal gas the restrictions of constant volume and constant pressure given in (VI, 8) and (VI, 9) may be eliminated, since both U and H are independent of pressure and volume and depend only on the temperature. Hence

$$dU = C_V\,dT \text{ and } dH = C_p\,dT \qquad \text{. . . (VI, 11)}$$

Substitution of these values into (VI, 10) gives

$$C_p \, dT = C_V \, dT + p \, dV \quad \dots \text{(VI, 12)}$$

For one mole of an ideal gas

$$pV = RT$$

If volume and temperature change at constant pressure, then

$$p \, dV = R \, dT$$

and (VI, 12) becomes

$$C_p \, dT = C_V \, dT + R \, dT$$
$$\therefore C_p = C_V + R$$

23. The nature of internal energy. The internal energy of a gas may be considered to consist of two components:

(a) E_0, the energy of the gas at 0 K, and
(b) E_T, the temperature-dependent energy.

Thus
$$U = E_0 + E_T$$

For a monatomic ideal gas at normal temperatures, E_T represents the translational energy of the atoms, which was shown in IV, **16** to be equal to $3RT/2$.

$$\therefore U = E_0 + 3RT/2$$
and
$$dU/dT = 3R/2$$

For polyatomic gases there will be contributions to E_T from translational, rotational, vibrational and electronic energies, the last two being significant only at high temperatures. Thus

$$U = E_0 + E_t + E_r + E_v + E_e$$

The contributions of these energies to the heat capacities and therefore to the internal energies have been discussed in IV, **16–24**.

24. Reversible expansion and maximum work. The work done when a gas expands against a constant opposing pressure p_{opp} is given by

$$Dw = p_{opp} \, dV \qquad (see\ \mathbf{7}\ above)$$

If $p_{opp} = p_{gas} - dp$, *i.e.* the opposing pressure is only infinitesimally smaller than the gas pressure, the expansion will take place reversibly and

$$Dw = (p_{gas} - dp)dV = p_{gas}dV$$

since the second-order terms $dpdV$ may be neglected. This reversible work is the *maximum* work which can be obtained from the system, and the actual work (the irreversible work) will always be less than this value, *i.e.* $Dw_{rev} > Dw_{irrev}$.

If the volume change is from V_1 to V_2 the reversible work will be given by

$$\int Dw_{rev} = w_{rev} = \int_{V_1}^{V_2} pdV$$

For an ideal gas $p = nRT/V$, and if the change takes place *isothermally*

$$w_{rev} = \int_{V_1}^{V_2} \frac{nRTdV}{V} = nRT\int_{V_1}^{V_2} \frac{dV}{V}$$

$$= nRT \ln(V_2/V_1) = nRT \ln(p_1/p_2)$$

25. Heat changes in reversible and irreversible processes. For a reversible process, (VI, 1) may be written

$$dU_{rev} = Dq_{rev} - Dw_{rev}$$

while for an irreversible process

$$dU_{irrev} = Dq_{irrev} - Dw_{irrev}$$

U is a function of state, therefore $dU_{rev} = dU_{irrev}$ and

Since
$$Dq_{rev} - Dw_{rev} = Dq_{irrev} - Dw_{irrev}$$
$$Dw_{rev} > Dw_{irrev}$$
$$\therefore Dq_{rev} > Dq_{irrev}$$

26. Reversible adiabatic expansion. For a reversible adiabatic expansion involving an amount n of an ideal gas

$$dU = -pdV = -nRTdV/V \quad . \quad . \quad . \quad (VI, 13)$$

From (VI, 11)

$$dU = nC_V dT \quad . \quad . \quad . \quad (VI, 14)$$

Comparison of (VI, 13) and (VI, 14) leads to

$$nC_V dT = -nRT dV/V$$

or

$$C_V dT/T = -R dV/V$$

Integrating between the limits T_1, V_1 and T_2, V_2, and realising that C_V is independent of temperature for an ideal gas, gives

$$C_V \int_{T_1}^{T_2} \frac{dT}{T} = -R \int_{V_1}^{V_2} \frac{dV}{V}$$

$$C_V \ln(T_2/T_1) = -R \ln(V_2/V_1) = R \ln(V_1/V_2) \ldots \text{(VI, 15)}$$

Now $R = C_p - C_V = C_V(\gamma - 1)$
where $\gamma = C_p/C_V$
Substituting for R in (VI, 15) gives

$$C_V \ln(T_2/T_1) = C_V(\gamma - 1) \ln(V_1/V_2)$$
$$\therefore \ln(T_2/T_1) = (\gamma - 1) \ln(V_1/V_2)$$

Taking antilogarithms, $\quad T_2/T_1 = (V_1/V_2)^{\gamma-1}$

or

$$T_1 V_1^{\gamma-1} = T_2 V_2^{\gamma-1} \ldots \text{(VI, 16)}$$

Since $T = pV/nR$, (VI, 16) may be rewritten

$$p_1 V_1^{\gamma} = p_2 V_2^{\gamma}$$

If the gas expands reversibly and adiabatically from p_2, V_2 to p_3, V_3 then

$$p_2 V_2^{\gamma} = p_3 V_3^{\gamma}$$

In general, $pV^{\gamma} = $ constant for a reversible adiabatic process. If $V_2 > V_1$, then since $\gamma - 1 > 0$, (VI, 16) indicates that $T_2 < T_1$—i.e. that an ideal gas will cool during a reversible adiabatic expansion.

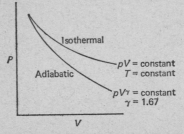

FIG. 18.—*Isothermal and adiabatic expansions for a monatomic ideal gas*

27. Comparison of an isothermal and an adiabatic expansion.
Pressure against volume curves for an isothermal (pV = constant) and an adiabatic expansion (pV^{γ} = constant) for a monatomic ideal gas ($\gamma = 1\cdot67$) are compared in Fig. 18. The curves for the adiabatic process are steeper than those for the isothermal process. In the latter process heat is absorbed to make up for the work done by the gas in expansion. In the adiabatic expansion the work done results in a decrease in the thermal energy of the gas, and hence the temperature falls.

APPLICATION OF THE FIRST LAW TO REAL GASES

28. The Joule experiment. For an ideal gas $(\partial U/\partial V)_T$ is zero by definition. The first attempt to measure $(\partial U/\partial V)_T$ for a real gas was made by Joule in 1844. He allowed air to expand from a filled into an evacuated container (Fig. 19) and attempted to measure the rise in temperature of the water surrounding the containers. Joule was unable to detect any

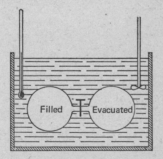

Fig. 19.—*The Joule experiment*

change in temperature, hence $q = 0$, and since the gas expanded into an evacuated space $w = 0$. From the first law $\Delta U = 0$ and U does not change with volume, *i.e.* $(\partial U/\partial V)_T = 0$. Unfortunately, Joule's technique was not sensitive enough to measure the small temperature change which does in fact take place.

29. The Joule–Thomson experiment. A more sensitive technique to measure $(\partial U/\partial V)_T$ was developed by Joule and

Thomson, and the apparatus, which was well insulated from the surroundings, is shown schematically in Fig. 20. The gas in the left hand chamber at pressure p_1 and initial volume V_1 is allowed to pass slowly through the porous plug into the right

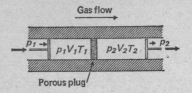

FIG. 20.—*The Joule–Thomson experiment*

hand chamber, which is at a lower pressure p_2, until the final volume in this chamber is V_2. The slow rate of flow of gas is essential to keep the kinetic energy resulting from the flow to a minimum. In contrast to Joule's experiment, the work performed is not zero.

Work done on the gas in the left hand chamber

$$= \int_{V_1}^{0} p_1 \mathrm{d}V = -p_1 V_1$$

Work done by the gas in the right hand chamber

$$= \int_{0}^{V_2} p_2 \mathrm{d}V = p_2 V_2$$

The net work $= p_2 V_2 - p_1 V_1$

Since there is no heat exchange with the surroundings, $q = 0$; hence

$$\Delta U = U_2 - U_1 = q - w = -w = p_1 V_1 - p_2 V_2$$

Rearranging $\quad U_2 + p_2 V_2 = U_1 + p_1 V_1$

$$\therefore H_2 = H_1, \text{ i.e. } \Delta H = 0$$

Thus the Joule–Thomson experiment takes place at constant enthalpy, in contrast to the Joule experiment, which occurs at constant internal energy.

Most gases (except hydrogen and helium) we found to undergo cooling on passing through the porous plug, the extent of cooling being dependent on the initial temperature and pressure.

30. The Joule–Thomson coefficient.

The change of the temperature of the gas with pressure on passing through the porous plug, $(\partial T/\partial p)_H$, is called the *Joule–Thomson coefficient* and is given the symbol μ_{JT}. This coefficient is positive when temperature and pressure both decrease, *i.e.* when the gas is cooled on passing through the porous plug.

Hydrogen and helium warm up on expansion at room temperature, but these gases can be caused to cool if they are initially below their *inversion temperatures*. These are the temperatures at which $\mu_{JT} = 0$ and the gases exhibit neither cooling nor heating. For hydrogen, μ_{JT} becomes positive below 192·7 K.

The Joule–Thomson effect is used extensively in the liquefaction of air and other gases.

31. Explanation of the Joule–Thomson effect.

Since $U + pV$, *i.e.* H remains constant, any change in pV during the process must be compensated for by a change in U. Fig. 12, which is actually a plot of pV/RT against p, but is identical in shape to one of pV against p, shows that for most gases pV initially *rises* with *decrease* in p at moderate pressures. This increase in pV results in a decrease in U which leads to a fall in temperature, *i.e.* $T_2 < T_1$. For hydrogen and helium, pV falls with decreasing p, resulting in an increase in U and $T_2 > T_1$.

32. $(\partial U/\partial V)_T$ for real gases.

At low pressure and temperature, $(\partial U/\partial V)_T$ is very close to zero for all gases, and increases with increasing pressure and temperature. $(\partial U/\partial V)_T$ tends to zero as the pressure tends to zero, supporting the postulate that $(\partial U/\partial V)_T = 0$ for ideal gases (*see* 20 above).

PROGRESS TEST 6

NOTE: In all problems the standard pressure is approximated to 101·3 kN m^{-2}.

1. Calculate the work done when one mole of water is vaporised at 373·2 K and 101·3 kN m^{-2}, given that the density of water under these conditions is $9·583 \times 10^2$ kg m^{-3}. Assume that water vapour behaves as an ideal gas. (7)

2. If the amount of heat required to vaporise one mole of water at 373·2 K and 101·3 kN m^{-2} is 40·64 kJ, calculate the change

in internal energy, using the value for the work done found in question 1 above. (14)

3. When one mole of liquid methanol is completely burned in oxygen to form liquid water and gaseous carbon dioxide, ΔU is $-724 \cdot 9$ kJ at $298 \cdot 2$ K. Calculate ΔH at this temperature. (18)

4. Calculate the work done when 10 dm^3 of an ideal gas initially at $298 \cdot 2$ K and $99 \cdot 98$ kN m^{-2} are expanded isothermally and reversibly to 20 dm^3. (24)

5. One mole of an ideal monatomic gas, initially at a temperature of 300 K and a pressure of 1013 kN m^{-2}, is expanded adiabatically and reversibly under a constant opposing pressure of $101 \cdot 3$ kN m^{-2}. Calculate the final gas temperature. What would be the final temperature if the expansion were carried out suddenly, *i.e.* irreversibly, in one step ? (26)

6. Explain what is meant by (a) an extensive property (2), (b) an adiabatic process (5), (c) an exact differential (10), (d) a thermodynamically reversible process (12).

7. State the first law of thermodynamics in (a) words, (b) symbols. (13)

8. For ideal gases derive the relationship between (a) C_p and C_V (22), (b) ΔH and ΔU (18), (c) p, V and γ for a reversible adiabatic expansion (26).

9. Define (a) an ideal gas (20), (b) the Joule–Thomson coefficient. (30)

CHAPTER VII

THERMOCHEMISTRY

Thermochemistry is a branch of thermodynamics concerned with the heat changes caused by chemical and physical changes. Processes accompanied by the evolution of heat (q negative) are called *exothermic*; those accompanied by the absorption of heat (q positive) are called *endothermic*.

THERMOCHEMICAL EQUATIONS

1. Thermochemical equations. A thermochemical equation shows the chemical or physical process which takes place, gives the state of aggregation of the reactants and products, and indicates the heat change which accompanies the process at the specified temperature. Since most processes carried out in the laboratory take place at constant pressure (the atmospheric pressure), the heat change can be equated with ΔH (*see* VI, **17**). Thus the reaction

$$C(\text{graph}) + O_2(g) = CO_2(g) \qquad \Delta H_{298} = -393 \cdot 5 \text{ kJ}$$
$$\text{. . . (VII, 1)}$$

indicates that when one mole of carbon in the form of graphite reacts with one mole of gaseous oxygen, one mole of gaseous carbon dioxide is formed and $393 \cdot 5$ kJ of heat are liberated at 298 K and $101 \cdot 325$ kN m^{-2} pressure. ΔH, the enthalpy change, is often called the *heat of reaction*.

ΔH refers to the difference in enthalpy between the final and initial states and is independent of how the process is carried out, since ΔH is a function of state, *i.e.*

$$\Delta H = H \text{ (products)} - H \text{ (reactants)}$$
$$= H(CO_2, g) - H(C, \text{graph}) - H(O_2, g)$$

It is not possible to measure $H(CO_2, g)$, etc., separately.

If (VII, 1) is written in the reverse manner, the sign of ΔH must also be reversed, *i.e.*

$$CO_2(g) = C(\text{graph}) + O_2(g) \qquad \Delta H_{298} = 393 \cdot 5 \text{ kJ}$$
$$\text{. . . (VII, 2)}$$

If the number of moles involved in thermochemical equations is multiplied by a factor, then ΔH must be multiplied by the same factor, *e.g.*

$$2CO_2(g) = 2C(graph) + 2O_2(g) \qquad \Delta H_{298} = 787 \cdot 0 \text{ kJ}$$

2. Standard states. Thermodynamic data are usually tabulated for substances in their *standard states*, which refer to the most stable form of the substance at a specified temperature (usually 298 K) and a pressure of $101 \cdot 325$ kN m^{-2}. Thus at 298 K the standard state of carbon is graphite, of sulphur, rhombic sulphur and of water, liquid water. The enthalpies of substances in their standard states are designated by the symbol $H^{\ominus}$, and $\Delta H_{298}^{\ominus}$ is the enthalpy change when the reactants and products are all in their standard states at 298 K.

3. Hess's law, or the law of constant heat summation. This law may be stated as follows: *the resultant heat change in a chemical reaction is the same whether the reaction takes place in one or several stages*. This is a direct consequence of the first law of thermodynamics, since both ΔH and ΔU for any reaction depend only on the initial and final states and not on any intermediate states. Hess's law permits the computation of heat changes for processes for which direct measurements are not available.

EXAMPLE 1: Calculate $\Delta H_{298}^{\ominus}$ for the synthesis of acetylene from its elements given:

$$C(graph) + O_2(g) = CO_2(g) \; \Delta H_{298}^{\ominus} = -393 \cdot 5 \text{ kJ} \; \ldots \text{ (VII, 3)}$$
$$H_2(g) + \tfrac{1}{2}O_2(g) = H_2O(l) \; \Delta H_{298}^{\ominus} = -285 \cdot 8 \text{ kJ} \; \ldots \text{ (VII, 4)}$$
$$C_2H_2(g) + \tfrac{5}{2}O_2(g) = 2CO_2(g) + H_2O(l) \; \Delta H_{298}^{\ominus} = -1300 \text{ kJ}$$
$$\ldots \text{ (VII, 5)}$$

Hess's law allows the addition and subtraction of equations along with their associated heat changes. The combination

$$2(\text{VII, 3}) + (\text{VII, 4}) - (\text{VII, 5})$$

gives (VII, 6), the desired reaction with its associated heat change.

$$2C(graph) + H_2(g) - C_2H_2(g) = 0 \qquad \Delta H_{298}^{\ominus} = 226 \cdot 8 \text{ kJ}$$
$$\ldots \text{ (VII, 6)}$$

Equation (VII, 6) may be rearranged to the more normal form

$$2C(\text{graph}) + H_2(g) = C_2H_2(g) \qquad \Delta H^{\ominus}_{298} = 226\cdot8 \text{ kJ}$$
$$\cdots \text{ (VII, 7)}$$

The above reaction cannot be performed directly, since reaction between C(graph) and $H_2(g)$ leads to the formation of hydrocarbons in addition to $C_2H_2(g)$.

HEATS OF REACTION

4. Heats of formation. $\Delta H^{\ominus}_{298}$ in Example 1 is for the formation of $C_2H_2(g)$ from its elements in their standard states, and is known as the *standard heat of formation*, $\Delta H^{\ominus}_f$. Thus we may write

$$\Delta H^{\ominus}_f (C_2H_2, g) = H^{\ominus}(C_2H_2, g) - 2H^{\ominus}(C, \text{graph})$$
$$- H^{\ominus}(H_2, g) \cdots \text{ (VII, 8)}$$

By adopting a convention with regard to the enthalpies of elements in their standard states, *relative* values of the enthalpies of formation of compounds may be obtained. Thus the enthalpy of an element in its standard state is *arbitrarily given the value zero*, and (VII, 8) becomes

$$\Delta H^{\ominus}_f(C_2H_2, g) = H^{\ominus}(C_2H_2, g) \cdots \text{ (VII, 9)}$$

The standard heat of formation of a compound may be considered to be its enthalpy at 298 K, and is the heat of reaction when one mole of a compound in its standard state is formed from its elements in their standard states at a temperature of 298 K. Standard heats of formation are extremely useful, since they allow calculations to be made of the heat changes in a wide variety of reactions. Consider the reaction

$$a\text{A} + b\text{B} = c\text{C} + d\text{D} \qquad \Delta H^{\ominus}_{298}$$
$$\Delta H^{\ominus}_{298} = cH^{\ominus}(\text{C}) + dH^{\ominus}(\text{D}) - aH^{\ominus}(\text{A}) - bH^{\ominus}(\text{B})$$

According to (VII, 9), the above equation may be rewritten in terms of heats of formation, *i.e.*

$$\Delta H^{\ominus}_{298} = c\Delta H^{\ominus}_f(\text{C}) + d\Delta H^{\ominus}_f(\text{D}) - a\Delta H^{\ominus}_f(\text{A}) - b\Delta H^{\ominus}_f(\text{B})$$
$$\cdots \text{ (VII, 10)}$$

EXAMPLE 2: Calculate the heat change in the reaction

$$2NH_3(g) + \tfrac{7}{2}O_2(g) = 2NO_2(g) + 3H_2O(l)\ \Delta H^{\ominus}_{298}$$

given the following standard heats of formation (in kJ mol^{-1}):

$$NH_3(g) - 46 \cdot 1;\ H_2O(l) - 285 \cdot 8;\ NO_2(g)\ 33 \cdot 2$$

$$\Delta H^{\ominus}_{298} = 2H^{\ominus}_f(NO_2,\ g) + 3\Delta H^{\ominus}_f(H_2O,\ l) - 2\Delta H^{\ominus}_f(NH_3,\ g)$$
$$-\tfrac{7}{2}\Delta H^{\ominus}_f(O_2,\ g)$$
$$= (2 \times 33 \cdot 2) - (3 \times 285 \cdot 8) + (2 \times 46 \cdot 1) - 0$$
$$= -698 \cdot 8\ \text{kJ mol}^{-1}.$$

5. Heats of combustion. The heat of combustion is the heat liberated when one mole of a substance is completely burned in oxygen. Heats of combustion may often be measured experimentally and are used in determining other heats of reaction. Example 1 is a case where heats of combustion were used to obtain a value for the standard heat of formation of acetylene. Another use is to measure the heat of reaction for the phase change of an element from one allotropic form to another. Thus the heat of transition of C(diamond) to C(graph) may be found from

$$C(\text{diamond}) + O_2(g) = CO_2(g)\ \Delta H^{\ominus}_{298} = -395 \cdot 4\ \text{kJ}$$
$$C(\text{graph})\quad + O_2(g) = CO_2(g)\ \Delta H^{\ominus}_{298} = -393 \cdot 5\ \text{kJ}$$

Subtracting the second reaction from the first gives

$$C(\text{diamond}) = C(\text{graph})\qquad \Delta H^{\ominus}_{298} = -1 \cdot 9\ \text{kJ}$$
$$\text{. . . (VII, 11)}$$

6. Heats of solution. The heat change accompanying the dissolution of one mole of a solute in a liquid is known as the *heat of a solution*, and depends on the final concentration of the solution, *e.g.*

$$H_2SO_4(l) + 2H_2O(l) = H_2SO_4(2H_2O)\ \Delta H_{298} = -41 \cdot 6\ \text{kJ}$$
$$H_2SO_4(l) + 10H_2O(l) = H_2SO_4(10H_2O)\ \Delta H_{298} = -67 \cdot 9\ \text{kJ}$$
$$H_2SO_4(l) + aq \qquad = H_2SO_4(aq)\ \Delta H_{298} = -96 \cdot 1\ \text{kJ}$$

where the symbol aq indicates that a very large amount of water is present, *i.e.* that the solution is infinitely dilute. At infinite dilution further addition of solvent gives no measurable heat change.

7. Heats of neutralisation. The heat release accompanying the neutralisation of strong acids and strong bases in dilute aqueous solution is remarkably constant. This is because strong acids and strong bases are fully dissociated into ions in aqueous solution (*see* XII, **15** and XIII, **2**), and the same process takes place in all such neutralisations, namely

$$H_3O^+(aq) + OH^-(aq) = 2H_2O(l) \qquad \Delta H_{298}^{\ominus} = -56 \cdot 0 \text{ kJ}$$
$$\qquad \qquad \qquad \qquad \ldots \text{ (VII, 12)}$$

NOTE: Although there are 2 moles of $H_2O(l)$ on the product side of (VII, 12), the heat of neutralisation is for the formation of 1 mole of $H_2O(l)$. This is because the reactant side of the equation contains 1 mole of $H_2O(l)$ intimately associated with the proton in the form $H_3O^+(aq)$. If the proton were represented by H^+, (VII, 12) would be written

$$H^+(aq) + OH^-(aq) = H_2O(l) \; \Delta H_{298}^{\ominus} = -56 \cdot 0 \text{ kJ}$$

The standard state for an ion in solution is that of unit *activity* (*see* VIII, **26–28**). For the present, the activity of an ion or other solute may be taken as an idealised concentration. As the concentration tends to zero the activity approaches the concentration, and in the limit of infinite dilution they are equal.

The heat of neutralisation is not constant for reactions involving weak acids and weak bases. Thus only $49 \cdot 8 \text{ kJ mol}^{-1}$ of heat are liberated in the neutralisation of acetic acid by ammonium hydroxide. This is because the heat required to ionise the weak acid and the weak base must be considered.

8. Heats of formation of ions. The reverse of (VII, 12) represents the formation of $H_3O^+(aq)$ and $OH^-(aq)$.

$$2H_2O(l) = H_3O^+(aq) + OH^-(aq) \qquad \Delta H_{298}^{\ominus} = 56 \cdot 0 \text{ kJ}$$
$$\qquad \qquad \qquad \qquad \ldots \text{ (VII, 13)}$$

The formation of ions in solution necessarily involves two or more ions, since there is no way of producing a single ionic species in solution. The heat change in (VII, 13) may be written in terms of standard heats of formation.

$$\Delta H_{298}^{\ominus} = \Delta H_f^{\ominus}(H_3O^+, aq) + \Delta H_f^{\ominus}(OH^-, aq) - \Delta H_f^{\ominus}(H_2O, l)$$

Hence $\Delta H_f^{\ominus}(H_3O^+, aq) - \Delta H_f^{\ominus}(OH^-, aq)$

$$= 56 \cdot 0 - 285 \cdot 8 = -229 \cdot 8 \text{ kJ} \qquad \ldots \text{ (VIII, 14)}$$

NOTE: Equation (VII, 14) involves the standard heat of formation of 1 mole of $H_2O(l)$ for the reasons given in the note in **7** above.

The heats of formation of individual ions cannot be found without some assumption with regard to the heat of formation of one of the ions. By convention $\Delta H_f^{\ominus}(H_3O^+, aq)$ at 298 K and unit activity is zero, *i.e.*

$$\frac{1}{2}H_2(g) + H_2O = H_3O^+(aq) + e^-(aq) \quad \Delta H_f^{\ominus}(H_3O^+, aq) = 0$$

Hence relative values of heats of formation of ions can be found. Equation (VII, 14) becomes

$$\Delta H_f^{\ominus}(OH^-, aq) = -229 \cdot 8 \text{ kJ}$$

Similar procedures have been used to evaluate the heats of formation of other ions in aqueous solution, and allow heat changes to be found for ionic reactions in aqueous solution.

9. Latent heats. The *latent* heat is the heat change associated with a change in physical state, such as fusion, vaporisation and alteration of crystal structure. These heat changes take place at constant temperature. Thus the heat absorbed in melting ice at $273 \cdot 16$ K causes no change in temperature until all the ice has melted (hence the term "latent," meaning hidden). An identical amount of heat is liberated when the same quantity of water freezes.

$$H_2O(l) = H_2O(s) \qquad \Delta H_{273}^{\ominus} = -6 \cdot 01 \text{ kJ}$$

The heat released in (VII, 11), the change from one allotropic form of carbon to another, is called the *latent heat of transition*.

10. Trouton's rule. This states that the latent heat of vaporisation per mole is proportional to the boiling point in K, *i.e.* $\Delta H_v = kT_b$. For a large number of liquids with very different boiling points and latent heats of vaporisation, $k \approx 90$ J K^{-1} mol^{-1}. Two classes of compounds show significant deviations from this rule. Thus liquids such as water, alcohols, etc., which are extensively hydrogen-bonded in the liquid (*see* XIX, **6–7**), but are essentially non-associated in the vapour have k values greater than 90 J K^{-1} mol^{-1}. The reason is that energy in addition to the latent heat of vaporisation must be supplied in order to break these bonds. The second class of compounds, the carboxylic acids, are also hydrogen-bonded in solution, resulting in the formation of dimers, which persist to a large extent in the vapour state. When allowance is made for this dimerisation the low values of k (59 J K^{-1} mol^{-1} for acetic acid) may be explained.

11. Other heats of reaction. In addition to the heats of reaction mentioned in 4–9 above, many other types exist. Typical of these are heats of sublimation, dissociation and atomisation. For a particular reaction there may be two ways of stating the heat release. Thus for the formation of gaseous sodium atoms from solid sodium (VII, 15), the heat change may be called the heat of sublimation ΔH_S or the heat of atomisation ΔH_A

$$Na(s) = Na(g) \qquad \Delta H_S \text{ or } \Delta H_A \quad . \quad . \quad . \quad (VII, 15)$$

12. Calorimetry. A calorimeter is used to measure heats of reaction directly, and consists of a reaction vessel which is immersed in a liquid in an insulated container. The temperature rise of the liquid caused by the heat release in an exothermic reaction is measured with a sensitive thermometer. The heat evolved in the reaction may be calculated from the temperature rise in two ways:

(a) By prior calibration of the calorimeter, using a reaction whose heat change is accurately known.
(b) By measuring the amount of electrical energy required to reproduce the temperature rise in the reaction being studied.

Heats of combustion are often measured in a bomb calorimeter, in which the substance is burned in an excess of oxygen in a pressurised container. In this apparatus the heat change is at constant volume (ΔU) and must be corrected to give the heat change at constant pressure (ΔH) (see VI, **18**).

BOND ENERGIES

Chemical reactions between covalent compounds involve the breaking and making of chemical bonds, and the energies involved in these processes must be reflected in the accompanying heat changes.

13. Bond energies. The energy involved in the breaking of a covalent bond is known as the *bond energy* and may be defined as the energy required to separate two covalently bonded atoms by an infinite distance in the gaseous state at $101 \cdot 325$ kN m^{-2} and the specified temperature.

14. Bond dissociation energies. The energy required to break a *given* bond in a *specific* compound is the *bond dissociation energy* (*see* XVII, 1). Thus, for example,

$$CH_4(g) = CH_3(g) + H(g) \qquad \Delta H_D[CH_3 - H] = 435 \text{ kJ}$$
$$C_2H_6(g) = C_2H_5(g) + H(g) \qquad \Delta H_D[C_2H_5 - H] = 402 \text{ kJ}$$
$$(CH_3)_3CH(g) = (CH_3)_3C + H(g) \quad \Delta H_D[(CH_3)_3C - H] = 377 \text{ kJ}$$

Bond dissociation energies are just special cases of heats of reaction. Each of the bond dissociation energies in the above reactions refer to the C–H bond, and although they are relatively constant, each is dependent on the exact environment of the bond.

15. Average bond energies. The approximate constancy of the bond dissociation energies of the C–H bond in a series of compounds leads to the concept of an *average bond energy*, $\Delta H_{\bar{D}}$, *i.e.* an average value of the bond dissociation energies of a given bond in a series of different dissociating species. In the above examples $\Delta H_{\bar{D}}[C–H]$ is 405 kJ mol^{-1}, while the value 413 kJ mol^{-1} is obtained from C–H bond energies taken from a wide selection of organic compounds.

NOTE: Ionic reactions cannot be treated in the above manner, since they invariably take place in solution, where the process of bond breakage and formation is greatly influenced by the solvent.

16. The use of average bond energies. The primary value of average bond energies lies in the calculation of heats of reactions involving compounds for which no enthalpy data are available.

EXAMPLE 3: Calculate the standard heat of formation of ethylene, given that the average bond energies of C=C and C–H are 610 and 413 kJ mol^{-1} respectively and the heats of formation of C(g) and H(g) are 716 and 218 kJ mol^{-1} respectively.

The dissociation of C_2H_4(g) into gaseous atoms requires an expenditure of energy equal to the average bond energies of one C=C and four C–H bonds. Hence the enthalpy change for the formation of C_2H_4(g) from its gaseous atoms is given by

$$2C(g) + 4H(g) = C_2H_4(g) \quad \Delta H_{298}$$
$$= -(\Delta H_{\bar{D}}[C=C] + 4\Delta H_{\bar{D}}[C–H]) \quad \text{. . . (VII, 15)}$$

The negative sign is required since average bond energies refer to the breaking of bonds, whereas in the above reaction we are concerned with bond formation. Hence

$$\Delta H_{298} = -(610 + 4 \times 413) = -2262 \text{ kJ mol}^{-1}$$

However, gaseous atoms of carbon and hydrogen are not the standard states of these elements, and the reactions required to produce these atoms are given below.

$$2C(\text{graph}) = 2C(g) \quad \Delta H_{298} = 2 \times 716 \text{ kJ} \ldots \text{(VII, 16)}$$
$$2H_2(g) \quad = 4H(g) \quad \Delta H_{298} = 4 \times 218 \text{ kJ} \ldots \text{(VII, 17)}$$

Addition of (VII, 15), (VII, 16) and (VII, 17) gives

$$2C(\text{graph}) + 2H_2(g) = C_2H_4(g)$$

and the accompanying heat change

$$\Delta H_f^{\ominus}(C_2H_4, g) = -2262 + 1432 + 872 = 42 \text{ kJ mol}^{-1}$$
$$\ldots \text{(VII, 18)}$$

The experimentally determined value of 52·3 kJ mol^{-1} is in reasonable agreement with the calculated value since the latter involves a small difference between two large numbers thus magnifying any errors.

THE TEMPERATURE DEPENDENCE OF HEATS OF REACTION

17. Kirchhoff's equation. The thermochemical data considered so far have all been at 298 K. However, heats of reaction are temperature-dependent, and since chemical reactions are often carried out at temperatures other than 298 K, it is important to have information on the variation of heats of reaction with temperature. The heat of a reaction may be written

$$\Delta H = H(\text{products}) - H(\text{reactants})$$

Differentiating both sides with respect to temperature at constant pressure

$$(\partial \Delta H / \partial T)_p = (\partial H(\text{products}) / \partial T)_p + (\partial H(\text{reactants}) / \partial T)_p$$

Since $(\partial H/\partial T)_p = C_p$ the above equation becomes

$$(\partial \Delta H/\partial T)_p = C_p(\text{products}) - C_p(\text{reactants}) = \Delta C_p \qquad \ldots \text{ (VII, 19)}$$

Equation (VII, 19) is known as *Kirchhoff's equation*. If the enthalpy limits ΔH_1 and ΔH_2 correspond to the temperature limits T_1 and T_2 integration of (VII, 19) gives

$$\int_{\Delta H_1}^{\Delta H_2} \mathrm{d}(\Delta H) = \Delta H_2 - \Delta H_1 = \int_{T_1}^{T_2} \Delta C_p \mathrm{d}T \quad \ldots \text{ (VII, 20)}$$

The integration can be performed only if ΔC_p is a constant, or its variation with temperature is known. If ΔC_p is a constant, as will be the case for an ideal gas and also for a real gas over a small temperature range, (VII, 20) reduces to

$$\Delta H_2 - \Delta H_1 = \Delta C_p(T_2 - T_1) \quad \ldots \text{ (VII, 21)}$$

Over a wide range of temperature an empirical equation of the form $C_p = a + bT + cT^2$ may be used for the heat capacities of the reactants and the products, where the coefficients a, b and c are determined experimentally.

For reactions at constant volume, (VII, 19) may be rewritten

$$(\partial \Delta U/\partial T)_V = \Delta C_V$$

18. Application of Kirchhoff's equation. ΔH may be calculated at any temperature, provided information is available on the value of ΔH at some other temperature and on the molar heat capacities of the reactions and products.

EXAMPLE 4: Calculate $\Delta H^{\ominus}_{1000}$ for the reaction

$$\tfrac{1}{2}N_2(g) + \tfrac{3}{2}H_2(g) = NH_3(g) \quad \Delta H^{\ominus}_{298} = -46 \cdot 19 \text{ kJ}$$

given the following mean molar heat capacities (in $J \, K^{-1} \, mol^{-1}$)
$C_p(H_2) = 28 \cdot 6$; $C_p(N_2) = 28 \cdot 8$; $C_p(NH_3) = 35 \cdot 9$.

$$\begin{aligned}
\Delta C_p &= C_p(NH_3) - \tfrac{1}{2}C_p(N_2) - \tfrac{3}{2}C_p(H_2) \\
&= 35 \cdot 9 - (28 \cdot 8/2) - (3 \times 28 \cdot 6/2) = -21 \cdot 4 \text{ J K}^{-1} \text{ mol}^{-1}
\end{aligned}$$

Hence from (VII, 21)

$$\begin{aligned}
\Delta H^{\ominus}_{1000} + 46 \cdot 190 &= -21 \cdot 4 \, (1000 - 298) = -15\,020 \\
\therefore \Delta H^{\ominus}_{1000} &= -61\,210 \text{ J mol}^{-1} = -61 \cdot 21 \text{ kJ mol}^{-1}
\end{aligned}$$

PROGRESS TEST 7

1. Given that $\Delta H^{\ominus}_{298}$ for the reaction

$$2H_2S(g) + SO_2(g) = 2H_2O(l) + 3S(\text{rhombic})$$

is $-234 \cdot 5$ kJ mol^{-1} and the standard heats of formation of $SO_2(g)$ and $H_2O(l)$ are $-296 \cdot 9$ and $-285 \cdot 9$ kJ mol^{-1} respectively, calculate the standard heat of formation of $H_2S(g)$. Also calculate $\Delta U^{\ominus}_{298}$ for the above reaction, assuming the gases to behave ideally. (VII, 4 and VI, 18)

2. Calculate $\Delta H^{\ominus}_{298}$ for the reaction

$$Na(s) + \tfrac{1}{2}Cl_2(g) + aq = Na^+(aq) + Cl^-(aq)$$

given the following standard heats of formation (in kJ mol^{-1}):

$$Na^+(aq) = -240 \cdot 6;\ Cl^-(aq) = -166 \cdot 9 \qquad (8)$$

3. The heat of combustion of gaseous ethane is -1560 kJ mol^{-1}. From this and the following data, calculate the mean C–C bond energy in ethane. Standard heats of formation (in kJ mol^{-1}): $CO_2(g) - 393 \cdot 5$; $H_2O(l - 285 \cdot 9$; $C(g)$ 716; $H(g)$ 218. The mean C–H bond energy $= 413$ kJ mol^{-1}. (15–16)

4. Calculate $\Delta H^{\ominus}_{1000}$ for the reaction

$$\tfrac{1}{2}N_2(g) + \tfrac{3}{2}H_2(g) = NH_3(g) \quad \Delta H^{\ominus}_{298} = -46 \cdot 19 \text{ kJ}$$

given $C_p(N_2) = 27 \cdot 2 + 4 \cdot 2 \times 10^{-3}T$
$\quad C_p(H_2) = 27 \cdot 2 + 3 \cdot 8 \times 10^{-3}T$
and $C_p(NH_3) = 33 \cdot 6 + 2 \cdot 9 \times 10^{-3}T + 21 \cdot 3 \times 10^{-6}T^2$. (17–18)

5. Define (a) standard state (2), (b) heat of formation (4), (c) heat of solution (6), (d) bond dissociation energy (14), (e) average bond energy (15).

6. State (a) Hess's law (3), (b) Trouton's rule (10).

7. Derive an expression for the variation of heat of reaction with temperature. (17)

CHAPTER VIII

THE SECOND AND THIRD LAWS OF THERMODYNAMICS

THE ENTROPY

1. The limitations of the first law. In Chapter VI we examined the changes brought about in a system by the absorption of heat and the subsequent performance of work. No restrictions were placed on the direction of change even though we know from experience that, under given conditions, the direction of change is fixed. Thus heat flows along a metal bar from a region of higher temperature to one of lower temperature, a gas passes from a region of higher pressure to one of lower pressure, and mixing aqueous solutions of barium chloride and sodium sulphate always produces a precipitate of barium sulphate. These are all *natural* processes and proceed spontaneously (*i.e.* of their own accord, without the aid of any external agency) in the direction indicated. The reverse processes do not occur spontaneously even though they would not be forbidden by the first law. All the above-mentioned processes tend to proceed spontaneously *in a direction which will lead to equilibrium.*

2. The need for a new criterion. It was originally thought that all spontaneous reactions were exothermic, as indeed many are. However, spontaneous endothermic reactions are also known. Thus ΔU and ΔH are not good criteria of spontaneity, and a new function of state is required which will change in a characteristic manner when a reaction proceeds spontaneously. A function which satisfies this condition is the *entropy, S.*

3. The definition of entropy. Entropy itself is not easy to define and it is more convenient initially to define a change in entropy. When a quantity of heat Dq_{rev} is absorbed in a reversible process at a temperature T, the change in entropy, dS, is given by

$$dS = \mathrm{D}q_{rev}/T \qquad . \ . \ . \text{ (VIII, 1)}$$

For a finite change at constant temperature,

$$\Delta S = S_2 - S_1 = \int \frac{\mathrm{D}q_{rev}}{T} = \frac{1}{T} \int \mathrm{D}q_{rev} = \frac{q_{rev}}{T} \ . \ . \ . \text{ (VIII, 2)}$$

When a system gains an amount of heat q_{rev} the surroundings must lose an amount of heat q_{rev}, and the entropy change of the surroundings will be

$$\Delta S_{surr} = -q_{rev}/T$$

The total entropy change of the system and the surroundings, ΔS_{net}, will be

$$\Delta S_{net} = \Delta S_{syst} + \Delta S_{surr} = (q_{rev}/T) - (q_{rev}/T) = 0$$

Since entropy is a function of state, dS is an exact differential (*see* VI, 10) and hence the integral of this function around a closed path will be zero, *i.e.* $\oint dS = 0$.

ENTROPY CHANGES IN REVERSIBLE PROCESSES

4. Entropy change for a phase transition. For a reversible change in phase at constant pressure

$$\Delta S_{syst} = q_{rev}/T = \Delta H_t/T$$

where ΔH_t is the latent heat of transition and T is the transition temperature. Thus for the change of one mole of ice to one mole of liquid water at 273·16 K, $\Delta H_t = 6010$ J mol⁻¹

$$\therefore \Delta S_{water} = 6010/273\cdot16 = 21\cdot99 \text{ J K}^{-1}\text{ mol}^{-1}$$

5. Entropy change for an ideal gas in terms of temperature and volume. The change in internal energy on supplying an ideal gas with a quantity of heat $\mathrm{D}q$ at constant pressure is given by

$$dU = \mathrm{D}q - pdV \quad \text{(only expansion work possible)}$$

If the heat change takes place reversibly, and remembering that for an ideal gas $dU = nC_V\,dT$ (*see* VI, 22), then the above equation can be rewritten

$$Dq_{rev} = nC_V\,dT + p\,dV$$

Now
$$dS_{gas} = Dq_{rev}/T = nC_V\,dT/T + p\,dV/T$$

For an ideal gas $p = nRT/V$, hence

$$dS_{gas} = nC_V\,dT/T + nR\,dV/V$$

If the temperature changes from T_1 to T_2 for a volume change from V_1 to V_2

$$\int dS_{gas} = \Delta S_{gas} = \int_{T_1}^{T_2} \frac{nC_V\,dT}{T} + \int_{V_1}^{V_2} \frac{nR\,dV}{V}$$

Since C_V is independent of temperature, for an ideal gas

$$\Delta S_{gas} = nC_V \ln(T_2/T_1) + nR \ln(V_2/V_1) \quad . \quad . \quad . \text{ (VIII, 3)}$$

The first term on the right hand side of (VIII, 3) gives the change in entropy for a change in temperature at constant volume, *i.e.* at constant volume

$$\Delta S_{gas} = nC_V \ln(T_2/T_1) \qquad . \quad . \quad . \text{ (VIII, 4)}$$

The second term gives the change in entropy for a change in volume at constant temperature, *i.e.* under isothermal conditions

$$\Delta S_{gas} = nR \ln(V_2/V_1) \quad . \quad . \quad . \text{ (VIII, 5)}$$

Equation (VIII, 3) indicates that the entropy of an ideal gas increases on warming and on expansion.

6. Entropy change for an ideal gas in terms of temperature and pressure. If the the temperature, volume and pressure of an ideal gas change, then

$$p_1V_1 = nRT_1 \quad \text{and} \quad p_2V_2 = nRT_2$$

It follows that

$$V_2/V_1 = p_1T_2/p_2T_1$$

Substituting this ratio into (VIII, 3) gives

$$\Delta S_{gas} = nC_V \ln(T_2/T_1) + nR \ln(T_2/T_1) + nR \ln(p_1/p_2)$$

Since $C_p = C_V + R$ the above equation may be simplified

$$\Delta S_{gas} = nC_p \ln(T_2/T_1) + nR \ln(p_1/p_2) \quad \ldots \text{(VIII, 6)}$$

The first term on the right hand side of (VIII, 6) gives the change in entropy due to a change in temperature at constant pressure, *i.e.* at constant pressure

$$\Delta S_{gas} = nC_p \ln(T_2/T_1) \quad \ldots \text{(VIII, 7)}$$

The second term gives the change in entropy due to a change in pressure at constant temperature, *i.e.* under isothermal conditions

$$\Delta S_{gas} = nR \ln(p_1/p_2) \quad \ldots \text{(VIII, 8)}$$

Equation (VIII, 6) indicates that the entropy of an ideal gas increases on warming and on reducing the pressure.

7. The net entropy change. In the above examples the entropy change of the surroundings will be equal and opposite to the entropy change of the system, and the net entropy will therefore be zero. The combination of system plus surroundings corresponds to an isolated system (*see* VI, 1). Hence for a reversible change in an isolated system $\Delta S = 0$ for a finite change and $dS = 0$ for an infinitesimal change.

ENTROPY CHANGES IN IRREVERSIBLE PROCESSES

Since entropy is a function of state, ΔS will be the same for reversible and irreversible processes between the same initial and final states. Thus in order to calculate the entropy change accompanying an irreversible process it is only necessary to find a reversible path between the same initial and final states.

8. Isothermal irreversible expansion of an ideal gas. The entropy change of the gas, ΔS_{gas}, will be given by (VIII, 5). However, ΔS_{surr} will differ from that obtained for the reversible process. Consider the case of the gas expanding into a vacuum with the performance of no work, *i.e.* $w = 0$. Since the change is isothermal, $\Delta U = 0$ (the change in internal energy of an ideal gas is solely a function of temperature), and from the first law $q = 0$. Therefore $\Delta S_{surr} = 0$

$$\therefore \Delta S_{net} = \Delta S_{gas} + \Delta S_{surr} = nR \ln(V_2/V_1)$$

Since $nR \ln(V_2/V_1) > 0$, the irreversible expansion is accompanied by an increase in the net entropy.

9. Flow of heat from a hot body to a cold body.

Suppose two bodies at temperatures T_1 and T_2 ($T_2 > T_1$), insulated from the surroundings, are brought together momentarily so that a small quantity of heat, q, flows directly between them. q is such that T_1 and T_2 are effectively unchanged by the transfer, and therefore there is no entropy change in the surroundings. The entropy change of the cold body is q/T_1, while that of the hot body is $-q/T_2$. The net entropy change will be

$$\Delta S_{net} = (q/T_1) - (q/T_2) > 0, \text{ since } T_2 > T_1$$

Again there is a net increase in entropy.

10. General statement.

The examples given in 4–9 above illustrate the completely general result that for a finite change in an *isolated* system

$$\Delta S_{net} \geqslant 0 \qquad \text{. . . (VIII, 8)}$$

and for an infinitesimal change

$$dS_{net} \geqslant 0 \qquad \text{. . . (VIII, 9)}$$

These equations form a mathematical statement of the *second law of thermodynamics*. Since reversible processes are always in a state of *equilibrium* (they are, in effect, a continuous sequence of equilibrium states), and irreversible processes correspond to spontaneous, *i.e.* observable, processes, the criterion of equilibrium is given by

$$dS_{net} = 0 \qquad \text{. . . (VIII, 10)}$$

while for a spontaneous process

$$dS_{net} \geqslant 0 \qquad \text{. . . (VIII, 11)}$$

11. Verbal statement of the second law.

In an isolated system the direction of change is such that the entropy increases to a maximum; at equilibrium the entropy is constant. The importance of the second law lies in its ability to determine whether or not a given process is likely to occur.

For an isolated system:

(a) dS_{net} *positive indicates that the process proceeds spontaneously in the forward direction.*

(b) dS_{net} *zero indicates that the process is in equilibrium.*

(c) dS_{net} *negative indicates that the process proceeds spontaneously in the reverse direction.*

The second law was initially deduced from the impossibility of certain thermal processes, and one often quoted version is that due to Clausius: *heat cannot be conveyed from a body at a lower temperature to another at a higher temperature without the expenditure of work.* This is merely an alternative way of stating that heat flows spontaneously from a body at a higher temperature to one at a lower temperature.

ENTROPY AND DISORDER

12. Matter on the molecular scale. Changes which are accompanied by an *increase* in entropy (*i.e.* spontaneous changes) all have one feature in common: they result in an *increase* in molecular disorder. Thus in fusion the system changes from the high degree of order of the crystal lattice to the more disordered state of the liquid. This increase in disorder is continued in vaporisation, where the change is from the liquid to the highly disordered gaseous state. Both fusion and vaporisation are accompanied by an increase in entropy. The increase in entropy on heating a substance can be related to the increase in the thermal motion of the molecules. Similar changes in disorder may be noted in chemical reactions. Thus the reaction

$$2BaO_2(s) = 2BaO(s) + O_2(g)$$

is accompanied by a large increase in entropy since there is the net generation of one mole of gaseous oxygen, which greatly increases the disorder on the right hand side of the equation.

13. Disorder and probability. A disordered system has a high probability of occurrence, whereas an ordered system has a low probability. The chance of shuffling a pack of cards into an ordered arrangement of the four suits is very small, while a disordered arrangement is highly probable. Since entropy and disorder are related, so are entropy and probability. Entropy increases as systems move from ordered (low probability) states to disordered (high probability) states. This relationship led Boltzmann to propose that entropy could be defined by the equation

$$S = k \ln \Omega \qquad \ldots \text{(VIII, 12)}$$

where k is Boltzmann's constant, and Ω represents the *thermodynamic probability*, *i.e.* the number of possible ways of distributing the molecules among the various positions and energy levels available which will lead to the observed state of the system.

14. The third law of thermodynamics. If there is only one way of performing the distribution, $\Omega = 1$ and hence $S = 0$. This situation might be expected at 0 K, when all atoms, ions and molecules will be in a state of lowest possible energy, and gives rise to the *third law of thermodynamics*. This states that *the entropy of all perfect crystalline substances may be taken as zero at* 0 K. Perfect crystalline substances are those in which all the atoms, ions or molecules are arranged in a flawless lattice. Imperfect crystals, glasses and supercooled liquids have residual entropies at 0 K.

ABSOLUTE ENTROPIES

15. Calculation of absolute entropies. At constant pressure, $\mathrm{D}q_{rev}$ in (VIII, 1) may be replaced by $\mathrm{d}H$ and rewritten

$$\mathrm{d}S = \mathrm{d}H/T$$

From (VI, 11), $\mathrm{d}H = C_p \mathrm{d}T$ and the above equation becomes

$$\mathrm{d}S = C_p \mathrm{d}T/T$$

The change in entropy on heating a solid from 0 K to T_1 K at constant pressure is given by

$$\Delta S = S_{T_1} - S_0 = \int_0^{T_1} \frac{C_p \mathrm{d}T}{T} \quad \ldots \text{(VIII, 13)}$$

If the solid is perfectly crystalline at 0 K, $S_0 = 0$ and (VIII, 13) becomes

$$S_{T_1} = \int_0^{T_1} \frac{C_p \mathrm{d}T}{T} \qquad \ldots \text{(VIII, 14)}$$

For substances which are gases at T_1 K, the entropies involved in any phase changes (such as fusion and vaporisation) must be added to (VIII, 14). For example, for a gas at 298 K

$$S_{298} = \int_0^{298} \frac{C_p \mathrm{d}T}{T} + \sum_{\substack{\text{all phase} \\ \text{changes}}} \frac{\Delta H}{T} \quad \cdots \quad (\text{VIII, 15})$$

If the substance is in its standard state, then S_{298} may be replaced by $S_{298}^{\ominus}$, the standard entropy. The entropies given in (VIII, 14) and (VIII, 15) are *absolute* entropies and not *changes in* entropy. If the solid is not perfectly crystalline at 0 K, the residual entropy must be taken into account when computing absolute entropies.

16. Standard entropies of solids, liquids and gases.

(*a*) *Solids.* The entropy of a solid at 298 K may be found by graphical integration of (VIII, 14) if the variation of C_p with temperature is known. Since values of C_p are not usually known below about 15 K, an extrapolation to 0 K is necessary (*see* Fig. 21).

(*b*) *Gases.* In this case (VIII, 15) must be used, where all the phase changes are included. Consider the case of a substance

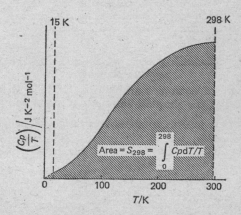

FIG. 21.—*Plot of C_p/T against T for the determination of the absolute entropy of a solid*

which is solid at 0 K, melts to a liquid at T_f K and vaporises at T_v K. The various contributions to $S^{\ominus}_{298}$ are shown below.

(i) $\displaystyle\int_0^{T_f} \frac{C_p \mathrm{d}T}{T}$ for the solid S_1

(ii) $\Delta H / T_f$ for fusion S_2

(iii) $\displaystyle\int_{T_f}^{T_v} \frac{C_p \mathrm{d}T}{T}$ for the liquid S_3

(iv) $\Delta H / T_v$ for vaporisation S_4

(v) $\displaystyle\int_{T_v}^{298} \frac{C_p \mathrm{d}T}{T}$ for the gas S_5

$$S^{\ominus}_{298} = S_1 + S_2 + S_3 + S_4 + S_5$$

The five contributions to $S^{\ominus}_{298}$ are shown pictorially in Fig. 22.

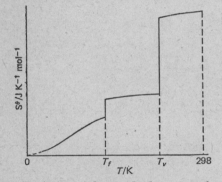

FIG. 22.—*Variation of the standard entropy of a substance with temperature*

Some standard entropies are shown in Table 11, where it can be seen that the simplest solids have the lowest entropies

TABLE 11   STANDARD ENTROPIES (IN $J\ K^{-1}\ mol^{-1}$)

Solids		Gases	
C(graph)	5·9	He	126·1
C(diamond)	2·5	H_2	130·6
Ba	67	O_2	205·0
$BaSO_4$	132·2	CO	197·6
Liquids		CO_2	213·6
H_2O	70·0	H_2O	188·7
Hg	76·0	CH_4	186·2
Br_2	152·3	C_6H_6	269·2

and that the entropy increases with increasing complexity of the molecule. In general, the entropy increases in the order solid < liquid < gas.

17. Use of standard entropies. Standard entropies of substances may be used to calculate the standard entropy changes in chemical reactions in just the same way as standard enthalpies of formation. There is, however, one important difference. *The standard entropy of an element is not arbitrarily given the value zero.*

THE GIBBS FREE ENERGY

18. The need for an additional criterion of spontaneity. Although dS is a reliable criterion it is not very convenient in practice, since account must be taken of the entropy change of the system *and* its surroundings. A better criterion would be expressed in terms of the properties of the system alone. A function of state which does this is the *Gibbs free energy* or simply the *free energy*, G, given by

$$G = H - TS \qquad . \; . \; . \; \text{(VIII, 16)}$$

For an infinitesimal isothermal change ($dT = 0$)

$$dG = dH - TdS \qquad . \; . \; . \; \text{(VIII, 17)}$$

while for a similar finite change

$$\Delta G = \Delta H - T\Delta S \qquad . \; . \; . \; \text{(VIII, 18)}$$

19. The criterion of spontaneity at constant temperature and pressure. At constant temperature and pressure dH and TdS may be replaced by Dq and Dq_{rev} respectively (*see* VI, **17** and VIII, **3**) and (VIII, 17) becomes

$$dG = Dq - Dq_{rev}$$

There are two possibilities:

(a) For a reversible process, *i.e.* one in equilibrium, $Dq = Dq_{rev}$

$$\therefore dG = 0$$

(b) For an irreversible (spontaneous process), $Dq < Dq_{rev}$ (*see* VI, **25**)

$$\therefore dG < 0$$

If dG *is negative, the process will proceed spontaneously in the forward direction.*

If dG *is zero, the process is in equilibrium.*

If dG *is positive, the process will proceed spontaneously in the reverse direction.*

In general, at constant temperature and pressure

$$\mathrm{d}G \leqslant 0$$

while for a finite change

$$\Delta G \leqslant 0$$

20. Free energy and equilibrium. Equation (VIII, 18) may be rewritten

$$-\Delta G = -\Delta H + T\Delta S$$

The driving force behind a process is the tendency for ΔG to diminish; the more negative ΔG, the greater the tendency to change. The tendency for a process to take place depends on two factors:

(a) $-\Delta H$, which represents the tendency of a system to achieve a state of minimum energy.

(b) $T\Delta S$, which represents the tendency of a system to achieve a state of maximum entropy (*i.e.* maximum disorder).

At equilibrium the two tendencies are just balanced and $\Delta G = 0$. It may not be possible for a system to achieve both minimum energy and maximum entropy, and a compromise is usually reached. Thus a system may move to a state of higher energy if this can be balanced by a sufficiently large increase in entropy.

At normal temperatures $\Delta H > T\Delta S$ and the majority of chemical reactions are exothermic. Since $T\Delta S$ increases with temperature more quickly than does ΔH, $T\Delta S$ becomes significant at high temperatures.

21. Free energy and useful work. If a system undergoes an infinitesimal change in volume at constant temperature and pressure, $\mathrm{d}H = \mathrm{d}U + p\mathrm{d}V$ (*see* VI, **18, 22**) and (VIII, 17) may be written

$$\mathrm{d}G = \mathrm{d}U + p\mathrm{d}V - T\mathrm{d}S \quad . \quad . \quad . \quad \text{(VIII, 19)}$$

For a reversible process in which types of work other than expansion may be performed

$$dU = Dq_{rev} - Dw_{rev} = TdS - Dw_{rev} \quad . \quad . \quad . \quad \text{(VIII, 20)}$$

Substitution of (VIII, 20) into (VIII, 19) gives

$$dG = -(Dw_{rev} - pdV)$$

or

$$-dG = Dw_{rev} - pdV = Dw_{useful}$$

For a finite change under similar conditions

$$-\Delta G = w_{useful}$$

Dw_{useful} is the useful or non-expansion work which can be obtained from a reversible reaction at constant temperature and pressure. The decrease in free energy during a process is equal to the useful work that may in principle be obtained from that process. One common form of useful work is electrical work, and an important way of measuring ΔG is to carry out the process in an electrochemical cell at constant temperature and pressure (*see* XIV, 12).

Equation (VIII, 18) may be stated in words

Useful or available work	=	Total available energy	−	unavailable energy
$-\Delta G$	=	$-\Delta H$	+	$T\Delta S$

22. Standard free energies of formation. These are analogous to standard heats of formation (*see* VII, 4). Thus $\Delta G_f^{\ominus}$ represents the change in free energy when one mole of a substance in its standard state is formed from its component elements in their standard states at $101 \cdot 325$ kN m^{-2} and the specified temperature. The standard free energy of formation of an element is zero by convention. Standard free energies of formation may be used to calculate standard free energy changes in a manner completely analogous to that used for standard heats of formation. Thus the analogous equation to (VII, 10) is

$$\Delta G_{298}^{\ominus} = c\Delta G_f^{\ominus}(\text{C}) + d\Delta G_f^{\ominus}(\text{D}) - a\Delta G_f^{\ominus}(\text{A}) - b\Delta G_f^{\ominus}(\text{B})$$

EXAMPLES OF FREE ENERGY CHANGES

23. Phase changes

EXAMPLE 1: Calculate $\Delta G^{\ominus}$ for the conversion of one mole of liquid water to one mole of ice at 273·16 K and 101·325 kN m^{-2}.

$$\Delta H_{273}^{\ominus} = \Delta H_{\text{fusion}}$$
$$\Delta S^{\ominus} = \Delta H_{\text{fusion}}/T \qquad \therefore T\Delta S = \Delta H_{\text{fusion}}$$
Now $\qquad \Delta G^{\ominus} = \Delta H^{\ominus} - T\Delta S^{\ominus}$
$$= \Delta H_{\text{fusion}} - \Delta H_{\text{fusion}} = 0$$

The result is expected, since ice and water are in equilibrium at 273·16 K and 101·325 kN m^{-2}.

The values of $\Delta H^{\ominus}$ and $\Delta S^{\ominus}$ for this phase change are known both above and below 273·16 K (Table 12). Thus at 263·16 K

TABLE 12

VALUES OF $\Delta G^{\ominus}$ FOR THE PHASE CHANGE $H_2O(l) = H_2O(s)$

T/K	$\Delta H^{\ominus}/$ J mol^{-1}	$\Delta S^{\ominus}/$ J K^{-1} mol^{-1}	$-T\Delta S^{\ominus}/$ J mol^{-1}	$\Delta G^{\ominus}/$ J mol^{-1}
263·16	−5616	20·5	5395	−221
273·16	−6005	22·0	6005	0
283·16	−6394	23·4	6626	232

$\Delta G^{\ominus}$ is negative, and hence the phase change from liquid water to ice is spontaneous. At 283·16 K $\Delta G^{\ominus}$ is positive and the change is not spontaneous.

24. Chemical reactions

EXAMPLE 2: Using the entropy data given in Table 11 and given that $\Delta H_f^{\ominus}(CO_2)$ is −393·5 kJ mol^{-1}, calculate $\Delta G_f^{\ominus}(CO_2)$.

The reaction is

$$C(\text{graph}) + O_2(g) = CO_2(g) \qquad \Delta H_{298}^{\ominus} = -393·5 \text{ kJ}$$
$$\Delta S^{\ominus} = S^{\ominus}(CO_2) - S^{\ominus}(O_2) - S^{\ominus}(C)$$
$$= 213·6 - 205 - 5·7 = 3·1 \text{ J K}^{-1} \text{ mol}^{-1}$$

Now $\qquad \Delta G^{\ominus} = \Delta H^{\ominus} - T\Delta S^{\ominus}$
$$\therefore \Delta G_f^{\ominus}(CO_2) = -393·5 - (298 \times 3·1 \times 10^{-3})$$
$$= -394·4 \text{ kJ mol}^{-1}$$

The reaction proceeds with a large decrease in free energy, which is in agreement with the fact that it is spontaneous. The fact that the reaction does not proceed at an observable rate can be explained from kinetic considerations (see XXII).

The sign of the free energy change indicates whether or not a reaction will proceed spontaneously, but gives no information with regard to the rate of the reaction.

25. Pressure changes for ideal gases. Since $H = U + pV$, (VIII, 16) may be written

$$G = U + pV - TS$$

For an infinitesimal change in G

$$dG = dU + pdV + Vdp - TdS - SdT \quad . . . \text{(VIII, 21)}$$

If only expansion work is performed $dU = Dq - pdV$ and (VIII, 21) becomes

$$dG = Dq + Vdp - TdS - SdT$$

For a reversible change $Dq = TdS$. Hence

$$dG = Vdp - SdT \quad . . . . \text{(VIII, 22)}$$

Furthermore, if the change takes place isothermally (VIII, 22) simplifies to

$$dG = Vdp \quad . . . \text{(VIII, 23)}$$

If the pressure change is from p_1 to p_2 (VIII, 23) may be integrated

$$\Delta G = G_2 - G_1 = \int_{p_1}^{p_2} Vdp \quad . . . \text{(VIII, 24)}$$

Since $V = RT/p$ for one mole of an ideal gas

$$G_2 - G_1 = RT\int_{p_1}^{p_2} \frac{dp}{p} = RT \ln (p_2/p_1) \quad . . . \text{(VIII, 25)}$$

If state 1 represents the standard state, then $p_1 = p^\ominus$ (= $101 \cdot 325 \, \text{kN m}^{-2}$) and $G_1 = G^\ominus$. Putting $p_2 = p$ and $G_2 = G$, (VIII, 25) becomes

$$G - G^\ominus = RT \ln (p/p^\ominus)$$
$$\text{or} \qquad G = G^\ominus + RT \ln (p/p^\ominus) \quad . . . \text{(VIII, 26)}$$

26. Pressure changes for real gases. For real gases, integration of (VIII, 24) would require a real gas equation of state such as van der Waals' equation (*see* V, 3). This would result in a complicated dependence of ΔG on pressure. It would be convenient if expressions of the type used in 25 above could be retained for real gases. This may be achieved by replacing pressure by a new function, the *fugacity*, f, and (VIII, 26) becomes

$$G = G^{\ominus} + RT \ln (f/f^{\ominus}) \quad . \quad . \quad . \text{(VIII, 27)}$$

where $f^{\ominus}$, the standard state of fugacity, is $101 \cdot 325$ kN m^{-2}. The ratio $f/f^{\ominus}$ is called the *activity*, a, and (VIII, 27) may be written

$$G = G^{\ominus} + RT \ln a \quad . \quad . \quad . \text{(VIII, 28)}$$

In the standard state $G = G^{\ominus}$, $RT \ln a = 0$ and therefore $a = 1$. *Thus the standard state of activity is unity.*

For an ideal gas $f = p$ and for all gases f approaches p as p tends to zero. The ratio f/p is called the activity coefficient, γ, and is unity for an ideal gas. It is important to realise that both a and γ are dimensionless quantities.

27. Concentration changes in solution. If a solute in solution obeys Henry's law (*see* X, 30), *i.e.* is ideal, the pressure in (VIII, 26) may be replaced by the concentration, c.

$$G = G^{\ominus} + RT \ln (c/c^{\ominus}) \quad . \quad . \quad . \text{(VIII, 29)}$$

where $c^{\ominus}$, the standard state of concentration, is 1 mol dm^{-3}. Because of the choice of standard state, c is numerically equal to $c/c^{\ominus}$ and (VIII, 29) becomes

$$G = G^{\ominus} + RT \ln c \quad . \quad . \quad . \text{(VIII, 30)}$$

For solutes which do not behave ideally the fugacity concept is again introduced, where the standard state of fugacity is 1 mol dm^{-3}. Hence

$$G = G^{\ominus} + RT \ln (f/f^{\ominus})$$

Again $a = f/f^{\ominus}$, but since $f^{\ominus} = 1$ mol dm^{-3}, $a = f$ and the above equation becomes

$$G = G^{\ominus} + RT \ln a$$

where the standard state is that of unit activity. The more dilute the solution the closer a approaches c. The activity coefficient $\gamma = a/c$ indicates the non-ideality of the solution and approaches unity as c tends to zero. Activity may be thought of as an effective concentration.

28. Other standard states.

(a) *Solvents.* For ideal solvents, *i.e.* those which obey Raoult's law (*see* X, 2)

$$G = G^{\ominus} + RT \ln x$$

where the standard state is that of unit mole fraction. For non-ideal solvents

$$G = G^{\ominus} + RT \ln a$$

where the standard state is that of unit activity.

(b) *Pure liquids and solids.* For pure liquids and solids G is essentially constant, *i.e.* $G = G^{\ominus}$, hence $a = 1$, and the standard state is that of unit activity.

29. Temperature changes. For a reversible change at constant pressure (VIII, 22) reduces to

$$dG = -S dT$$

If the subscripts $_1$ and $_2$ refer to the initial and final states

$$dG_1 = -S_1 dT \text{ and } dG_2 = -S_2 dT$$

Hence

$$d(G_2 - G_1) = -(S_2 - S_1) dT$$

or

$$d\Delta G = -\Delta S dT$$

Since the change is at constant pressure, the above equation may be written in terms of partial derivatives, *i.e.*

$$(\partial \Delta G / \partial T)_p = -\Delta S = (\Delta G - \Delta H)/T \quad \text{. . . (VIII, 31)}$$

This is known as the *Gibbs–Helmholtz equation.* A more useful form of this equation may be obtained if we note that

$$\left[\frac{\partial(\Delta G / T)}{\partial T} \right]_p = \frac{T(\partial \Delta G / \partial T)_p - \Delta G}{T^2} \quad \text{. . . (VIII, 32)}$$

Substitution of (VIII, 31) into (VIII, 32) gives

$$\left[\frac{\partial(\Delta G / T)}{\partial T} \right]_p = \frac{T(\Delta G - \Delta H)/T - \Delta G}{T^2} = \frac{-\Delta H}{T^2}$$
$$\text{. . . (VIII, 33)}$$

If ΔH is independent of temperature, integration of (VIII, 33) shows that a plot of $\Delta G / T$ against $1/T$ will yield a straight line of slope ΔH.

THE EQUILIBRIUM CONSTANT

30. The relationship between ΔG and K. The free energy change for the reaction

$$a\text{A} + b\text{B} = c\text{C} + d\text{D}$$

is given by

$$\Delta G = cG(\text{C}) + dG(\text{D}) - aG(\text{A}) - bG(\text{B})$$

The general relationship between G and a for non-ideal systems is $G = G^{\ominus} + RT \ln a$, and substitution of this relationship into the above equation yields

$$\Delta G = c(G^{\ominus}(\text{C}) + RT \ln a_C) + d(G^{\ominus}(\text{D}) + RT \ln a_D) - \\ a(G^{\ominus}(\text{A}) + RT \ln a_A) - b(G^{\ominus}(\text{B}) + RT \ln a_B)$$

Hence $\Delta G = \Delta G^{\ominus} + RT \ln \dfrac{(a_C)^c (a_D)^d}{(a_A)^a (a_B)^b}$　　. . . (VIII, 34)

where $\Delta G^{\ominus} = cG^{\ominus}(\text{C}) + dG^{\ominus}(\text{D}) - aG^{\ominus}(\text{A}) - bG^{\ominus}(\text{B})$.

Equation (VIII, 34) is known as the *van't Hoff reaction isotherm*.

If the reaction is allowed to proceed to equilibrium, $\Delta G = 0$ and

$$\Delta G^{\ominus} = -RT \ln \left[\frac{(a_C)^c (a_D)^d}{(a_A)^a (a_B)^b} \right]_{\text{equil}}$$

The ratio of the equilibrium activities is given the symbol K_A and is commonly called the *thermodynamic equilibrium constant in terms of activities*. This equilibrium constant is exact and is independent of any assumption of the behaviour of the participants. Hence

$$\Delta G^{\ominus} = -RT \ln K_A \quad . . . \text{(VIII, 35)}$$

31. K_A and spontaneity. If K_A is less than one, $\ln K_A$ is negative and hence $\Delta G^{\ominus}$ is positive. The smaller K_A, the more positive $\Delta G^{\ominus}$ and the less we may expect the reaction to proceed from left to right. A positive value of $\Delta G^{\ominus}$ does not mean that the reaction is impossible, only that it does not proceed spontaneously in the direction indicated under the specified experimental conditions. If K_A is a large positive quantity, $\Delta G^{\ominus}$ is very negative and the reaction proceeds until a large concentration of products is built up and the equilibrium state is reached. It can readily be seen that $\Delta G^{\ominus}$ can vary from a positive to a negative quantity, depending on the values of a_A, a_B, etc.

32. $\Delta G^{\ominus}$ and other equilibrium constants. For ideal systems the van't Hoff reaction isotherm may be derived in terms of pressures, concentrations or mole fractions. Thus for ideal gases (VIII, 35) may be written

$$\Delta G^{\ominus} = -RT \ln K_p \text{ and } K_p = \left[\frac{(p_C/p_C^{\ominus})^c(p_D/p_D^{\ominus})^d}{(p_A/p_A^{\ominus})^a(p_B/p_B^{\ominus})^b}\right]_{\text{equil}}$$
$$\cdots \text{ (VIII, 36)}$$

whilst for ideal solutions

$$(a)\ \Delta G^{\ominus} = -RT \ln K_c \text{ and } K_c = \left[\frac{(c_C/c_C^{\ominus})^c(c_D/c_D^{\ominus})^d}{(c_A/c_A^{\ominus})^a(c_B/c_B^{\ominus})^b}\right]_{\text{equil}}$$
$$\cdots \text{ (VIII, 37)}$$

$$(b)\ \Delta G^{\ominus} = -RT \ln K_x \text{ and } K_x = \left[\frac{(x_C/x_C^{\ominus})^c(x_D/x_D^{\ominus})^d}{(x_A/x_A^{\ominus})^a(x_B/x_B^{\ominus})^b}\right]_{\text{equil}}$$
$$\cdots \text{ (VIII, 38)}$$

where $x^{\ominus}$ is the standard state of mole fraction, which is unity.

33. The variation of K_p and K_c with temperature. Equation (VIII, 33) may be written in terms of standard states:

$$\left[\frac{\partial(\Delta G^{\ominus}/T)}{\partial T}\right]_p = -\frac{\Delta H^{\ominus}}{T^2}$$

Substitution of (VIII, 36) into the above equation gives

$$\left[\frac{\partial(-RT \ln K_p/T)}{\partial T}\right]_p = -\frac{\Delta H^{\ominus}}{T^2}$$

$$\therefore R\left(\frac{\partial \ln K_p}{\partial T}\right)_p = \frac{\Delta H^{\ominus}}{T^2}$$

or
$$\frac{d \ln K_p}{dT} = \frac{\Delta H^{\ominus}}{RT^2} \cdots \text{ (VIII, 39)}$$

This equation is known as the *van't Hoff reaction isochore*. If $\Delta H^{\ominus}$ is considered to be constant over a narrow range of temperature, (VIII, 39) may be rearranged and integrated

$$\int_{K_1}^{K_2} d \ln K_p = \frac{\Delta H^{\ominus}}{R}\int_{T_1}^{T_2} \frac{dT}{T^2}$$

$$\therefore \ln\frac{(K_p)_2}{(K_p)_1} = -\frac{\Delta H^{\ominus}}{R}\left(\frac{1}{T_2} - \frac{1}{T_1}\right) \cdots \text{ (VIII, 40)}$$

Hence $\Delta H^{\ominus}$ may be found from a knowledge of K_p at two different temperatures.

If the variation of K_c with temperature is required, $\Delta H^{\ominus}$ is replaced by $\Delta U^{\ominus}$, since the heat change is now at constant volume and (VIII, 39) becomes

$$\frac{\mathrm{d} \ln K_c}{\mathrm{d} T} = \frac{\Delta U^{\ominus}}{RT^2} \qquad \text{. . . (VIII, 41)}$$

34. Le Chatelier's principle (*see* also XI, 6). This principle states that whenever a constraint is placed on a system in a state of dynamic equilibrium the system will always react in a direction which will tend to remove the constraint. The van't Hoff isotherm is in accord with this principle, since for an endothermic reaction ($\Delta H^{\ominus}$ positive), increase in temperature will increase K and will therefore favour the forward reaction. Conversely, if the reaction is exothermic ($\Delta H^{\ominus}$ negative), then as the temperature is increased K will decrease and the backward reaction will be favoured.

THE APPLICATION OF FREE ENERGY CHANGES TO PHASE EQUILIBRIA

35. The Clapeyron equation. Consider two phases of a single substance in equilibrium at constant temperature and pressure

$$\text{phase A} \rightleftharpoons \text{phase B}$$

Under these conditions

$$\Delta G = G_B - G_A = 0$$
$$\therefore \ G_A = G_B \qquad \text{. . . (VIII, 42)}$$

i.e. the molar free energies are equal. Suppose that the temperature and pressure are altered infinitesimally. For the system to be in equilibrium under these conditions

$$G_A + \mathrm{d} G_A = G_B + \mathrm{d} G_B$$

where $\mathrm{d} G_A$ and $\mathrm{d} G_B$ are the corresponding infinitesimal free energy changes. From (VIII, 42) it follows that

$$\mathrm{d} G_A = \mathrm{d} G_B$$

For simultaneous changes in temperature and pressure (VIII, 22) is appropriate. Thus

$$V_A \mathrm{d}p - S_A \mathrm{d}T = V_B \mathrm{d}p - S_B \mathrm{d}T \quad . . . \text{(VIII, 43)}$$

where V_A, V_B and S_A, S_B are the molar volumes and molar entropies respectively in the two phases. Rearranging (VIII, 43)

$$(V_B - V_A)\mathrm{d}p = (S_B - S_A)\mathrm{d}T$$
$$\therefore \frac{dp}{\mathrm{d}T} = \frac{S_B - S_A}{V_B - V_A} = \frac{\Delta S}{\Delta V} \quad . . . \text{(VIII, 44)}$$

For a reversible change in phase at constant pressure $\Delta S = \Delta H_t/T$ (see 4 above) and substitution of this relationship into (VIII, 44) yields the *Clapeyron equation*.

$$\frac{dp}{dT} = \frac{\Delta H_t}{T \Delta V} \quad . . . \text{(VIII, 45)}$$

36. The Clausius–Clapeyron equation. If the phase change is between a gaseous and a condensed (liquid or solid) phase the Clapeyron equation may be simplified:

$$\Delta V = V_B - V_A = V_{\text{gas}} - V_{\text{cond}} \approx V_{\text{gas}}$$

since the molar volume of a gas will be much greater than that of the condensed phase. Furthermore, if the gas is assumed to be ideal $V_{\text{gas}} = RT/p$.

The Clapeyron equation becomes

$$\frac{\mathrm{d}p}{\mathrm{d}T} = \frac{\Delta H_t}{T(RT/p)} = \frac{\Delta H_t p}{RT^2}$$

or $\quad \dfrac{\mathrm{d}p}{p} = \mathrm{d}\ln p = \dfrac{\Delta H_t}{R} \cdot \dfrac{\mathrm{d}T}{T^2} \quad . . . \text{(VIII, 46)}$

This is known as the Clausius–Clapeyron equation. If ΔH_t is constant, integration of (VIII, 46) between the pressure limits p_1 and p_2 and the corresponding temperature limits T_1 and T_2 leads to an equation analogous to (VIII, 40), *i.e.*

$$\ln\!\left(\frac{p_2}{p_1}\right) = -\frac{\Delta H_t}{R}\!\left(\frac{1}{T_2} - \frac{1}{T_1}\right)$$

Thus ΔH_t may be found from a knowledge of the vapour pressures at two temperatures (see also IX, 5).

SUMMARY OF METHODS OF FINDING $\Delta H^\ominus$, $\Delta S^\ominus$ AND $\Delta G^\ominus$

37. $\Delta H^\ominus$

(a) By calorimetry (VI, 12).

(b) From equilibrium constants,

$$\mathrm{d} \ln K_p = \frac{\Delta H^\ominus}{R} \cdot \frac{\mathrm{d}T}{T^2} \qquad \text{(VIII, 33)}$$

(c) From electrochemical cell measurements,

$$\Delta H^\ominus = nF[T(\partial E^\ominus/\partial T)p - E^\ominus] \qquad \text{(XIV, 13)}$$

38. $\Delta S^\ominus$

(a) From heat capacity data and the third law (VIII, 15–16).

(b) From cell measurements, $\Delta S^\ominus = nF(\partial E^\ominus/\partial T)_p$ (XIV, 12).

39. $\Delta G^\ominus$

(a) From equilibrium constants, $\Delta G^\ominus = -RT \ln K$ (VIII, 30, 32).

(b) From cell measurements, $\Delta G^\ominus = -nFE^\ominus$ (XIV, 12).

(c) From $\Delta H^\ominus$ and $\Delta S^\ominus$, using the relationship $\Delta G^\ominus = \Delta H^\ominus - T\Delta S^\ominus$.

EXTRACTION OF METALS BY HIGH-TEMPERATURE PROCESSES

40. The Ellingham diagram. An example of the use of thermodynamic principles in an industrial process is provided by the high-temperature extraction of metals from their ores. A large number of commercially important metals occur as oxides, and we will consider the thermodynamic aspects of the reduction of these oxides. The reduction of a metallic oxide by another element is, in effect, the competition of both elements for oxygen. The affinity of an element for oxygen is measured by the standard free energy of formation of its oxide. In the high-temperature reduction processes the variation of the standard free energy of formation of the oxide with temperature is clearly important. Fig. 23 shows the variation of $\Delta G^\ominus$ with temperature for the formation of some oxides.

This so-called *Ellingham diagram* gives linear plots, the slopes of which are defined by (VIII, 31), *i.e.*

$$(\partial \Delta G^{\ominus}/\partial T)_p = -\Delta S^{\ominus}$$

The oxidation reactions represented by the Ellingham diagram all involve *one mole of oxygen* at $101 \cdot 325$ kN m^{-2}.

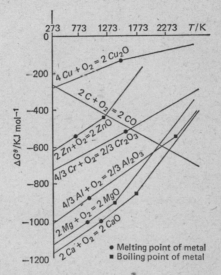

FIG. 23.—*Ellingham diagram for oxide formation*

The line representing the reaction

$$2Mg + O_2 = 2MgO$$

shows abrupt changes at the melting point (923 K) and at the boiling point (1393 K) of magnesium. Such changes in slope are due to the entropy increases on fusion and on vaporisation, the latter being the most significant.

41. The use of the Ellingham diagram. Since $\Delta G^{\ominus}$ is a function of state, the oxidation reactions and their associated standard free energy changes may be added and subtracted to yield the desired reaction. Thus the standard free energy change for the reduction of MgO(s) by C(graph) may be found by subtraction of (VIII, 48) from (VIII, 47)

$$2C(graph) + O_2(g) = 2CO(g) \quad \Delta G^{\ominus}(CO) \quad . \quad . \quad . \quad (VIII, 47)$$

$$2Mg(s) \quad + O_2(g) = 2MgO(s) \quad \Delta G^{\ominus}(MgO) . \quad . \quad . \quad (VIII, 48)$$

$$2MgO(s) + 2C(graph) = 2Mg(s) + 2CO(g) \quad \Delta G^{\ominus}$$

and $\Delta G^{\ominus} = \Delta G^{\ominus}(CO) - \Delta G^{\ominus}(MgO)$

The value of $\Delta G^{\ominus}$ for this reduction at any desired temperature may be found from the values of $\Delta G^{\ominus}(CO)$ and $\Delta G^{\ominus}(MgO)$. From the Ellingham diagram it can be seen that $\Delta G^{\ominus}(CO) - \Delta G^{\ominus}(MgO)$ is positive below 2100 K and negative above this temperature. Thus the reduction of MgO(s) by C(graph) is thermodynamically feasible only above 2100 K, when the CO line lies below the MgO line.

The most stable oxides are those at the bottom of the Ellingham diagram, *i.e.* those with the most negative standard free energies of formation. The metal which forms the most stable oxide is a potential reducing agent for a less stable oxide.

42. The use of carbon as a reducing agent. In most oxidation reactions, one mole of gaseous oxygen is consumed, and this contributes a large negative term to $\Delta S^{\ominus}$ resulting in the majority of lines sloping *upwards* from left to right. This indicates that the majority of oxides become less stable at elevated temperatures. However, in reaction (VIII, 47) an extra mole of gas is produced, making $\Delta S^{\ominus}$ large and positive, and the CO line therefore slopes *downwards* from left to right and cuts across the other oxide lines. This means that C(graph) will reduce almost any oxide provided that the temperature is high enough, and explains the effectiveness of carbon as a general reducing agent for metal oxides.

PROGRESS TEST 8

1. Calculate the entropy change when one mole of steam is condensed at 373 K, the water is cooled to 273 K and finally frozen to ice at 273 K.

Latent heat of vaporisation of water
$$= 2257 \cdot 5 \text{ J g}^{-1} \text{ at } 373 \text{ K.}$$
Latent heat of fusion of water $= 333 \cdot 5$ J g^{-1} at 273 K.
$$c_p(\text{H}_2\text{O, l}) = 4 \cdot 18 \text{ J K}^{-1} \text{ g}^{-1} \qquad \textbf{(4–5)}$$

2. Calculate the entropy change when one mole of an ideal monatomic gas at 273 K and 101·3 kN m^{-2} is heated to 1000 K and compressed so that its final pressure is 1013 kN m^{-2}. **(6)**

3. The crystal lattice of solid N_2O at 0 K is not perfectly ordered, and each N_2O molecule may be orientated in one of two ways with respect to the remainder of the lattice. Calculate the residual entropy per mole of N_2O at 0 K. **(13)**

4. From the following data determine the thermodynamic feasibility of the reaction at 298 K.

$$\text{C}_2\text{H}_4(\text{g}) + \text{H}_2(\text{g}) = \text{C}_2\text{H}_6(\text{g})$$

Standard heats of combustion (in kJ mole^{-1}): $\text{C}_2\text{H}_4(\text{g})$ — 1410; $\text{H}_2(\text{g})$ — 286; $\text{C}_2\text{H}_6(\text{g})$ — 1560. Standard entropies (in J K^{-1} mol^{-1}): $\text{C}_2\text{H}_4(\text{g})$ 219·4; $\text{H}_2(\text{g})$ 130·6; $\text{C}_2\text{H}_6(\text{g})$ 229·5. (VII, **5**; VIII, **17–18**)

5. At 1500 K and 101·3 kN m^{-2} water vapour is dissociated into hydrogen and oxygen to the extent of 2×10^{-2} %. Calculate the standard free energy change. **(30, 32)**

6. At 10 130 kN m^{-2} the following results were obtained for the synthesis of ammonia from nitrogen and hydrogen:

T/K	Equilibrium percentage of ammonia
823	6·7
1023	1·5

Calculate K_p at each of these temperatures and hence find the enthalpy change for the synthesis. **(33)**

7. From the following data calculate graphically the molar latent heat of vaporisation, assuming that water vapour behaves as an ideal gas.

T/K	p/mmHg
273	4·6
293	17·5
313	55·3
333	149·4
353	335·1
373	760·0

(35–36)

8. Explain, with examples, what is meant by a spontaneous process. **(1)**

9. Indicate the physical significance of entropy. **(12)**

10. Show how dS and dG may be used as criteria of spontaneity and explain why the latter is preferred. (**4–10, 18, 19**)

11. Derive (*a*) the Gibbs–Helmholtz equation (**29**), (*b*) the relationship between $\Delta G^{\ominus}$ and K (**30**), (*c*) the van't Hoff reaction isochore (**33**), (*d*) the Clausius–Clapeyron equation (**36**).

12. State the third law of thermodynamics and explain its importance with regard to absolute entropies. (**14–17**)

13. Explain the usefulness of the Ellingham diagram in the high-temperature reduction of metal oxides. (**40–42**)

PART FOUR

EQUILIBRIA

The next three chapters deal with the types of equilibria of interest to the chemist.

In Chapter IX the concept of the equilibrium state is introduced and illustrated by discussion of one-component systems in terms of the phase rule. A simple two-component system is also discussed. It is important to realise that the characteristics of the equilibrium state discussed in this chapter apply to *all* equilibria.

The first part of Chapter X contains a discussion of solutions and their colligative properties as examples of multi-component, one-phase systems. In the second part of this chapter, systems are discussed in which partial miscibility of the components leads to the appearance of more than one phase.

Chapter XI contains a discussion of chemical equilibria, the use of the equilibrium constant being illustrated by specimen calculations. Much of the second part of this chapter, devoted to heterogeneous chemical equilibria, could logically have been presented in Chapter IX. However, the use of the equilibrium constant allows quantitative predictions to be made about the behaviour of such systems which is not possible using the phase rule.

Ionic and oxidation–reduction equilibria, though governed by exactly analogous laws to the equilibria studied in Chapters IX–XI, are sufficiently important to warrant separate consideration, and are to be found in Part Five, along with allied electrochemical material.

CHAPTER IX

PHASE EQUILIBRIA

THE EQUILIBRIUM STATE

1. Phase equilibria. In the present chapter we shall be primarily concerned with systems in which one phase is converted to another, and in this context we may say that the system is in equilibrium if a number of phases exist together and there is no net conversion of one phase to another. The properties of the equilibrium state are summarised in the next section, with particular reference to phase equilibria. However, it is extremely important to realise that these properties are characteristic of *all* equilibria.

2. Properties of the equilibrium state (*see also* VIII).

(*a*) For any system at equilibrium under constant temperature and pressure, $\Delta G = \Delta H - T\Delta S = 0$ (*see* VIII, 20), *i.e.* equilibrium is attained when the tendency of a system to reach a state of minimum energy is balanced by its tendency to seek a state of maximum disorder.

(*b*) Equilibrium is dynamic and is reached when the rates of two opposing processes become equal. If the system is disturbed from equilibrium by some change in the external conditions, it will react in such a manner as to restore equilibrium.

(*c*) The properties of the equilibrium state are independent of the path by which equilibrium was reached.

3. The balance between ΔH and ΔS. The balance between the tendencies towards minimum energy and maximum disorder has been discussed in VIII, 20.

4. The liquid–vapour equilibrium—the equilibrium vapour pressure (*see* also 5 below). The dynamic character of the equilibrium state (2 (*b*) and (*c*) above) is nicely illustrated by consideration of the equilibrium between a liquid and its vapour in a closed container at constant temperature.

Evaporation takes place at a constant rate because, even though the average potential energy is less in the liquid than in the gas, there is always a certain fraction of molecules at the surface with energies greater than the cohesive energy of the liquid. However, as the number of molecules in the vapour increases, the number colliding with and re-entering the liquid surface also increases, and equilibrium is established when the rates of these two opposing processes become equal. The pressure at which equilibrium is established is called the *equilibrium vapour pressure.*

5. Effect of temperature and pressure on the vapour pressure.

If the temperature of the system is increased the rate of evaporation increases and equilibrium is established at a higher vapour pressure. This increase in vapour pressure can be predicted from Le Chatelier's principle (*see* VIII, 34). Since ΔH must be positive for the transition from liquid to vapour, increase in temperature will shift the equilibrium over to the vapour side, *i.e.* the vapour pressure will increase. The variation of vapour pressure with temperature is given by the Clausius–Clapeyron equation, (VIII, 46). The integrated form of this equation

$$\lg p = (-\Delta H_v / 2 \cdot 303 \, RT) + \text{constant}$$

enables heats of vaporisation to be determined from vapour pressure measurements at various temperatures (*see* VIII, 36).

If the pressure in the system is decreased at constant temperature, then the number of molecules colliding with the liquid surface decreases whilst the rate of evaporation remains constant, so that there is a net conversion of liquid to vapour until the equilibrium vapour pressure is re-established. Likewise, if the pressure is increased a net conversion of vapour to liquid occurs until the equilibrium vapour pressure is re-established.

If the vessel is open to the atmosphere the molecules in the vapour are swept away so that condensation does not occur, and evaporation continues at a constant rate until no further liquid remains.

6. The phenomenon of boiling.

When a liquid boils, evaporation takes place from all parts of the liquid rather than just from the surface, *i.e.* bubbles of vapour form within the liquid. Since the formation of a stable bubble requires that the pressure inside the bubble is equal to the external pressure, a liquid does not boil until its vapour pressure is equal to the external pressure.

THE PHASE RULE

7. Statement of the phase rule. The conditions of equilibrium between a number of phases are generally expressed in the form of a *phase diagram*. The general law governing such equilibria is known as the *phase rule* (J. Willard Gibbs, 1875–8) and may be expressed in the form

$$F = C - P + 2$$

where F is the number of degrees of freedom, C the number of components and P the number of phases in a heterogeneous system. These terms are precisely defined in **8–10** below.

8. Definition of phase. The term "phase" has been defined in VI, 1 and its use is illustrated below.

Any gas or mixture of gases always constitutes a single phase, as does a homogeneous liquid or solid solution.

A mixture of water and ether constitutes a two-phase system, since two liquid layers are obtained (one containing a saturated solution of ether in water, the other containing a saturated solution of water in ether), which are separated from each other by a definite boundary surface.

Under normal conditions water constitutes a three-phase system—ice, liquid water, water vapour. In fact six different crystalline forms of ice have been found at very high pressures, each of which constitutes a separate phase, since it is clearly separated from the other crystalline forms by the crystal boundary.

9. Definition of component. The number of components in a system is the minimum number of *independent* chemical constitutuents necessary to describe the composition of each phase present, either directly or in the form of a chemical equation.

NOTE: This refers only to equations with respect to which the system is in equilibrium.

The composition of each of the three phases present in water can be expressed in terms of one component—water—even though the molecular arrangement is different in the three phases.

The decomposition of calcium carbonate

$$CaCO_3 \rightleftharpoons CaO + CO_2$$

involves three phases but only two components, since, if the amounts of $CaCO_3$ and CO_2 are specified, the amount of CaO is automatically fixed.

A problem may arise within this definition if the system is not allowed sufficient time to reach equilibrium. Thus if the amounts of CaO and $CaCO_3$ are measured at a series of pressures in a time shorter than that required for the equilibrium to shift under the change in conditions, then the equilibrium is effectively non-existent and the system contains three components.

10. Definition of degree of freedom. The number of degrees of freedom is the number of *independent intensive* variables (*e.g.* pressure, temperature, concentration, density) which must be fixed in order to completely define the system at equilibrium. The use of this term is clarified in the examples which make up the rest of this chapter.

APPLICATION OF THE PHASE RULE TO WATER

As outlined in **8–9** above, under normal conditions water is a three-phase, one-component system. The phase rule is used to predict the general form of the temperature against pressure phase diagram, as illustrated in **11–13** below.

NOTE: Any pair of intensive variables could be used, *e.g.* T against p, refractive index against p, etc.

11. One-phase regions. Where only a single phase is present

$$F = 1 - 1 + 2 = 2$$

that is, each single phase has two degrees of freedom, so that both temperature and pressure must be specified in order to define the condition of the phase. Since two independent variables are generally necessary to specify a point in an area, each phase must occupy a definite area in the diagram. Since there are three phases, there will be three such areas.

12. Two-phase regions. Where two phases are present in equilibrium

$$F = 1 - 2 + 2 = 1$$

that is, two phases in equilibrium have one degree of freedom, so that only one of the two variables temperature and pressure need be specified in order to define the system. Since it is necessary to specify only one variable, two phases in equilibrium must be represented by a line on the diagram. Since there are three phases, there will be three such lines.

13. Three-phase regions. Where three phases are present in equilibrium

$$F = 1 - 3 + 2 = 0$$

that is, if three phases co-exist in equilibrium, no variables need to be specified in order to define the system—the statement that three phases exist in equilibrium is sufficient. Since no variables need be specified, this state must be represented by the point on the diagram at which the three lines intersect. Such a point is known as a *triple point*. Since there are three phases, there can be only one triple point.

THE PHASE DIAGRAM FOR WATER

14. General description. The phase diagram for water is shown in Fig. 24. The slopes of the lines may be obtained from the Clapeyron equation for the solid–liquid equilibrium and

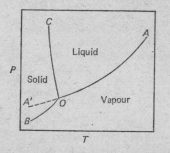

FIG. 24.—*Phase diagram for water*

from the Clausius–Clapeyron equation for the liquid–vapour and solid–vapour equilibria (*see* VIII, **35, 36**).

OA and OB represent the conditions for equilibrium between liquid–vapour and solid–vapour respectively, *i.e.* they represent the variation of the vapour pressures of the solid and liquid with temperature. OA continues to the critical temperature, where liquid and vapour become indistinguishable, and OB continues down to 0 K.

OC represents the conditions for equilibrium between ice and water, *i.e.* OC shows the effect of pressure on the melting point of ice. The fact that OC slopes to the left indicates that the melting point of ice decreases with increasing pressure. The only other common substances showing this anomalous effect are antimony and bismuth.

15. The triple point. OA, OB and OC meet at the *triple point* O, where all three phases are in equilibrium. The triple point for water occurs at $0.0098°C$ at an equilibrium vapour pressure of 611 N m^{-2} (4.58 mm Hg).

NOTE: Notice the difference from the normal melting point of ice at 0°C, where solid and liquid are in equilibrium under a pressure of 101.325 kN m^{-2} (760 mm Hg).

If the vapour is cooled at a pressure below 611 N m^{-2} direct transformation of vapour to solid occurs without the intermediate formation of liquid. The reverse process (sublimation) is an important purification method, *e.g.* iodine, triple point 114.15°C at 1.20 kN m^{-2} (90 mm Hg), can be purified by heating the solid at 114.15°C keeping the vapour pressure below 1.20 kN m^{-2} by condensing the vapour on to a cold surface.

Substances in which the vapour reaches atmospheric pressure at a temperature below the triple point do not show normal melting points. Carbon dioxide, for example, has a vapour pressure of 101.325 kN m^{-2} at -78.1°C, though it can be liquefied at -56.6°C under a pressure of 518 kN m^{-2}, *i.e.* triple point -56.6°C at 518 kN m^{-2}.

16. Metastable equilibrium. The vapour pressure curve of the liquid can be continued past the triple point to A', as shown by the broken line, *i.e.* the liquid can be supercooled. The liquid–vapour system along OA' is said to be in a state of *metastable equilibrium*, where metastable is used to describe a specific equilibrium which is not the most stable equilibrium under the given conditions.

NOTE: It is important to realise that the phase rule applies equally to all equilibria and does not distinguish between stable and metastable states.

The metastable liquid always has a greater vapour pressure than the stable solid at the same temperature, indicating its tendency to change spontaneously to the solid on addition of an ice crystal (seeding).

POLYMORPHISM

The occurrence of the same substance in more than one crystalline form is known as *polymorphism*. In the case of elements the term *allotropy* is used. There are two types of polymorphism—*enantiotropy* and *monotropy*.

17. Definition of enantiotropy. Two crystalline forms of a substance are said to be enantiotropic if each has a definite range of stability and, at a given pressure, the change from one form to another occurs at a definite temperature in either direction.

18. The transition point. The temperature at which the enantiotropic change takes place is called the *transition point* or *transition temperature*. Any deviation from the transition point results in complete transformation to the stable form. Transition points may be determined utilising the abrupt changes in density, vapour pressure or electrical properties which occur when an enantiotropic change takes place. Also, since the transition point is the *only* temperature at which the two solid phases co-exist, a cooling curve for the system will be characterised by a constant temperature region corresponding to the transition point and lasting until the enantiotropic change is completed.

19. Phase diagram for an enantiotropic substance. The phase diagram of a hypothetical substance having two enantiomers α and β, which are stable below and above the transition point respectively, is shown in Fig. 25. For simplicity the lines representing the α solid–β solid and the solid–liquid equilibria have been omitted.

AB and BC, the vapour pressure curves of the α and β solids respectively, meet at the transition point B. CD is the vapour pressure curve of the liquid and C the triple point for β–liquid–vapour.

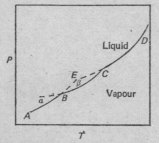

Fig. 25.—*Phase diagram for a substance having two enantiomers, α and β*

If the system is heated slowly from A then, as expected, the vapour pressure of the α form increases along AB. At B the temperature remains constant until all the α form has been converted to the β form, and the vapour pressure then increases along BC. At C the β form melts, and the vapour pressure of the liquid then increases along CD.

If the system is heated quickly the α form may continue in a metastable state along BE. At E the metastable triple point, the α form melts to metastable liquid whose vapour pressure then increases along ECD. This explains why sulphur apparently melts at 114·5°C if heated quickly (metastable triple point of rhombic form) and at 119·25°C if heated slowly (stable triple point of monoclinic form).

20. Definition of monotropy. Monotropy occurs when one crystalline form is stable and the other metastable over the entire range of their existence. Monotropic changes take place in one direction only—metastable to stable.

21. Phase diagram for a monotropic substance. In Fig. 26 the α form is stable and the β form metastable. The line representing the solid–liquid equilibrium has again been omitted.

AB and *DE* are the vapour pressure curves of the stable α and metastable β forms respectively. *BC* is the vapour pressure curve of the liquid. *B* is the stable triple point for α–liquid–vapour and *E* the metastable triple point for β–liquid–vapour. *AB* and *DE* meet at *F*, the hypothetical transition point—hypothetical because it lies above the melting points of both the α and β forms so that the solids do not exist at *F*.

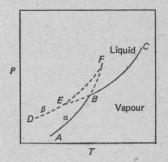

FIG. 26.—*Phase diagram for a monotropic substance*

The β form can be obtained by quickly cooling the liquid, which may then pass along *CB* to *E*, where crystals of the β form separate out. The β form will always tend to change spontaneously to the α form, though the change may take place very slowly, *e.g.* white to violet phosphorus at room temperature. The direct transformation from violet to white phosphorus is not possible.

NOTE: The so-called red phosphorus is probably finely divided violet phosphorus containing some white phosphorus.

It is interesting to note that white phosphorus exists in two enantiotropic forms, with a transition point at −76·9°C at atmospheric pressure.

APPLICATION OF THE PHASE RULE TO SULPHUR

22. The phase diagram—general considerations. The phase diagram for sulphur, shown in Fig. 27, clearly illustrates most of the types of equilibria discussed above.

The system contains one component and four phases:

rhombic and monoclinic solids, liquid and vapour (abbreviated to r, m, l and v in the discussion below).

NOTE: Actually the liquid is an equilibrium mixture of three types of sulphur molecule — $\lambda(S_8)$, $\mu(S_6)$, $\pi(S_4)$.

23. One-phase regions.

Each single phase is represented by an area. Since there are four phases (r, m, l and v), there will be four such areas. The slopes of the lines BF and CF are such that the monoclinic form can exist only in the triangular region BCF.

24. Two-phase regions.

Two phases in equilibrium are represented by a line. Since there are six ways of choosing two phases from four, there should be six lines in Fig. 27, corresponding to the equilibria: $r–m$(BF), $r–l$(EFG, EF metastable), $r–v$ (ABE, BE metastable), $m–l$ (CF), $m–v$ (BC), $l–v$ (ECD, EC metastable). By rapid heating, rhombic sulphur can be converted directly to the liquid without going through the monoclinic form, and the phase diagram would then consist of three lines, AE, EG and ED.

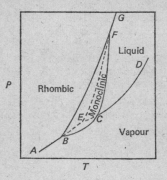

FIG. 27.—*Phase diagram for sulphur*

25. Three-phase regions.

Three phases in equilibrium are represented by a triple point on the diagram. Since there are four ways of choosing three phases from four, there should be four triple points in Fig. 27, corresponding to the equilibria: $r–m–l$ (F), $r–m–v$ (B), $r–l–v$ (E, metastable), $m–l–v$ (C).

26. Four-phase regions. For four phases in equilibrium

$$F = 1 - 4 + 2 = -1$$

Since the concept of a negative number of degrees of freedom is meaningless, there can be no set of conditions under which all four phases r–m–l–v coexist in equilibrium.

TWO-COMPONENT SYSTEMS

27. The "reduced" phase rule. For a two-component system the phase rule shows that the maximum possible number of degrees of freedom is three. A complete graphical representation of the system would therefore require a three-dimensional phase diagram, *i.e.* a solid model. For this reason it is usual to fix one of the degrees of freedom so that the system can be represented on a normal two-dimensional graph. In such a case the phase rule reduces to

$$F = C - P + 1$$

When dealing with solids or liquids it is usual to fix the pressure (at $101 \cdot 325$ kN m^{-2}), since its effect on the system is small. The two independent variables normally used are temperature and composition.

APPLICATION OF THE PHASE RULE TO THE PHENOL-DIPHENYLAMINE SYSTEM

28. The phase diagram: general considerations. This is an example of the simplest type of two-component system, in which the two components are completely miscible in the liquid phase but completely immiscible as solids. The phase diagram is shown in Fig. 28 and described in **29–31** below.

29. One-phase regions. In the area above ACB there is but a single phase

$$\therefore F = 2 - 1 + 1 = 2$$

and both temperature and composition must be specified to define the system.

30. Two-phase regions. In the areas ACD and BCE there are two phases, solid phenol–liquid and solid diphenylamine–liquid, respectively. It is probably easiest to visualise AC as

representing the effect of addition of diphenylamine on the
freezing point of phenol and BC as representing the effect of
addition of phenol on the freezing point of diphenylamine

$$F = 2 - 2 + 1 = 1$$

so that at any temperature the composition of the liquid in
equilibrium with pure solid is fixed and vice versa.

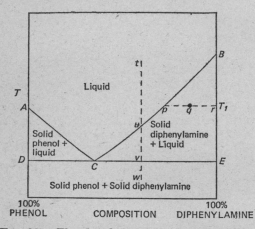

FIG. 28.—*The phenol–diphenylamine phase diagram*

(a) *Tie lines.* The composition of the liquid in equilibrium
with pure solid at a given temperature (say T_1) is established
by drawing a horizontal *tie line* (pr in Fig. 28) to cut BC or AC;
p then gives the composition of the liquid.

(b) *The lever rule.* At any point q along pr the composition
of the liquid is fixed at p and the relative amounts of liquid and
pure solid present may be obtained from the *lever rule*, *i.e.* by
considering pr as a lever with its fulcrum at q. Thus at q

Amount of liquid composition p × length pq = Amount of
pure solid × length qr

$$\therefore \frac{\text{Amount of liquid composition } p}{\text{Amount of pure solid}} = \frac{qr}{pq} \qquad . \quad . \quad . \text{ (IX, 1)}$$

31. Three-phase regions—the eutectic point. At point C all
three phases are in equilibrium

$$\therefore F = 2 - 3 + 1 = 0$$

This point is called the *eutectic point* and is analogous to the triple point in one-component systems. At the eutectic point the composition of the solid is the same as that of the liquid from which it separates. This equality of composition, together with the fact that the eutectic occurs at a definite temperature, led to the early belief that compound formation was taking place at the eutectic point. However, microscopic examination shows that the eutectic is a fine-grained mixture of the two solid phases, and altering the fixed pressure alters the composition of the eutectic.

NOTE: Below *DCE* two solid phases are present

$$\therefore F = 2 - 2 + 1 = 1$$

so that only the temperature need be fixed to define the system. This is fairly obvious on a physical basis, since the composition of each phase is fixed—pure phenol and pure diphenylamine.

32. Thermal analysis. The form of the phase diagram, and in particular the eutectic point, are normally determined by the method of thermal analysis. As an example we may consider cooling a system of composition given by the line *tuvw*.

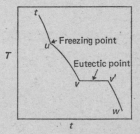

FIG. 29.—*Cooling curve for a phenol–diphenylamine mixture of composition tuvw*

If liquid of composition *t* is cooled, no solid separates until point *u* is reached. Here solid diphenylamine is deposited, giving up its latent heat and causing a check on the rate of cooling. The remaining solution contains less diphenylamine, and its composition and temperature move down the line *BC* towards *C*. Between *u* and *v* the composition of the liquid is given by the tie line to *BC*. At *v* the eutectic temperature is reached and solid phenol also begins to separate. Since the phase rule shows that there is only one temperature at which pure diphenylamine, pure phenol and liquid can be in equilibrium, the temperature of the system now remains constant until all the liquid has solidified. The temperature then begins to fall again along *vw*.

The cooling curve for this system is shown in Fig. 29, where the letters have the same significance as in Fig. 28. Obviously, the closer *tw* is to the eutectic composition, the shorter will be *uv* and the longer *vv'*. When *tw* has the eutectic composition, the first solid to separate out is the eutectic, and the cooling curve shows only one break.

33. Mixed melting points. It can be seen from Fig. 28 that any mixture of phenol and diphenylamine melts at a lower temperature than pure diphenylamine. The technique of the mixed melting point, in which the identity of a substance is checked by adding to it a small sample of a known substance believed to have the same identity, is based on this fact. For substances which are completely miscible as liquids, but completely immiscible as solids, the addition of the second substance will lower the melting point unless the two substances are identical. If solid solutions or compounds are formed, the melting point may be either raised or lowered.

PROGRESS TEST 9

1. Nitrogen and hydrogen react to form ammonia according to the equation

$$N_2(g) + 3H_2(g) \rightleftharpoons 2NH_3(g)$$

How many components are present in the systems (a) $N_2(g)$ + $H_2(g)$, (b) $NH_3(g)$? **(9)**

2. A certain substance exists in two solid modifications (both of which are more dense than the liquid), liquid and gas. How many equilibria are possible when (a) $P = 1$, (b) $P = 2$, (c) $P = 3$, (d) $P = 4$? How many degrees of freedom are there in each case? **(23–26)**

3. Thermal analysis of the bismuth–cadmium system shows two temperatures (t_1 and t_2) at which breaks occur in the cooling curve. The data for several compositions are shown below:

Percentage Bi	100	90	70	60	50	30	10	0
$t_1/°C$	271	230	170	144	193	257	300	321
$t_2/°C$	–	144	144	144	144	144	144	–

Construct the phase diagram, labelling all regions. **(28–31)**

4. From the phase diagram drawn in the previous question, calculate the composition of the liquid phase and the relative amount of solid in equilibrium with this liquid for the following mixtures: (a) 10% Bi at 275°C, (b) 90% Bi at 200°C, (c) 60% Bi at 144°C. State clearly which solid is formed in each case. (29)

5. Define the following terms: (a) equilibrium vapour pressure (4), (b) boiling point (6), (c) triple point (13), (d) polymorphism (16), (e) enantiotropy (17), (f) monotropy (20).

6. State the phase rule and explain the meaning of all terms used. (7)

7. Draw and explain the phase diagrams for (a) water (14–16), (b) sulphur (22–26).

8. Draw the phase diagrams for (a) an enantiotropic substance (19), (b) a monotropic substance (21).

9. Draw the phase diagram for a two-component system in which the two components are completely miscible in the liquid phase but completely immiscible as solids. Use the diagram to explain (a) tie lines, (b) the lever rule, (c) thermal analysis. (27–32)

CHAPTER X

SOLUTIONS

1. Definition. A solution is any homogeneous phase containing more than one component. The distinction between solvent and solute is purely verbal, the term solvent being used to describe the component present in excess.

NOTE: A solution is distinguished from a finely grained mechanical mixture, since a true solution is homogeneous right down to the molecular level.

IDEAL SOLUTIONS

2. Raoult's law. This states that the vapour pressure of any component in a solution varies linearly with the concentration of the component. Algebraically

$$p_i = p_i^{\bullet} x_i \qquad \ldots \text{(X, 1)}$$

where $p_i^{\bullet}$ and p_i are the vapour pressures exerted by pure component i and by component i in a solution in which its mole fraction is x_i respectively.

NOTE: The superscript $\bullet$ is used to indicate the property of a pure substance.

3. Raoult's law and the ideal solution. Raoult's law provides a limiting description of the behaviour of solutions in much the same way as the ideal gas law provides a limiting description of gases. Ideal solutions are those which obey Raoult's law over the entire range of composition.

In an ideal solution of two components A and B, all cohesive forces (A–A, A–B and B–B) must be identical, so that the escaping tendency of an A or B molecule is independent of whether it is surrounded by A molecules, B molecules or varying proportions of A and B molecules. In such a case the escaping tendency, as measured by the vapour pressure, of each component in the solution will be the same as that in the pure component, and Raoult's law will be obeyed. This is likely to be achieved only if A and B are of similar size and related constitution.

NOTE: The thermodynamic properties of ideal solutions were discussed in VIII, 28.

COLLIGATIVE PROPERTIES

Four characteristic effects occur when an *involatile* solute is dissolved in a solvent to form an ideal solution (*see also* III, 18): the vapour pressure and freezing point are lowered relative to the pure solvent, the boiling point is raised, and the solution exerts an osmotic pressure when separated from pure solvent by a semi-permeable membrane. Such effects are known as *colligative properties* and depend only on the number of solute molecules present.

4. Vapour pressure lowering. The vapour above a solution containing an *involatile* solute consists of pure solvent. If the solvent and solute are labelled 1 and 2 respectively, Raoult's law becomes

$$p_1 = p_1^{\bullet} x_1 \qquad \qquad \ldots \text{(X, 1)}$$
$$\therefore p_1^{\bullet} - p_1 = \Delta p = p_1^{\bullet} - p_1^{\bullet} x_1$$
$$\therefore \Delta p = p_1^{\bullet}(1 - x_1) = p_1^{\bullet} x_2 \quad \ldots \text{(X, 2)}$$

where Δp is the lowering of vapour pressure and x_2 the mole fraction of the solute.

NOTE: When several solutes are dissolved in a single solvent, the vapour pressure lowering is given by

$$\Delta p = p_1^{\bullet}(x_2 + x_3 + x_4 + \ldots)$$

Equation (X, 2) is utilised in the determination of the molar mass and hence the relative molecular mass of the solute as follows. For a very dilute solution (such solutions being the only ones to which Raoult's law is strictly applicable)

$$x_2 = n_2/(n_1 + n_2) \approx n_2/n_1 \text{ (since } n_1 \gg n_2)$$
$$\therefore \Delta p = \frac{n_2}{n_1}.p_1^{\bullet} = \frac{m_2 M_1}{m_1 M_2} \times p_1^{\bullet}$$
$$\therefore M_2 = \frac{m_2 M_1}{m_1} \times \frac{p_1^{\bullet}}{\Delta p} \qquad \qquad \ldots \text{(X, 3)}$$

where m_2, m_1 and M_1, M_2 are the masses and the molar masses of the solute and solvent respectively. $p_1^\bullet$, m_2, m_1 and M_1 are generally known, and Δp may be determined by some form of differential manometry.

5. Elevation of boiling point. Since the vapour pressure of any solution is always less than that of the pure solvent (*see* (X, 1) and (X, 2)), the solution must be heated to a higher temperature than the pure solvent before its vapour pressure reaches atmospheric. In Fig. 30, T_0 is the boiling point of the

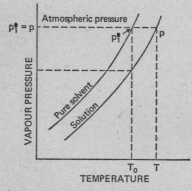

Fig. 30.—*Relationship between the lowering of vapour pressure and the elevation of boiling*

pure solvent and T that of the solution. It can be seen that $p_1^\bullet$, the vapour pressure of the pure solvent at T_0 and p, the vapour pressure of the solution at T are identical. p_1 is the vapour pressure of the solution at T_0.

For the *solution*, the Clausius–Clapeyron equation (*see* VIII, 36) is

$$\ln\left(\frac{p_2}{p_1}\right) = \frac{\Delta H_v}{R}\left(\frac{T - T_0}{TT_0}\right)$$

where ΔH_v is the latent heat of vaporisation. p may be replaced by $p_1^\bullet$

$$\ln\left(\frac{p_1^\bullet}{p_1}\right) = \frac{\Delta H_v}{R}\left(\frac{T - T_0}{TT_0}\right) \quad \cdot \quad \cdot \quad \cdot \quad (\text{X}, 4)$$

In a very dilute solution the difference between T and T_0 will be sufficiently small that we may write $TT_0 = T_0^2$ and $T - T_0 = \Delta T$. The above equation then becomes

$$\ln\left(\frac{p_1^\bullet}{p_1}\right) = \frac{\Delta H_v}{R} \times \frac{\Delta T}{T_0^2}$$

and from (X, 1)
$$p_1/p_1^\bullet = x_1$$

$$\therefore -\ln x_1 = \frac{\Delta H_v}{R} \times \frac{\Delta T}{T_0^2}$$

But $-\ln x_1 = -\ln(1 - x_2)$ and expanding $-\ln(1 - x_2)$ as a power series

$$-\ln(1 - x_2) = x_2 + (x_2^2/2) + (x_2^3/3) + \ldots$$

Since the solution is dilute, x_2 is small and terms other than the first can be neglected.

$$\therefore x_2 = \frac{\Delta H_v}{R} \times \frac{\Delta T}{T_0^2} \qquad \ldots \text{ (X, 5)}$$

NOTE: Equation (X, 5) reduces to (III, 4), since ΔH_v, R and T_0^2 are all constants, i.e. $\Delta T = K'x_2$.

Equation (X, 5) is usually approximated as follows:
In a dilute solution $x_2 \approx n_2/n_1 = m_2 M_1/m_1 M_2$

$$\therefore \Delta T = \left(\frac{RT_0^2 M_1}{\Delta H_v}\right) \frac{m_2}{M_2 m_1}$$

But $m_2/M_2 m_1$ is simply the amount of solute per kg of solvent and we can therefore set $m_2/M_2 m_1 = m$, where m is the molal concentration, i.e. the amount of solute per kg of solvent.

$$\therefore \Delta T = \left(\frac{RT_0^2 M_1}{\Delta H_v}\right) m = K_b m \quad \ldots \text{ (X, 6)}$$

K_b is called the *ebullioscopic constant* and is equal to the rise in boiling point for a 1-molal solution. Equation (X, 6) forms the basis of relative molecular mass determinations from the boiling point elevation.

Solvents with high relative molecular masses are chosen wherever possible, since these will have higher values of K_b (see (X, 6)) and thus of the boiling point elevation. Disadvantages of the method are the fact that T_0 (and therefore

K_b) is pressure-dependent, and that organic solutes may exert an appreciable pressure or may decompose at the boiling point of the solution.

6. Depression of freezing point. The depression of freezing point may be related to the solute concentration by a similar method (provided that pure solvent freezes out rather than solute or a solid solution). In Fig. 31 $p_1^{\bullet}$ is the vapour pressure

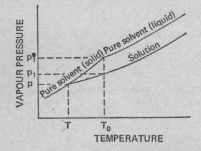

FIG. 31.—*Relationship between the lowering of vapour pressure and the depression of freezing*

of the pure solid solvent and of the pure liquid solvent at T_0 (the freezing point of the pure solvent) and p is the vapour pressure of the pure solid solvent and of the solution at T (the freezing point of the solution). p_1 is the vapour pressure of the solution at T_0. For dilute solutions, $p \approx p_1$, and the lowering of vapour pressure $p_1^{\bullet} - p \approx p_1^{\bullet} - p_1$.

The Clausius–Clapeyron equation may be applied to the pure solid solvent and to the dilute solution. For the solid

$$\ln\left(\frac{p_1^{\bullet}}{p}\right) = \ln p_1^{\bullet} - \ln p = \frac{\Delta H_s}{R}\left(\frac{T_0 - T}{T T_0}\right) \quad . \; . \; . \; \text{(X, 7)}$$

where ΔH_s is the latent heat of sublimation of the solvent. For the dilute solution

$$\ln\left(\frac{p_1}{p}\right) = \ln p_1 - \ln p = \frac{\Delta H_v}{R}\left(\frac{T_0 - T}{T T_0}\right) \quad . \; . \; . \; \text{(X, 8)}$$

where ΔH_v is the latent heat of vaporisation of the solvent.

Subtraction of (X, 8) from (X, 7) gives

$$\ln p_1^{\bullet} - \ln p_1 = \ln\left(\frac{p_1^{\bullet}}{p_1}\right) = \frac{\Delta H_s - \Delta H_v}{R}\left(\frac{T_0 - T}{TT_0}\right)$$

At a given temperature the same amount of heat is required to convert one mole of solid directly to vapour (ΔH_s), as is necessary first to melt the solid (ΔH_f) and then to vaporise it (ΔH_v), i.e. $\Delta H_s = \Delta H_f + \Delta H_v$.

$$\therefore \ln\left(\frac{p_1^{\bullet}}{p}\right) = \frac{\Delta H_f}{R}\left(\frac{T_0 - T}{TT_0}\right)$$

This equation is analogous to (X, 4), and leads to

$$\Delta T = \frac{RT_0^2}{\Delta H_f}x_2 \qquad \ldots \text{(X, 9)}$$

and for dilute solutions $\Delta T \approx K_f\text{m}$ $\qquad \ldots$ (X, 10)

where K_f is the *cryoscopic constant*, and is analogous to the ebullioscopic constant. Equation (X, 9) forms the basis of relative molecular mass determinations from freezing point depression. The advantage of the freezing point method is that the solute is less likely to exert an appreciable vapour pressure or to decompose at the freezing point of the solution.

7. Solubility. In a saturated solution, equilibrium takes place between pure solid *solute* and the solution. This is analogous to the equilibrium taking place on freezing between pure solid *solvent* and solution. Since we have already mentioned that the terms "solute" and "solvent" are interchangeable, the solubility of the solute in an ideal solution should depend only on the properties of the solute, *i.e.* as in equation (X, 4)

$$\ln x_2 = \frac{-\Delta H_f}{R}\left(\frac{T - T_0}{TT_0}\right) \quad \ldots \text{(X, 11)}$$

where ΔH_f and T_0 are the latent heat of fusion and freezing point of the *solute*. Equation (X, 11) is an expression of the *ideal law of solubility*. According to this law, the solubility of a substance should be the same in all solvents with which it forms ideal solutions.

8. Osmotic pressure. When a solution of a solute is separated from the pure solvent by a semi-permeable membrane (*i.e.* a membrane permeable to solvent but not to solute) solvent flows through the membrane into the solution. If the solution and the pure solvent are present in a confined space, this movement of solvent sets up an *osmotic pressure*. The definition of osmotic pressure may be seen by a consideration of Fig. 32,

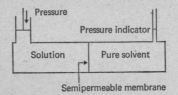

FIG. 32—*Osmotic pressure of a solution*

where a chamber is divided into two parts by a semi-permeable membrane. The one on the right contains pure solvent and that on the left the solution. A piston allows pressure to be applied to the solution to counteract the flow of pure solvent into the solution. The osmotic pressure is defined as the pressure required to prevent osmosis of the solvent into the solution in a confined space.

9. The van't Hoff equation for osmotic pressure. Osmosis occurs because the free energy of the solvent in the solution is less than the free energy of the pure solvent. The solvent moves spontaneously in the direction of the lower free energy, *i.e.* from pure solvent to solution. If it is assumed that the solvent vapour behaves as an ideal gas, then the lowering of free energy per mole of solvent caused by the addition of solute is given by (VIII, 26), which may be written in the form

$$G - G^{\bullet} = \Delta G_A = RT \ln (p/p^{\bullet}) = RT \ln x_1$$

The process of osmosis may be balanced by applying an external pressure to the solution, which has the effect of increasing its free energy. The dependence of free energy on pressure at constant temperature is given by (VIII, 23), *i.e.*

$$\Delta G_B = V_1 \Delta p \quad \text{for a finite change.}$$

Since liquids are quite incompressible, V_1, the volume of one mole of the solvent in the solution, can be considered to be independent of pressure. Now Δp is simply the external pressure which must be applied to prevent osmosis, *i.e.* it is the osmotic pressure, π.

$$\therefore \Delta G_B = \pi V_1$$

At equilibrium the two processes must be balanced, *i.e.* $\Delta G_A + \Delta G_B = 0$. Therefore

$$\pi V_1 + RT \ln x_1 = 0$$
or
$$\pi V_1 = -RT \ln x_1$$

Now $-\ln x_2 = -\ln (1 - x_2) \approx x_2$ for dilute solutions (*see* 5 above)

$$\therefore \pi V_1 = RT x_2 \qquad . . . (X, 12)$$

Since $x_2 \approx n_2/n_1$ for dilute solutions, (X, 12) may be written

$$\pi V_1 n_1 = n_2 RT$$

Now $V_1 n_1$ is the total volume of the solvent containing n_2 mol of solute and for dilute solutions is essentially the volume, V, of the solution

$$\therefore \pi V = n_2 RT \qquad . . . (X, 13)$$

The formal analogy between the van't Hoff equation, (X, 13), and the ideal gas law should be noted.

Equation (X, 13) may be written in terms of concentrations:

$$\pi = n_2 RT/V = cRT$$

An equation which is found to be more satisfactory at higher concentrations is

$$\pi = mRT \qquad . . . (X, 14)$$

10. The mechanism of osmosis. Several theories have been advanced to explain the mechanism of osmosis, which probably differs from membrane to membrane.

(*a*) *The molecular sieve theory.* This theory pictures the membrane as a sieve through which small solvent molecules can pass while larger solute molecules cannot. Solvent molecules will pass from a region of higher solvent concentration to one of lower solvent concentration.

(*b*) *The vapour pressure theory.* This suggests that the membrane consists of small capillaries, the walls of which are not wetted by the solvent, which therefore cannot enter the capillaries. However, the solvent vapour is able to diffuse down the capillary tubes and the solvent distills from a region of higher vapour pressure (pure solvent) to one of lower vapour pressure (solution).

(c) *The membrane solution theory.* The solvent may be soluble in the membrane material while the solute is not, and hence the solvent may be transferred directly from the pure solvent into the solution.

11. Osmotic pressure and relative molecular masses of polymers.

From (X, 13) it can be seen that for a 1 M solution at 27°C

$$\pi = n_2 RT/V = 1 \times 8 \cdot 314 \times 300/10^{-3} = 2 \cdot 494 \text{ MN m}^{-2}$$

This pressure, being equivalent to a hydrostatic head of about 170 m, shows that osmotic pressure is a very large effect indeed.

Using (X, 12), (X, 13) or (X, 14) osmotic pressure measurements have become a very important means of measuring the relative molecular masses of natural and artificial polymers, which, because of their very high relative molecular masses, can usually be obtained only as very dilute solutions.

VOLATILE SOLUTES

As mentioned in IX, graphical representation of a two-component system requires a three-dimensional phase diagram. Since this is inconvenient, the graphs exhibited below have been obtained by holding one variable constant and plotting the behaviour of the other two.

12. Raoult's law.

For an ideal solution containing two volatile components, 1 and 2, Raoult's law becomes

$$p_1 = p_1^{\bullet} x_1 \quad p_2 = p_2^{\bullet} x_2 \quad \ldots \quad (X, 1)$$

and, provided the vapour behaves as an ideal gas, the total pressure p is given by

$$p = p_1 + p_2 = p_1^{\bullet} x_1 + p_2^{\bullet} x_2 \quad \ldots \quad (X, 15)$$

It should be noted that the linear variation of vapour pressure implied by (X, 1) and (X, 15) is with respect to the composition of the liquid rather than the vapour. The vapour is richer in the more volatile component as shown in the next section.

13. Vapour pressure against composition diagrams (constant temperature). The behaviour of solutions of ethyl iodide in ethyl bromide is very nearly ideal. Their vapour pressures at 16·7°C are 60·2 kN m^{-2} and 21·6 kN m^{-2} respectively, and the graph of vapour pressure against composition is shown in Fig.

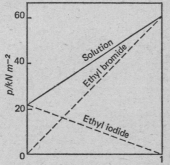

Fig. 33.—*Vapour pressure against composition diagram (constant temperature) for ideal solutions of ethyl bromide and ethyl iodide*

33. Here the broken lines show the partial vapour pressures of each of the components and the solid line shows the total vapour pressure of the solution.

The composition of the vapour from a solution of given composition may be calculated from Raoult's law, as follows. Consider a mixture containing 0·5 mol fraction of each component at 16·7°C (ethyl bromide ≡ 1, ethyl iodide ≡ 2):

$$p_1^{\bullet} = 60 \cdot 2 \text{ kN m}^{-2} \qquad p_2^{\bullet} = 21 \cdot 6 \text{ kN m}^{-2}$$
$$p_1 = p_1^{\bullet} x_1 = 60 \cdot 2 \times 0 \cdot 5 = 30 \cdot 1 \text{ kN m}^{-2}$$
$$p_2 = p_2^{\bullet} x_2 = 21 \cdot 6 \times 0 \cdot 5 = 10 \cdot 8 \text{ kN m}^{-2}$$
$$\therefore p = p_1 + p_2 = 40 \cdot 9 \text{ kN m}^{-2}$$

The composition of the vapour is obtained using Dalton's law:

$$x_1 = p_1/p = 30 \cdot 1/40 \cdot 9 = 0 \cdot 736$$
$$x_2 = p_2/p = 10 \cdot 8/40 \cdot 9 = 0 \cdot 264$$

i.e. the vapour is richer in the more volatile component, ethyl bromide.

If this vapour is condensed the vapour pressures of the components in the resulting solution will be:

$$p_1 = p_1^{\circ} x_1 = 60\cdot2 \times 0\cdot736 = 44\cdot3 \text{ kN m}^{-2}$$
$$p_2 = p_2^{\circ} x_2 = 21\cdot6 \times 0\cdot264 = 5\cdot70 \text{ kN m}^{-2}$$
$$\therefore p = p_1 + p_2 = 50\cdot0 \text{ kN m}^{-2}$$

and the composition of the vapour will be:

$$x_1 = p_1/p = 44\cdot3/50\cdot0 = 0\cdot886$$
$$x_2 = p_2/p = 5\cdot7/50\cdot0 = 0\cdot114$$

i.e. the vapour is again richer in the more volatile component.

Such a procedure forms the basis of the technique of *fractional distillation*. In practice it is more convenient to perform the distillation at constant pressure rather than at constant temperature, and the various steps in the distillation are better illustrated on a boiling point against composition diagram as shown in the next section.

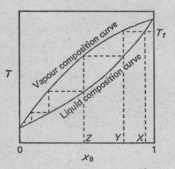

Fig. 34.—*Boiling point against composition diagram (constant pressure) for ideal solutions of A and B*

14. Boiling point against composition diagrams (constant pressure). The boiling point against composition diagram for an ideal binary solution in which component A is more volatile than component B is shown in Fig. 34. The upper curve represents the composition of the vapour and the lower curve the composition of the liquid. At any given temperature the vapour is richer in the more volatile component.

Liquid of composition X begins to boil at T_1. The vapour initially collected has composition Y, *i.e.* is richer in the more volatile component A. The boiling point of the residual liquid will rise, since it becomes richer in B. If the vapour is condensed and reboiled the vapour initially collected will have composition Z, *i.e.* still richer in A. This procedure can be continued until pure A is obtained in the distillate. In practice the fractions collected will cover a range of compositions and X, Y, Z refer to the average compositions within these ranges.

The labour involved in separating A and B on a batch basis as above would be prohibitive, and the process of successive vaporisation and condensation is normally carried out in a fractionating column. A simple fractionating column which nicely illustrates the above ideas is the "bubble cap" column.

15. The "bubble cap" column. This simple fractionating column consists of a series of plates arranged above the boiling liquid so that the vapour has to bubble through a film of liquid on each plate before passing up the column to the next plate.

The mode of operation of the column is as follows. Vapour from the flask bubbles through a film of condensed liquid on the first plate. Since the liquid on the plate is cooler than that in the flask, partial condensation occurs and the vapour leaving the plate is enriched in the more volatile component. A similar enrichment takes place on the next plate, and so on until ideally the distillate consists of pure A, and pure B is left in the flask.

Each partial condensation, corresponding to an equilibrium between liquid and vapour on one of the plates, is known as a *theoretical plate*, and the efficiency of any fractionating column is measured by its number of theoretical plates. In order to separate liquids having very close boiling points a large number of theoretical plates may be required, and this is achieved in the laboratory by packing the column with glass beads or helices.

NEGATIVE DEVIATIONS FROM IDEALITY

The process of solution is always accompanied by the absorption or evolution of heat, unless the components form an ideal solution. Where evolution of heat takes place the resulting solution shows a negative deviation from ideality, *i.e.*

the partial vapour pressure of each component in the solution is less than that predicted by Raoult's law.

16. Reasons for negative deviations. The evolution of heat accompanying the process of solution shows that the solution represents a state of lower energy than either of the two components separately, *i.e.* the solution is more stable than either of the pure components. Thus the escaping tendency of the molecules, as measured by the vapour pressures of the components, will be less in the solution than in either of the pure components.

Solutions showing negative deviations are obtained when some bonding exists in the solution which is not possible in either of the pure components. For example, mixtures of chloroform and acetone show a negative deviation due to hydrogen bonding $[Cl_3C - H \cdots O = C(CH_3)_2]$, which is not possible in either pure chloroform or pure acetone (*see* XVII, 6–7).

17. Vapour pressure against composition diagrams for solutions showing a negative deviation. Fig. 35 shows the vapour

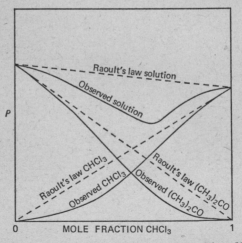

Fig. 35.—*Vapour pressure against composition diagram (constant temperature) for acetone–chloroform solutions, showing negative deviations from ideality*

pressure against composition diagram for a solution showing a negative deviation from ideality (*e.g.* chloroform–acetone solutions at constant temperature). The broken lines indicate the variation of vapour pressure predicted by Raoult's law and the solid curves the experimentally observed variation.

It can be seen that the individual partial vapour pressure curves fall below the Raoult's law line and that the total vapour pressure exhibits a minimum which lies below the vapour pressure of either of the pure components.

18. Raoult's law and Henry's law. In the region where the mole fraction of acetone is close to unity (*i.e.* where acetone can be considered as the solvent) Fig. 35 shows that its partial vapour pressure closely approximates the Raoult's law line. In this same region chloroform can be considered as the solute, and it can be seen that its vapour pressure varies approximately linearly, but with a slope different from the Raoult's law line. In fact the slope approximately agrees with that predicted by Henry's law, *i.e.* $p_{CHCl_3} = K_{CHCl_3} x_{CHCl_3}$, where K_{CHCl_3} is a constant not equal to $p^{\bullet}_{CHCl_3}$. These two facts may be summed up by saying that the solvent in a dilute solution obeys Raoult's law, whilst the solute obeys Henry's law.

NOTE: If the solution is ideal, $K = p^{\bullet}$ and Henry's law reduces to Raoult's law. Henry's law is dealt with in rather more detail in connection with the solubility of gases (*see* **29–32** below).

19. Boiling point against composition diagrams for solutions showing a negative deviation. Fig. 36 shows the boiling point

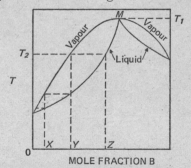

FIG. 36.—*Boiling point against composition diagram (constant pressure) for solutions showing negative deviations from ideality*

against composition diagram for a solution of A and B showing a negative deviation from ideality (*e.g.* water–hydrochloric acid solutions at constant pressure). The minimum in the vapour pressure in Fig. 35 is replaced by a maximum M in the boiling point, and at M the liquid and vapour have the same composition. The mixture of this composition is called the *maximum boiling azeotrope*.

20. Fractional distillation of solutions showing a negative deviation. If a mixture having composition M in Fig. 36 is distilled, the vapour formed at T_1 has exactly the same composition as the liquid and no separation of the components is achieved.

If a mixture of composition Z is heated, the vapour initially formed at T_2 has composition Y, *i.e.* richer in component A, so that the residue in the flask is richer in component B. If this vapour is then condensed and re-distilled, the vapour initially formed will have composition X, *i.e.* still richer in component A.

Carrying out this procedure in a fractionating column should result finally in the separation of pure A in the distillate, leaving the maximum boiling azeotrope in the flask.

By an exactly similar procedure, it can be predicted that if the initial mixture had a composition to the right of M, fractionation would have resulted in the separation of pure B in the distillate, leaving the maximum boiling azeotrope in the flask.

In general, such systems are separable by fractionation into the maximum boiling azeotrope (flask) and the pure component in excess of the azeotropic composition (distillate).

POSITIVE DEVIATIONS FROM IDEALITY

Where the process of solution is accompanied by the absorption of heat, the resulting solution shows a positive deviation from ideality, *i.e.* the partial vapour pressure of each component in the solution is greater than that predicted by Raoult's law.

21. Reasons for positive deviations. The absorption of heat shows that the solution represents a state of higher energy than either of the pure components separately, *i.e.* the solution is less stable than either of the pure components. Thus the escaping tendency of the molecules, as measured by the vapour pressures of the components, will be greater in the solution than in either of the pure components.

Solutions showing positive deviations are usually obtained on mixing a highly polar compound A with a non-polar compound B. In the solution the strong dipole–dipole stabilisation between A molecules is partially replaced by the much weaker interaction (dipole–induced dipole) between A and B (*see* XVIII, 8).

22. Vapour pressure against composition diagram for solutions showing a positive deviation. Fig. 37 shows the vapour pressure against composition diagram for a solution showing a

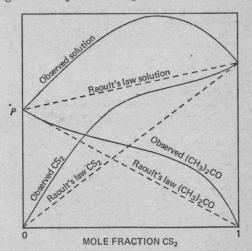

FIG. 37.—*Vapour pressure against composition diagram (constant temperature) for acetone–carbon disulphide solutions, showing positive deviations from ideality*

positive deviation from ideality (*e.g.* carbon disulphide–acetone solutions at constant temperature). The broken lines indicate the variation of vapour pressure predicted by Raoult's law, and the solid lines the experimentally observed variation.

The individual partial pressure curves fall well above the

Raoult's law line and the total vapour pressure exhibits a maximum which lies above the vapour pressure of either of the pure components.

23. Boiling point against composition diagram for solutions showing a positive deviation. Fig. 38 shows the boiling point against composition diagram for a solution of A and B showing

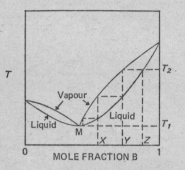

Fig. 38.—*Boiling point against composition diagram (constant pressure) for solutions showing positive deviations from ideality*

a positive deviation from ideality (*e.g.* water–ethanol solutions at constant pressure). The maximum in vapour pressure is replaced by a minimum M in the boiling point. The mixture of composition M is called the *minimum boiling azeotrope*.

24. Fractional distillation of solutions showing a positive deviation. By the procedure of 20 it can be shown that such systems are separable into the minimum boiling azeotrope (distillate) and the pure component in excess of the azeotropic composition (flask).

PARTIAL MISCIBILITY

25. Partial miscibility and Raoult's law. The formation of an ideal solution is always accompanied by a decrease in free energy, since in the equation governing the solution equilibrium

$$\Delta G = \Delta H - T\Delta S \quad . \quad . \quad . \quad \text{(VIII, 10)}$$

$\Delta H = 0$, and any mixing process is always accompanied by an increase in disorder. Such components will be miscible in all proportions.

If the solution shows a sufficiently large positive deviation from Raoult's law (*i.e.* the process of solution is accompanied by considerable absorption of heat) then the ΔH and ΔS terms in (VIII, 10) tend to drive the system in opposite directions and the components may not be miscible in all proportions. As successive portions of one component are added to the other, a limiting solubility is reached, beyond which two distinct liquid phases are formed.

26. Systems showing an upper critical solution (or consolute) temperature.

Normally, increasing the temperature tends to promote solubility, and the two-phase system present at low temperatures frequently forms a single phase at higher temperatures. The behaviour of such a system (*e.g.* phenol–water) may be shown on a temperature against composition diagram at constant pressure as in Fig. 39. Outside the loop

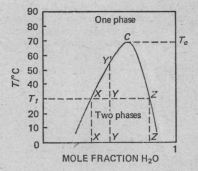

FIG. 39.—*Temperature against composition diagram (constant pressure) for phenol–water solutions, showing an upper critical solution temperature*

a single liquid phase is present, whilst inside there are two phases. At temperature T_1 a mixture of composition Y consists of two phases, called *conjugate solutions*, whose compositions are obtained from the tie line at T_1 as X and Z. The relative amounts of the conjugate solutions are given by the lever rule

$$\frac{\text{Amount of solution composition } X}{\text{Amount of solution composition } Z} = \frac{YZ}{XY}$$

As the temperature is increased along YY' the relative amount of the phenol-rich phase decreases while that of the water-rich phase increases, and at Y' the phenol-rich phase disappears entirely.

This gradual disappearance of one phase on heating is characteristic of all compositions except that corresponding to the maximum of the curve C. If a two-phase mixture of this composition is heated, the compositions of the phases approach each other until, at the temperature T_C corresponding to the maximum, the compositions become identical and a single phase remains. C is called the *critical composition* and T_C the *upper critical solution (or consolute) temperature*.

27. Systems showing a lower critical solution temperature.

In such systems the two phases present at high temperatures give a single phase on cooling.

This phenomenon is less common (*e.g.* water–triethylamine) and is apparently caused by large negative deviations from Raoult's law, the components tending to form loosely bound compounds. Such compound formation is enhanced at low temperatures, but the thermal dissociation induced by raising

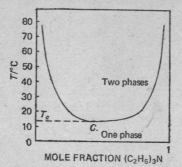

Fig. 40.—*Temperature against composition diagram (constant pressure) for water–triethylamine solutions, showing a lower critical solution temperature*

the temperature decreases the mutual solubility of the components.

The temperature against composition diagram for such a system is shown in Fig. 40, where C again represents the *critical composition* and T_c the *lower critical solution temperature*.

28. Systems showing an upper and a lower critical solution temperature. Such systems are more common at high pressures, but mixtures of water and nicotine show both an upper and a lower critical solution temperature at atmospheric pressure (*see* Fig. 41).

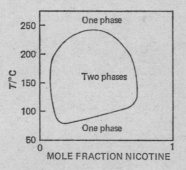

FIG. 41.—*Temperature against composition diagram (constant pressure) for nicotine–water solutions, showing both a lower and an upper critical solution temperature*

29. Steam distillation. The process of steam distillation can be applied to any mixture of water with an immiscible liquid. This technique is most useful for organic substances which decompose before reaching their boiling points, since it results in distillation at a temperature lower than the boiling point of either component. The calculation of the amounts of the two components present in the distillate is illustrated in the following example.

EXAMPLE 1: Consider the steam distillation of a mixture of water (A) and aniline (B). Since the two components are immiscible, each exerts its own vapour pressure independently of the other, so that the total pressure p is given by

$$p = p_A^\circ + p_B^\circ$$

The system will boil when p reaches $101 \cdot 325$ kN m^{-2}, which it does in this system at $98 \cdot 4°$C, when the vapour pressures of water and aniline are $95 \cdot 64$ kN m^{-2} and $5 \cdot 69$ kN m^{-2} respectively.

If x'_A and x'_B are the mole fractions in the vapour, then

$$p_A^\bullet / p_B^\bullet = x_A' p / x_B' p = x_A' / x_B'$$

Now $x_A' = n_A/(n_A + n_B)$ and $x_B' = n_B/(n_A + n_B)$, where n_A and n_B are the amounts of water and aniline in a given volume of the vapour. Consequently

$$p_A^\bullet / p_B^\bullet = n_A / n_B$$

Since $n = m/M$, this equation becomes

$$p_A^\bullet / p_B^\bullet = m_A M_B / m_B M_A$$
or
$$m_A / m_B = p_A^\bullet M_A / p_B^\bullet M_B$$

Thus for each kilogramme of water in the distillate the mass of aniline, m_B is given by

$$1/m_B = 95 \cdot 64 \times 0 \cdot 018 / 5 \cdot 69 \times 0 \cdot 094$$
Hence
$$m_B = 0 \cdot 30 \text{ kg}$$

i.e. there are $0 \cdot 30$ kg of aniline to every 1 kg of water in the distillate.

SOLUBILITY OF GASES IN LIQUIDS

30. Henry's law. The solubility of gases in liquids is governed by Henry's law, which may be expressed in the equivalent forms

$$m = Kp \qquad \cdot \ \cdot \ \cdot \ \text{(X, 15)}$$

where m is the mass of gas dissolved by a given amount of solvent under a pressure p, at constant temperature, and

$$p_2 = K' x_2 \qquad \cdot \ \cdot \ \cdot \ \text{(X, 16)}$$

where the symbols have their usual significance. For an ideal solution the constant K' is equal to $p_2^\bullet$, the vapour pressure of the pure liquefied gas, and (X, 1) reduces to Raoult's law for the solute

$$p_2 = p_2^\bullet x_2 \qquad \cdot \ \cdot \ \cdot \ \text{(X, 17)}$$

31. Solubility and nature of the gas. Equation (X, 15) suggests that the solubility of the gas in a solvent with which it forms an ideal solution should be dependent only on the nature of the gas. The lower the vapour pressure of the pure liquefied gas, the greater should be its solubility, *i.e.* easily liquefiable gases should be the most soluble.

32. The "salting out" effect. Gases are generally less soluble in aqueous solutions of electrolytes than in pure water, and this effect becomes more pronounced as the size of the ions decreases and their charge increases. The reason is that preferential solvation of the charged ions by the highly polar water molecules leaves a smaller amount of "free water" to dissolve the gas (*see* XX, **13**). Some large organic molecules, *e.g.* sugars, have the same effect, owing to their high polarisability.

33. Deviations from Henry's law. There are two common types of deviation:

(*a*) Those caused by failure of the gas to obey the ideal gas law. Such deviations are expected to occur at low temperatures and high pressures, particularly for the more easily liquefiable gases.

(*b*) Those caused by compound formation between the solute and solvent, or by dissociation or association of the solute in solution, *e.g.* solutions of ammonia in water dissociate according to the equation

$$NH_3 + H_2O \rightleftharpoons NH_4{}^+ + OH^-$$

This particular type of deviation is usually easy to allow for.

PROGRESS TEST 10

1. A solution of sodium chloride (relative molecular mass 58·5) and a solution of sucrose are placed side by side in a closed apparatus. Water distills from one solution to the other until equilibrium is reached, and on analysis the solutions are found to contain 1% sodium chloride and 10·5% sucrose. Find the relative molecular mass of sucrose. (4)

2. A solution of 2 g of a hydrocarbon (containing 93·76% C) in 100 g of benzene lowers the freezing point by 0·8°C. Find the empirical formula of the hydrocarbon. K_f for benzene is 5·12 K mol^{-1} kg. (6)

3. 2 g of a certain compound in 100 g of mesitylene (relative molecular mass 120) raises the boiling point from 164°C to 169°C and lowers the freezing point from −52°C to −55·6°C. ΔH_v for mesitylene is 29·5 kJ mol^{-1}. Find the relative molecular mass of the compound and ΔH_f for mesitylene. (5–6)

4. At 27°C the osmotic pressure of a solution of 1 g of insulin in 100 cm^3 of water was found to be 4·29 kN m^{-2}. Calculate the relative molecular mass of the protein. (9)

5. A mixture of benzene and toluene behaves as an ideal solution, the partial pressures of benzene and toluene at 20°C being 10·00 kN m^{-2} and 2·92 kN m^{-2} respectively. Calculate the composition of the vapour initially formed from a mixture containing 0·5 mole fraction of each component. If this vapour is condensed and then allowed to re-evaporate, calculate the composition of the initial fraction of the vapour thus formed. (13)

6. State (a) Raoult's law in words and symbols (2), (b) Henry's law in words and symbols (30). What is the relationship between the two laws? (18, 30)

7. Deduce the relationship between the molar mass of an involatile solute and (a) the lowering of vapour pressure of the solution (4), (b) the elevation of boiling point of the solution (5), (c) the osmotic pressure of the solution (9).

8. Outline the theories that have been advanced to explain the mechanism of osmosis. (10)

9. Draw the vapour pressure against composition diagrams (constant temperature) for two volatile compounds which form (a) ideal solutions (13), (b) solutions showing a negative deviation from ideality (17), (c) solutions showing a positive deviation from ideality (22).

10. Explain, with the aid of diagrams, the principles behind fractional distillation. What is an azeotropic mixture? (14, 15, 20, 24)

11. Discuss, with examples, the phenomenon of partial miscibility. (25–28)

CHAPTER XI

HOMOGENEOUS AND HETEROGENEOUS CHEMICAL EQUILIBRIA

THE EQUILIBRIUM STATE

1. Chemical equilibrium. The properties of the equilibrium state outlined in VIII, 2. and IX, 2 may be illustrated for chemical equilibria by the dissociation of calcium carbonate:

$$CaCO_3 \text{ (g)} \rightleftharpoons CaO \text{ (g)} + CO_2 \text{ (g)}$$

(*a*) The balance between the tendencies towards minimum energy and maximum entropy occurs since the right hand side of the equation represents a state of high entropy (because a gas is formed), whereas the left hand side represents a state of low energy (because the reaction is endothermic).

(*b*) Equilibrium occurs when the rate of dissociation of $CaCO_3$ becomes equal to its rate of formation from CaO and CO_2. This can be proved by allowing the system to come to equilibrium in a closed vessel and then connecting to it a source of CO_2, "labelled" with the radioactive ^{14}C isotope, at the equilibrium temperature and pressure. After a time the $CaCO_3$ will be found to contain the same proportion of ^{14}C as the CO_2.

2. The equilibrium constant K. As shown in VIII, 3, for a reaction

$$aA + bB \rightleftharpoons cC + dD \quad . \quad . \quad . \quad \text{(XI, 1)}$$

at equilibrium under constant pressure and temperature

$$\Delta G = \Delta G^\ominus - RT \ln K_A = 0$$

where K_A is the thermodynamic equilibrium constant in terms of activities and is given by

$$K_A = \left\{ \frac{(a_C)^c (a_D)^d}{(a_A)^a (a_B)^b} \right\}_{equil} \quad . \quad . \quad \text{(XI, 2)}$$

NOTE: In accordance with standard practice, the subscript$_{equil}$ will in future be omitted, since it is generally obvious that equilibrium conditions apply.

K_A is independent of activity but dependent on temperature. A large value of K_A indicates that the formation of products is favoured, a small value that the formation of reactants is favoured. Most apparently irreversible reactions can be considered to have a position of equilibrium. However, if K_A is very large, the reaction will apparently go to completion, whereas if K_A is very small the reaction will apparently not occur (see also VIII, 31).

If (XI, 1) had been written

$$cC + dD \rightleftharpoons aA + bB$$

then

$$K_A = \frac{(a_A)^a (a_B)^b}{(a_C)^c (a_D)^d}$$

i.e. K_A (forward reaction) $= 1/K_A$ (reverse reaction)

$$\text{. . . (XI, 3)}$$

If the activities in (XI, 2) are not the equilibrium values, then $\Delta G \neq 0$. If the activities of the products are too small relative to those of the reactants, ΔG is negative for the forward reaction. This reaction therefore proceeds at a faster rate than the reverse reaction until the activities reach their equilibrium values, ΔG becomes zero, and all *net* reaction ceases.

3. The equilibrium constants K_x, K_c and K_p. The systems studied in the present chapter are assumed to approximate sufficiently closely to ideality to enable activities to be replaced by mole fractions (K_x), concentrations (K_c) or pressures (K_p). These equilibrium constants have been defined in VIII, 32. Since they involve ratios of terms for each reactant, e.g. $p/p^\ominus$, $c/c^\ominus$, etc., they are dimensionless (x is by definition dimensionless; see VIII, 26) and hence have the same value for any given reaction.

Difficulties arise when the standard state is not included in the equation for the equilibrium constant. Since the standard state of concentration $c^\ominus$ is 1 mol dm^{-3}, c is numerically equal to $c/c^\ominus$ (see also VIII, 27) and (VIII, 37) may be written in the form

$$K_c = \frac{(c_C)^c (c_D)^d}{(c_A)^a (c_B)^b} \qquad \text{. . . (XI, 4)}$$

which is not strictly correct.

Similarly if $p^\ominus = 1$ atm, then K_p may be written as

$$K_p = \frac{(p_C)^c (p_D)^d}{(p_A)^a (p_B)^b} \qquad \text{. . . (XI, 5)}$$

This treatment cannot be applied if the standard state of pressure is 101·325 kN m^{-2} and here it is essential to use the form $p/p^\ominus$.

The equilibrium constants given by (XI, 4) and (XI, 5) are not equal. The relationship between them can readily be found by substituting $p = nRT/V = cRT$ into (XI, 5) and is

$$K_p = K_c (RT)^{\Delta n}$$

where $\Delta n = (c + d) - (a + b)$, *i.e.* the change in the amounts of gaseous reagents.

4. The dynamic character of equilibrium. Equilibrium is reached when the rates of the forward and reverse reactions in (XI, 1) are equal. If experiments indicate that the rate of the forward reaction is $k_f [A]^a [B]^b$ and that of the reverse reaction is $k_b [C]^c [D]^d$, where k_f and k_b are the respective rate constants (*see* XXI, 3) and square brackets rather than c's are used to indicate concentrations, then at equilibrium

$$k_f [A]^a [B]^b = k_b [C]^c [D]^d$$
$$\therefore K_c = \frac{k_f}{k_b} = \frac{[C]^c [D]^d}{[A]^a [B]^b}$$

This kinetic approach to equilibrium constants, although extensively used, needs great care, since rate laws cannot be deduced from the stoichiometric coefficients of the reactants and must be determined experimentally (*see* XXI, 2–4).

Note that the value of K_c itself measures the ratio of the concentrations of products to reactants, but gives no information about the rate of approach to equilibrium. Thus a particular value of K_c may be the ratio of two very small rate constants, in which case the approach to equilibrium will be slow, or the ratio of two large rate constants, in which case the approach to equilibrium will be rapid.

This non-thermodynamic approach to equilibrium constants does not involve the use of standard states. Since concentration units are used, the numerical value of the equilibrium constant will be the same regardless of the method of derivation (*see* **3** above). For convenience, the standard state of concentration will be omitted in the remainder of the book. Under these conditions K_c will have units, unless $(c + d)$ $= (a + b)$ in (XI, 4). Where pressure units are used, the standard state of $101 \cdot 325$ kN m^{-2} will be included.

5. Measurement of equilibrium constants.

There are two general methods of measuring equilibrium constants:

(*a*) Direct measurement of the composition of the equilibrium mixture without disturbing the position of equilibrium. This method is not always applicable. For example, a mixture of hydrogen and oxygen can be left indefinitely at room temperature with no discernible formation of water. The partial pressures of hydrogen and oxygen may take any value because the system has not reached equilibrium.

Because of the wide disparity in the rates of approach of different reactions to equilibrium it is of paramount importance to ensure that equilibrium has truly been attained. Precautions to be taken include approaching equilibrium both from the side of the products and from the side of the reactants, from temperatures and pressures above and below the desired values, and using catalysts to speed the rate of attainment of equilibrium.

(*b*) Using the relation $\Delta G^\ominus = -RT \ln K$, where $\Delta G^\ominus$ may be determined by the methods outlined in VIII, **39** (*b*) and (*c*).

Detailed procedures for the determination of equilibrium constants by method (*a*) are given in **7–13** below, where examples of various equilibria are considered.

6. Le Chatelier's principle.

Examples of the application of this principle, which was stated in VIII, **34**, to various equilibria are:

(*a*) $$SO_2(g) + \tfrac{1}{2}O_2(g) \rightleftharpoons SO_3(g)$$

If excess oxygen is added to the system at equilibrium, Le Chatelier's principle predicts that the yield of SO_3 will increase in order to consume the excess oxygen. This could have been predicted from the form of the equilibrium constant, since immediately after the excess oxygen has been added

$$[SO_3]/[SO_2][O_2]^{\frac{1}{2}} < K_c$$

and must therefore increase to re-establish the equality.

(*b*) $$2NO_2(g) \rightleftharpoons N_2O_4(g)$$

If the pressure in the system at equilibrium is increased, Le Chatelier's principle predicts that the reaction will produce more N_2O_4, since this decreases the volume of the system.

This could also have been predicted from the form of the equilibrium constant. For example, doubling the pressure corresponds to doubling all the concentrations. Thus the denominator in the expression for the equilibrium constant increases by a factor of 4, while the numerator increases only by a factor of 2.

$$[N_2O_4]/[NO_2]^2 < K_c$$

and $[N_2O_4]$ must therefore increase to re-establish the equality.

(c) The effect of an increase in temperature is to increase the yield of products in an endothermic reaction or the reactants in an exothermic reaction. This has been discussed in VIII, 34.

It should be noted that, whatever the dictates of Le Chatelier's principle, an increase in temperature invariably speeds up the rate of attainment of equilibrium and vice versa. It may therefore not be economic to increase the yield of an exothermic reaction by decreasing the temperature, since this may decrease the rate of attainment of equilibrium to an unacceptably low value.

EXAMPLES OF HOMOGENEOUS CHEMICAL EQUILIBRIA

The derivation and use of the equilibrium constant are illustrated in the calculations attached to the following sections.

7. The hydrogen–iodine reaction.

$$H_2(g) + I_2(g) \rightleftharpoons 2HI(g)$$

This reaction has been studied by sealing known amounts of HI or of H_2 and I_2 into fused quartz bulbs and heating in a thermostat until equilibrium is reached. The equilibrium is then "frozen" by breaking the bulbs below ice-water and the mixtures analysed for iodine by titration with sodium thiosulphate solution.

EXAMPLE 1: 0·02 g of hydrogen and 1·28 g of iodine are allowed to come to equilibrium at 460°C. On analysis the equilibrium mixture is found to contain 0·0021 mol of iodine. Calculate the equilibrium constant at this temperature.

The initial amounts of hydrogen and iodine are:

$$n_{H_2} = 0.02 \times 10^{-3}/2 \times 10^{-3} = 0.01 \text{ mol}$$
$$n_{I_2} = 1.28 \times 10^{-3}/128 \times 10^{-3} = 0.01 \text{ mol}$$

Thus if x mol of iodine have reacted,

$$H_2(g) + \quad I_2(g) \rightleftharpoons 2HI(g)$$

Initially:	0.01 mol	0.01 mol	0 mol
At equilibrium:	(0.01 − x) mol	(0.01 − x) mol	2x mol

$$\therefore K_c = [HI]^2/[H_2][I_2] = 4x^2/(0.01 - x)^2$$

But $\quad 0.01 - x = 0.0021$

$$\therefore x = 0.0079$$
$$\therefore K_c = 4(0.0079)^2/(0.0021)^2 = 56.6$$

EXAMPLE 2: Using the equilibrium constant calculated above, find the fraction of iodine reacting if the initial mole ratio of hydrogen to iodine is 3:1.

Suppose that initially y mol of I_2 are present and that a fraction α reacts. Then,

$$H_2(g) + \quad I_2(g) \rightleftharpoons 2HI(g)$$

Initially:	3y mol	y mol	0 mol
At equilibrium:	y(3 − α) mol	y(1 − α) mol	2yα mol

$$\therefore K_c = \frac{4y^2\alpha^2}{y^2(3 - \alpha)(1 - \alpha)}$$

Rearranging this equation to obtain α in terms of K_c

$$\alpha = \frac{4K_c \pm [16 K_c^2 - 12 K_c(K_c - 4)]^{\frac{1}{2}}}{2(K_c - 4)}$$

The positive sign can obviously be discarded, since α cannot be greater than unity.

$$\therefore \alpha = 0.93$$

Notice that in the previous example α was 0.79. The presence of excess H_2 has increased the yield of HI, as would be predicted by Le Chatelier's principle.

8. The dissociation of dinitrogen tetroxide.

$$N_2O_4(g) \rightleftharpoons 2NO_2(g)$$

This is an example of an equilibrium in which there is a change in the number of gaseous molecules. The course of the reaction is normally followed by measuring the density or the apparent relative molecular mass of the equilibrium mixture, as shown in Progress Test II, 11.

EXAMPLE 3: Calculate the degree of dissociation of dinitrogen tetroxide at total pressures of (a) $101 \cdot 3$ kN m^{-2}, (b) $25 \cdot 3$ kN m^{-2}. K_p for this reaction is $0 \cdot 14$.

If the initial amount of N_2O_4 is a, the degree of dissociation α, and total equilibrium pressure p:

$$N_2O_4(g) \rightleftharpoons 2NO_2(g)$$

Initially: a mol   0 mol   total a mol
At equilibrium: $a(1 - \alpha)$ mol $2a\alpha$ mol   total $a(1 + \alpha)$mol

The mole fractions present at equilibrium are:

$$x_{N_2O_4} = a(1 - \alpha)/a(1 + \alpha), \quad x_{NO_2} = 2a\alpha/a(1 + \alpha)$$

By Dalton's law the equilibrium partial pressures are:

$$p_{N_2O_4} = p(1 - \alpha)/(1 + \alpha), \quad p_{NO_2} = p2\alpha/(1 + \alpha)$$

$$\therefore K_p = \frac{(p_{NO_2}/p^{\ominus}_{NO_2})^2}{(p_{N_2O_4}/p^{\ominus}_{N_2O_4})} = \frac{\{2\alpha/(1 + \alpha)\}^2\{p/p^{\ominus}_{NO_2}\}^2}{\{(1 - \alpha)/(1 + \alpha)\}\{p/p^{\ominus}_{N_2O_4}\}}$$

Now $p^{\ominus}_{NO_2} = p^{\ominus}_{N_2O_4} = p^{\ominus} = 101 \cdot 3$ kN m^{-2}

$$\therefore K_p = \frac{4\alpha^2}{1 - \alpha^2} \times \frac{p}{p^{\ominus}}$$

Hence $\alpha = \left\{ \left(\frac{p^{\ominus}K_p}{p} \right) \Big/ \left(4 + \frac{p^{\ominus}K_p}{p} \right) \right\}^{\frac{1}{2}}$

(a) Substituting $p = 101 \cdot 3$ kN m^{-2}, $K_p = 0 \cdot 14$: $\alpha = 0 \cdot 19$
(b) Substituting $p = 25 \cdot 3$ kN m^{-2}, $K_p = 0 \cdot 14$: $\alpha = 0 \cdot 40$

i.e. the degree of dissociation increases from 19% at $101 \cdot 3$ kN m^{-2} to 40% at $25 \cdot 3$ kN m^{-2}, as would be predicted qualitatively by Le Chatelier's principle.

9. The formation of ammonia

$$N_2(g) + 3H_2(g) \rightleftharpoons 2NH_3(g)$$

This reaction has been investigated by passing mixtures of H_2 and N_2 of known composition through an iron tube lined with a finely divided iron catalyst. The equilibrium mixture may be bubbled through water, the concentration of ammonia obtained by titration with acid, and the other gases by difference.

EXAMPLE 4: The above experiment, which was carried out at 350°C and a total pressure of 1013 kN m⁻² with a mole ratio of H_2 to N_2 of 3:1, resulted in the formation of 0·0735 mole fraction of NH_3. Calculate K_p for the forward and reverse reactions.

$$p_{NH_3} = x_{NH_3}p = 0.0735 \times 1013 = 74.5 \text{ kN m}^{-2}$$

Since total pressure $p = 1013$ kN m⁻²

$$p_{H_2} + p_{N_2} = 1013 - 74.5 = 938.5 \text{ kN m}^{-2}$$

But $p_{H_2} = 3p_{N_2}$

$$\therefore p_{N_2} = 234.6 \text{ kN m}^{-2} \text{ and } p_{H_2} = 703.9 \text{ kN m}^{-2}$$

For the forward reaction, $K_p = \dfrac{(p_{NH_3}/p_{NH_3}^{\ominus})^2}{(p_{N_2}/p_{N_2}^{\ominus})(p_{H_2}/p_{H_2}^{\ominus})^3}$

Since $p_{NH_3}^{\ominus} = p_{N_2}^{\ominus} = p_{H_2}^{\ominus} = p^{\ominus} = 101.3$ kN m⁻²

$$K_p = \frac{74.5 \times p^{\ominus 2}}{234.6 \times 703.9^3} = 6.8 \times 10^{-3}$$

For the reverse reaction, $K_p = 1/(6.8 \times 10^{-4}) = 1.47 \times 10^3$
If the equation for the formation of NH_3 had been written

$$\tfrac{1}{2}N_2(g) + \tfrac{3}{2}H_2(g) \rightleftharpoons NH_3(g)$$
$$K_p' = \frac{(p_{NH_3}/p_{NH_3}^{\ominus})}{(p_{N_2}/p_{N_2}^{\ominus})^{\frac{1}{2}}(p_{H_2}/p_{H_2}^{\ominus})^{\frac{3}{2}}} = K_p^{\frac{1}{2}}$$

Hence $\quad K_p' = (6.8 \times 10^{-4})^{\frac{1}{2}} = 2.6 \times 10^{-2}$

Thus the value of the equilibrium constant for a specific reaction at a given temperature depends on the stoichiometric coefficients of the reactants and products, and these should always be clearly stated.

10. The esterification of acetic acid.
This is an example of an equilibrium occurring in a homogeneous liquid. Mixtures of the reagents at known concentrations are thermostated at the required temperature for several hours in a sealed tube. The tube is then broken under ice-water to freeze the equilibrium, and the acetic acid titrated with standard alkali, using phenolphthalein as indicator.

$$CH_3COOH + C_2H_5OH \rightleftharpoons CH_3COOC_2H_5 + H_2O$$

$$K_c = \frac{[CH_3COOC_2H_5][H_2O]}{[CH_3COOH][C_2H_5OH]} \quad \cdot \cdot \cdot \text{ (XI, 7)}$$

If this reaction is carried out using water as solvent the expression for the equilibrium constant may be simplified. Thus if 0·1 M acetic acid reacts with 0·1 M ethanol in water under conditions such that the reaction goes very nearly to completion, 0·1 M water is produced. The initial concentration of water is $1000/18 = 55·5$ M and the equilibrium concentration is therefore 55·6 M, *i.e.* while the concentrations of the other reagents have changed by almost 100% that of the water has changed by less than 0·2%. The concentration of water may therefore be treated as a constant and (XI, 7) reduces to

$$\frac{K_c}{[\text{H}_2\text{O}]} = K_c' = \frac{[\text{CH}_3\text{COOC}_2\text{H}_5]}{[\text{CH}_3\text{COOH}][\text{C}_2\text{H}_5\text{OH}]} \quad \cdot \cdot \cdot \text{(XI, 8)}$$

It is important to note that even though the solvent concentration does not now appear explicitly in the expression for the equilibrium constant, the nature and extent of such solution reactions are as much influenced by the solvent as by the solutes. Also, K_c' will have units, (concentration)$^{-1}$.

If the above reaction is carried out in a solvent other than water, the concentration of water must appear in the expression for the equilibrium constant, as in (XI, 7).

11. The dissociation of calcium carbonate:

$$\text{CaCO}_3(\text{s}) \rightleftharpoons \text{CaO}(\text{s}) + \text{CO}_2(\text{g})$$
$$K_c = \frac{[\text{CaO}][\text{CO}_2]}{[\text{CaCO}_3]} \quad \cdot \cdot \cdot \text{(XI, 9)}$$

The concentrations of pure liquids and pure solids are taken as constant and are usually included in the equilibrium constant. This is because so long as the liquid or solid remains pure its concentration will stay the same (assuming constant temperature).

NOTE: As mentioned in VIII, **28** (*b*), the activity of a pure liquid or pure solid is unity.

If CO_2 is assumed to behave ideally then (XI, 9) becomes

$$K_p = p_{\text{CO}_2}/p_{\text{CO}_2}^{\ominus} \quad \cdot \cdot \cdot \text{(XI, 10)}$$

i.e. at any given temperature the pressure of CO_2 is constant (since K_p and $p_{\text{CO}_2}^{\ominus}$ are constant) and is usually referred to as the *dissociation pressure*.

The reaction proceeds from left to right with an increase in volume so that Le Chatelier's principle predicts that an increase in pressure will result in a decreased yield of CO_2.

It is important to realise that (XI, 10) only holds if all the constituents are present. If, for example, equilibrium is approached by starting with $CaO(s)$ and excess $CO_2(g)$ and gradually increasing the pressure, CaO will be used up in the formation of $CaCO_3$ until finally only $CaCO_3(s)$ and $CO_2(g)$ remain. p_{CO_2} can then take any value and will not be related to K_p.

12. Salt hydrates. The equilibria occurring among the hydrates of cupric sulphate are:

$$CuSO_4.5H_2O(s) \rightleftharpoons CuSO_4.3H_2O(s) + 2H_2O(g)$$
$$CuSO_4.3H_2O(s) \rightleftharpoons CuSO_4.H_2O(s) + 2H_2O(g)$$
$$CuSO_4.H_2O(s) \rightleftharpoons CuSO_4(s) + H_2O(g)$$

If $H_2O(g)$ is assumed to behave ideally, then for each hydrate pair the equilibrium constant is given by

$$K_p = (p_{H_2O}/p_{H_2O}^{\ominus})^n$$

where n is the number of water molecules produced. Thus, for any hydrate pair, $p^n_{H_2O}$, and therefore p_{H_2O}, is constant, i.e. the vapour pressure above any hydrate pair is constant. The variation of vapour pressure above any hydrate pair given by an equation of the form

$$d \ln p = \frac{\Delta H}{nR} \frac{dT}{T^2} \qquad (see \text{ VIII, 36})$$

The phase diagram for this system is shown in Fig. 42. The horizontal lines represent the values of n over which the pressure is constant, and the vertical lines the ranges of pressure over which the pure solid phases (containing five, three, one and no molecules of water respectively) are stable. Below 0·11 kN m^{-2} no hydrate is formed. On reaching 0·11 kN m^{-2} the pressure will remain constant until all the anhydrous salt has been converted to monohydrate, which remains stable until a pressure of 7·5 kN m^{-2} is reached. Here the pressure again remains constant until all the monohydrate has been converted to trihydrate. The trihydrate is stable up to 10·4 kN m^{-2}, when the pentahydrate is formed. This remains stable until the vapour pressure of the saturated solution is reached, when the solid begins to dissolve, forming a saturated solution. On dehydration the reverse behaviour occurs.

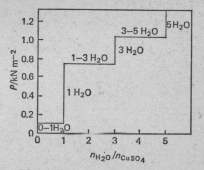

FIG. 42.—*Phase diagram for the system $CuSO_4$–H_2O*

If a particular hydrate has a vapour pressure less than that of the water vapour of the air, a higher hydrate will be formed. If the vapour pressure of the highest hydrate is less than that of the water vapour of the air, the solid will dissolve to form an aqueous solution. This phenomenon is known as *deliquescence*.

The reverse process—*efflorescence*—occurs if the vapour pressure of a particular hydrate is greater than that of the water vapour of the air.

13. The distribution law. Some substances are soluble in two immiscible solvents. Thus, for example, if a solution of iodine in water is shaken with carbon tetrachloride, the iodine distributes between the two solvents and at equilibrium at any given temperature

$$K_D = [I_2]_{H_2O}/[I_2]_{CCl_4}$$

where K_D is a constant known as the *distribution coefficient*.

This equation can readily be deduced from thermodynamic principles. If a single substance is distributed between two immiscible phases, then at equilibrium the molar free energies of the substance will be the same in both phases. If the phases are labelled A and B, then (VIII, 42) is appropriate, *i.e.*

$$G_A = G_B$$

$$\therefore\ G_A^\ominus + RT \ln a_A = G_B^\ominus + RT \ln a_B \quad (see\ VIII,\ 27)$$

or $$\ln(a_A/a_B) = (G_B^\ominus - G_A^\ominus)/RT$$

At constant temperature all the terms on the right hand side of the equation are constant

$$\therefore \ln (a_A/a_B) = \text{constant}$$
$$\text{or} \quad a_A/a_B = \text{constant}$$

This is the exact form of the distribution law, but, for dilute solutions, activities may be replaced by concentrations.

In some cases the substance exists in different states of molecular association in the two solvents, *e.g.* benzoic acid exists as hydrogen-bonded dimers in benzene, but as single molecules in water.

EXAMPLE 5: A molecule, A, exists totally as monomer in solvent 1 and totally as aggregates in solvent 2. Show how the state of aggregation may be obtained from measurements of the total concentration of A in solvent 1 and in solvent 2.

The equilibria are:

$$(A)_1 \underset{}{\overset{K_D}{\rightleftharpoons}} (A)_2 \underset{}{\overset{K_A}{\rightleftharpoons}} \frac{1}{n}(A_n)_2$$

$$\therefore K_D = [A]_2/[A]_1 \text{ and } K_A = [A_n]_2^{1/n}/[A]_2$$
$$\therefore K_D K_A = K = [A_n]_2^{1/n}/[A]_1$$

$$\text{or} \quad \lg[A]_1 + \lg K = \frac{1}{n}\lg[A_n]_2$$

n may therefore be determined from the slope of a plot of lg (total concentration of A in solvent 1) against lg (total concentration of A in solvent 2).

Some applications of the technique of distribution are the extraction of organic substances from water with an immiscible organic solvent, the distinction between bromides and iodides in analysis by reduction to bromine or iodine and shaking with carbon tetrachloride (iodine in carbon tetrachloride is violet, bromine is red), and the determination of the formulae of complex ions and associated molecules as shown above.

PROGRESS TEST 11

1. Find K_p and $\Delta G^{\ominus}$ for the following reaction at 1200°C:

(a) $CO_2(g) + H_2(g) \rightleftharpoons CO(g) + H_2O(g)$

given that for the reactions at 1200°C:

(b) $CO_2(g) \rightleftharpoons CO(g) + \frac{1}{2}O_2(g)$ $K_p = 1.14 \times 10^{-5}$
(c) $H_2(g) + \frac{1}{2}O_2(g) \rightleftharpoons H_2O(g)$ $K_p = 2.5 \times 10^5$

(2–3)

2. A mixture contains $12 \cdot 2$ kN m^{-2} of H_2, $20 \cdot 3$ kN m^{-2} of I_2 and $114 \cdot 6$ kN m^{-2} of HI. Will there be any net reaction and, if so, how much HI will be consumed or formed? K_p for the equilibrium is $55 \cdot 3$. (7)

3. For initial pressures of carbon monoxide and chlorine of $36 \cdot 48$ kN m^{-2} and $24 \cdot 32$ kN m^{-2} respectively, the total equilibrium pressure in the reaction

$$CO(g) + Cl_2(g) \rightleftharpoons COCl_2(g)$$

was $41 \cdot 54$ kN m^{-2}. Find K_p for this reaction. (6, 7)

4. Using the value of K_p found in the preceding section:

(a) Calculate the degree of dissociation of $COCl_2$ at the same temperature and $101 \cdot 3$ kN m^{-2} pressure.

(b) Calculate the degree of dissociation if the equilibrium mixture is diluted so that the total pressure remains at $101 \cdot 3$ kN m^{-2} but the partial pressure of the inert diluent is 76 kN m^{-2}.

(c) Comment on your results in terms of Le Chatelier's principle. (6–7)

5. Antimony pentachloride dissociates according to the equation

$$SbCl_5(g) \rightleftharpoons SbCl_3(g) + Cl_2(g)$$

At 128°C the apparent relative molecular mass of the equilibrium mixture (obtained from vapour density measurements) is 276. Calculate the degree of dissociation at this temperature. (8)

6. For the reaction

$$\tfrac{1}{2}N_2(g) + \tfrac{3}{2}H_2(g) \rightleftharpoons NH_3(g)$$

$\Delta H_f^\ominus = -51 \cdot 4$ kJ mol^{-1} and is approximately constant between 350°C and 500°C. K_p (350°C) $= 2 \cdot 66 \times 10^{-2}$. Find K_p (500°C). (VIII, 34)

7. When 1 mol of acetic acid was treated with $0 \cdot 33$ mol of ethanol, $0 \cdot 293$ mol of ethyl acetate was formed. Find K_x. If the measurement of the amount of ester was in error and the actual amount was $0 \cdot 296$ mol, find the effect on the equilibrium constant. (10)

8. Solid ammonium hydrosulphide is placed in a flask containing $101 \cdot 3$ kN m^{-2} pressure of ammonia gas. The reaction occurring is

$$NH_4HS(s) \rightleftharpoons NH_3(g) + H_2S(g)$$

for which $K_p = 0 \cdot 11$ at 25°C. Find the partial pressures of NH_3 and H_2S at this temperature. (11)

9. The distribution coefficient of mercuric bromide between water and benzene is $0 \cdot 9$. What mass would be extracted by

shaking with 100 cm³ of a water solution containing 5 g dm⁻³ of mercuric bromide with 100 cm³ of benzene (a) in one portion, (b) in successive 50 cm³ portions? **(13)**

10. Explain the difference between the thermodynamic and the dynamic (or kinetic) approaches to equilibrium constants. **(3–4)**

11. Indicate how equilibrium constants can be measured. **(5)**

12. Draw and explain the phase diagram for the system $CuSO_4 - H_2O$. **(12)**

13. Derive the exact form of the distribution law. **(13)**

PART FIVE

IONS IN SOLUTION

The first part of Chapter XII describes the electrolysis of electrolyte solutions; this is followed by an account of electrolytic conductivity. The Arrhenius theory is applied to the ionisation of weak electrolytes, while strong electrolytes are discussed in terms of the Debye–Hückel–Onsager theory. The migration of ions is considered in terms of transport numbers, ionic conductivities and ionic mobilities, and equations relating these three quantities are derived. The chapter closes with a discussion of a number of applications of conductivity measurements.

Chapter XIII gives an account of the properties of acids and bases in aqueous solution, and emphasis is placed on weak acids, weak bases and their salts. An account of the properties of buffer solutions is followed by a discussion of heterogeneous ionic equilibria.

In Chapter XIV the feasibility of oxidation–reduction reactions is discussed in terms of electrode potentials. The properties of single electrodes and cells are described, and the cell e.m.f. is related to the thermodynamic properties of the cell reaction and to the concentrations of the cell constituents. The scale of standard electrode potentials is derived and the conventions appropriate to their use are given. Finally the principles of potentiometric titrations are discussed, with several examples.

CHAPTER XII

ELECTROLYTES IN SOLUTION

Solids and liquids which are able to conduct an electric current may be divided into two categories:

(*a*) *Electronic conductors* such as metals and certain solid salts (*e.g.* CdS), where conduction takes place by migration of *electrons* under an applied potential.

(*b*) *Electrolytic conductors or electrolytes*, where conduction takes place by migration of *ions*. Examples include fused salts, and strong and weak electrolytes in aqueous solution.

ELECTROLYTES

This discussion will be restricted to acids, bases and salts dissolved in water.

1. The electrolytic cell. The effect of the passage of current through an electrolyte may be observed in an *electrolytic cell* consisting of two suitable conductors, called *electrodes*, im-

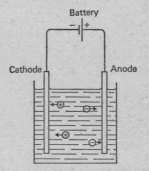

FIG. 43.—*The electrolytic cell*

mersed in the electrolyte solution. These electrodes are connected to an external source of direct electrical current such as a battery (*see* Fig. 43). The electrode connected to the negative side of the battery is called the *cathode*, while the other electrode is the *anode*. Under the applied field, the positive ions (*cations*) migrate to the cathode, and the negative ions (*anions*) to the anode. In addition to this migration of ions, chemical reactions occur at the electrodes which constitute the phenomenon of *electrolysis*.

2. The electrolysis of a silver nitrate solution.

Consider an electrolytic cell consisting of two silver electrodes immersed in a solution of silver nitrate. The ions present will be Ag^+, NO_3^- and H_3O^+, OH^- (from the slight dissociation of water; *see* XIII, **33**). Under an applied field the Ag^+ and H_3O^+ ions migrate to the cathode, where the Ag^+ ions are preferentially discharged (*see* **3** below) by the acquisition of an electron (the cathode, being negatively charged, has an excess of electrons) and are deposited on the cathode as metallic silver.

$$Ag^+ + e^- = Ag(s) \quad \ldots \text{(XII, 1)}$$

Although both the NO_3^- and OH^- ions travel to the anode, neither is discharged, and the reaction which does take place is the dissolution of $Ag(s)$ to give Ag^+, liberating electrons to the external circuit.

$$Ag(s) = Ag^+ + e^- \quad \ldots \text{(XII, 2)}$$

Reaction (XII, 2) is just the reverse of (XII, 1).

The net result is the transfer of an electron from the cathode to the anode. Since the *removal* of electrons from a chemical species is called *oxidation* and the *addition* is called *reduction*, oxidation is effected at the anode, and reduction at the cathode. (XII, 1) still takes place but (XII, 2) is replaced by

$$\tfrac{1}{2}H_2O = \tfrac{1}{2}O + H^+ + e^- \quad \ldots \text{(XII, 3)}$$

The oxygen atoms formed at the anode combine together to form gaseous oxygen, which is released. The migration of NO_3^- ions to the anode region, together with the H^+ ions (or H_3O^+ ions; *see* XIII, **33**) produced in (XII, 3), results in the formation of HNO_3 in the anode region.

3. Electrode processes.

From the above example it can be seen that the electrode processes are dependent on the sign and composition of the electrodes. They also depend on the nature of the ions in solution and the applied voltage. The electrode reactions which tend to occur are those which are thermodynamically favoured, *i.e.* those for which $\Delta G < 0$ (*see* VIII, **19**). If two or more reactions are possible, the reaction involving the most negative value of ΔG will occur.

4. Migration of ions. Since the passage of electricity through an electrolyte solution results in the migration of ions to the appropriate electrodes, matter as well as charge is transferred from one electrode to the other. The formation of HNO_3 in the anode region in the above example is evidence of this migration.

Thus current is conducted around the circuit by:

(*a*) Electron flow through the external metallic conductor.

(*b*) Electrode reactions which permit the electrons to enter or leave the solution at the electrodes.

(*c*) Migration of ions, which permits electrons to flow through the solution.

FARADAY'S LAWS OF ELECTROLYSIS

5. Statement of the laws. After extensive investigations of many types of electrolytic reaction, Faraday (1833) formulated his laws of electrolysis, which may be combined in the following statement: the amount n_B of a substance, B, formed as a result of the electrode reaction

$$A + ze^- = B \qquad . \; . \; . \; (XII, 4)$$

after the passage of an electric current I, for a time t, is given by

$$n_B = It/zF \qquad . \; . \; . \; (XII, 5)$$

where F, the proportionality constant, is known as the *Faraday constant* and is the quantity of electricity associated with 1 mol of electrons, *i.e.* $F = eL = 9{\cdot}648\,7 \times 10^4$ C mol^{-1} (approximated to $9{\cdot}649 \times 10^4$ C mol^{-1}).

For reaction (XII, 4) the passage of $9{\cdot}649 \times 10^4$ C results in the formation of $1/z$ mol of B. It is important to state the electrode reaction precisely. Thus for the electrode reaction

$$Cu^{2+} + 2e^- = Cu(s)$$

the passage of $9.648\ 7 \times 10^4$ C results in the formation of $1/2$ mol of copper, *i.e.* $0.031\ 77$ kg, while for the electrode reaction

$$Cu^+ + e^- = Cu(s)$$

an identical quantity of electricity forms 1 mol of copper, *i.e.* $0.063\ 54$ kg.

The physical quantity known as the equivalent mass is used extensively to represent the mass of substance associated with the transfer of 1 mol of electrons in a given reaction. From (XII, 4) the equivalent mass of B is equal to the molar mass divided by z. Neither the physical quantity nor its unit, written "equiv," will be used in this book.

EXAMPLE 1: Calculate the mass of zinc produced by the electrolysis of a 1 M solution of zinc sulphate using a current of 0.1 A for thirty minutes.

The electrode reaction is

$$Zn^{2+} + 2e^- = Zn(s)$$

From (XII, 4)
$$n_{Zn} = (0.1 \times 30 \times 60)/(2 \times 9.649 \times 10^4) = 9.33 \times 10^{-4}\text{mol}$$

The mass of zinc is given by

$$m_{Zn} = n_{Zn}\,M_{Zn} = 9.33 \times 10^{-4} \times 65.37 \times 10^{-3}$$
$$= 6.10 \times 10^{-5}\ \text{kg}$$

6. Coulometry. Equation (XII, 5) may be used to determine the quantity of electricity passing through a circuit by measuring the chemical changes produced in a suitable electrolytic cell, or *coulometer*, inserted into the circuit. A frequently used coulometer is the silver coulometer, which consists of two silver electrodes immersed in a solution of silver nitrate. At the end of the electrolysis the gain in mass of the cathode is determined. From (XII, 1) and (XII, 2), 9.649×10^4 C causes the deposition or removal of 1 mol $= 0.107\ 87$ kg of silver, and hence the quantity of electricity passed may be found.

Other coulometers include:

(a) The iodine coulometer, in which the iodine liberated from an electrolysed potassium iodide solution is determined by titration.

(b) The copper coulometer, in which the mass of copper deposited or removed from copper electrodes in copper sulphate solution is measured.

ELECTROLYTIC CONDUCTION

7. Conductance and conductivity. Electrolytic conductors, like electronic conductors, obey Ohm's law. Thus the current, I, flowing through an electrolyte solution is proportional to the potential difference, V, across the electrolyte

$$I = V/R \qquad \text{. . . (XII, 6)}$$

where R is the resistance of the solution, and is given by

$$R = \rho l/A \qquad \text{. . . (XII, 7)}$$

where l is the length of solution between electrodes of cross-sectional area A, and ρ, the proportionality constant, is known as the *resistivity*.

For electrolyte solutions, (XII, 6) is usually written in the form

$$I = GV \qquad \text{. . . (XII, 8)}$$

where $G = 1/R$ and is called the *conductance*, while the analogous formula to (XII, 7) is

$$G = \kappa A/l \qquad \text{. . . (XII, 9)}$$

For electrolyte solutions l and A are usually measured in cm and cm^2 respectively and hence κ, the *conductivity*, is the conductance of a solution between electrodes 1 cm^2 in area, 1 cm apart, *i.e.* 1 cm^3 of solution, and has the units Ω^{-1} cm^{-1}.

8. Measurement of κ. κ can in principle be obtained from measured values of R ($= 1/G$), l and A.

(a) *Measurement of R.* The resistance of an electrolyte solution is measured in a conductance cell, using a Wheatstone bridge circuit (Fig. 44). Alternating rather than direct current is used, since the latter produces chemical reactions at the electrodes which invalidate Ohm's law, the resistance depending on the applied voltage. This effect is due to polarisation in the conductance cell and is absent when alternating current is used, since the electrode reactions occurring in one half-cycle are reversed in the following half-cycle and the products of electrolysis do not build up. A frequency of a few thousand hertz is used and a set of earphones serves as a convenient detector. The variable capacitance in parallel with R (*see*

Fig. 44) serves to balance out the capacitance of the cell and improve the balance point. At balance

$$R_{cell}/R_1 = R_2/R_3$$

and hence R_{cell} may be found.

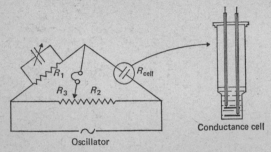

FIG. 44.—*Wheatstone bridge circuit for the determination of the resistance of electrolyte solutions*

(b) *Measurement of l and A.* Since the ratio of these two quantities l/A is a constant for any given cell (the *cell constant*), it is not necessary to determine each separately. Instead the resistance is measured with the cell containing a solution of known conductivity (invariably potassium chloride) and the cell constant is calculated from (XII, 7). Once the cell constant is known, κ can be obtained for any electrolyte solution from the measured value of R.

9. Molar conductivity. The value of κ depends on concentration, since 1 cm³ of different solutions could contain different amounts of electrolyte. It is useful to consider the conductivity with regard to 1 mol of electrolyte, the *molar conductivity*, Λ, defined by

$$\Lambda = \kappa V = \kappa/c \qquad . . . \text{(XII, 10)}$$

where V is the volume of solution containing 1 mol of electrolyte and c is the concentration.

NOTE: In this instance "molar" means divided by concentration (*see* I, 5).

The units of V and c are usually cm³ and mol cm⁻³ respectively. Hence Λ has the units Ω^{-1} cm² mol⁻¹.

Since an electric current represents the transfer of charge, it is more significant to compare electrolyte solutions which contain the same amount of charge rather than those which contain the same amount of substance. In general for an electrolyte $A_{v_+}B_{v_-}$ which produces v_+ mol of positive charge and v_- mol of negative charge, the quantity Λ/v should be used to compare molar conductivities. Thus the quantities $\Lambda(NaCl)$ and $\Lambda(\frac{1}{2}ZnCl_2)$ should be used to compare the molar conductivities of sodium chloride and zinc chloride since 1 mol of NaCl and 1 mol of $\frac{1}{2}ZnCl_2$ contain the same number of charges. Λ/v has been called the *equivalent conductance* and is the conductivity associated with 1 equivalent mass of an electrolyte.

10. The dependence of molar conductivities on concentration.

Kohlrausch found that for certain electrolyte solutions the relationship between the molar conductivity and the concentration was given by the empirical equation

$$\Lambda = A - Bc^{\frac{1}{2}} \quad \cdots \quad (XII, 12)$$

where A and B are constants.

NOTE: This relationship was originally formulated in terms of equivalent conductances.

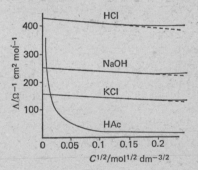

FIG. 45.—*Molar conductivities for electrolytes at 25°C*

The dependence of Λ on $c^{\frac{1}{2}}$ for several electrolytes is shown in Fig. 45. It is clear from this graph that electrolytes may be divided into two classes:

(*a*) *Strong electrolytes* which give essentially linear plots and have high Λ values at moderate concentrations.

(b) *Weak electrolytes* which show a marked increase in Λ with decreasing concentration at low concentrations.

11. Molar conductivities at infinite dilution.

The almost linear curves for strong electrolytes may be extrapolated to zero concentration to yield Λ^∞, the molar conductivity at infinite dilution. This represents the conductivity when the ions are so far apart that there is no interaction between them. The curved lines obtained for weak electrolytes do not allow accurate extrapolation to zero concentration. However, Λ^∞ for weak electrolytes may be obtained by application of *Kohlrausch's law of independent migration of ions*. Kohlrausch concluded that, at infinite dilution, cations and anions behave independently, so that Λ^∞ for an electrolyte is the sum of the molar conductivities of the individual ions. If 1 mol of an electrolyte, $A_{v_+}B_{v_-}$, is capable of producing v_+ mol of cations and v_- mol of anions in solution then

$$\Lambda^\infty(A_{v_+}B_{v_-}) = v_+\lambda_+^\infty + v_-\lambda_-^\infty$$

e.g.
$$\Lambda^\infty(\tfrac{1}{2}ZnCl_2) = \lambda^\infty(\tfrac{1}{2}Zn^{2+}) + \lambda^\infty(Cl^-)$$
$$\Lambda^\infty(ZnCl_2) = \lambda^\infty(Zn^{2+}) + 2\lambda^\infty(Cl^-) = 2\Lambda^\infty(\tfrac{1}{2}ZnCl_2)$$

where λ^∞ represents the molar conductivity of an ion at infinite dilution.

The validity of this law may be seen by reference to Table 13.

TABLE 13

TABLE 13 Λ^∞ VALUES FOR PAIRS OF STRONG ELECTROLYTES AT 25°C (Ω^{-1} cm^2 mol^{-1})

	KCl	149·9	KNO$_3$	145·0	KOH	71·5
	LiCl	115·0	LiNO$_3$	110·1	LiOH	36·7
Difference		34·9		34·9		34·8

Thus for pairs of strong electrolytes having a common ion there is a constant difference between the Λ^∞ values for each pair, which is equal to

$$\Lambda^\infty(KX) - \Lambda^\infty(LiX)$$
$$= \lambda^\infty(K^+) + \lambda^\infty(X^-) - \lambda^\infty(Li^+) - \lambda^\infty(X^-)$$
$$= \lambda^\infty(K^+) - \lambda^\infty(Li^+) = 34·9\,\Omega^{-1}\,cm^2\,mol^{-1}$$

It is possible to obtain Λ^∞ values for weak electrolytes by the addition and subtraction of Λ^∞ values for strong electrolytes. For a weak electrolyte such as acetic acid,

$$\Lambda^\infty(\text{HAc}) = \Lambda^\infty(\text{NaAc}) + \Lambda^\infty(\text{HCl}) - \Lambda^\infty(\text{NaCl})$$
$$= \lambda^\infty(\text{Na}^+) + \lambda^\infty(\text{Ac}^-) + \lambda^\infty(\text{H}^+) + \lambda^\infty(\text{Cl}^-)$$
$$- \lambda^\infty(\text{Na}^+) - \lambda^\infty(\text{Cl}^-)$$
$$= \lambda^\infty(\text{H}^+) + \lambda^\infty(\text{Ac}^-)$$

At 25°C,
$$\Lambda^\infty(\text{HAc}) = 91\cdot0 + 426\cdot2 - 126\cdot5 = 390\cdot7\Omega^{-1}\,\text{cm}^2\,\text{mol}^{-1}$$

THE ARRHENIUS THEORY OF IONISATION

12. The degree of ionisation. An explanation for the observed variation of Λ with concentration was proposed by Arrhenius as due to partial ionisation of the electrolyte. The more dilute the solution, the greater the ionisation, the greater the relative proportion of ions present, and hence the greater the molar conductivity. Arrhenius was able to calculate the degree of ionisation of an electrolyte from conductivity data on the assumption that at infinite dilution all electrolytes are completely ionised. Thus α, the degree of ionisation, is given by

$$\alpha = \Lambda/\Lambda^\infty \qquad \text{. . . (XII, 13)}$$

According to this equation, the degree of ionisation of any electrolyte is unity at infinite dilution.

Arrhenius made no distinction between strong and weak electrolytes even though Fig. 45 indicates very different behaviour in the two cases. It is now known that the Arrhenius theory is applicable only to weak electrolytes.

13. Weak electrolytes. The large increase in Λ with decreasing concentration for a weak electrolyte, MA, is satisfactorily explained in terms of the equilibrium

$$\text{MA} \rightleftharpoons \text{M}^+ + \text{A}^-$$

As the concentration decreases, the equilibrium shifts to the right and relatively more ions are available to carry the current. If the concentration of the electrolyte is cM, then at equilibrium

$$[\text{MA}] = c - \alpha c = c(1 - \alpha)$$
$$[\text{M}^+] = [\text{A}^-] = c\alpha$$

Thus K_c, the ionisation constant, is given by

$$K_c = [M^+][A^-]/[MA] = \alpha^2 c/(1 - \alpha) = \Lambda^2 c/(\Lambda^\infty \Lambda^\infty - \Lambda)$$
$$\text{. . . (XII, 14)}$$

Equation (XII, 14) constitutes the *Ostwald dilution law*. For acetic acid K_c is constant to within 2% over the concentration range 3×10^{-5} to 2×10^{-1} M, and provides strong evidence for the idea of incomplete dissociation.

14. Supporting evidence from colligative properties. The observed colligative properties of electrolyte solutions are always greater than those expected from the apparent concentration. Van't Hoff recognised that the colligative properties of solutions of electrolytes and non-electrolytes could be conveniently compared by writing (X, 6), (X, 10) and (X, 14) in the form:

$$\Delta T = iK_v \mathrm{m}$$
$$\Delta T = iK_f \mathrm{m}$$
$$\pi = iRT \mathrm{m}$$

where i, the *van't Hoff factor*, is greater than unity for electrolytes and may be interpreted as the ratio of the actual number of particles present to the number obtained assuming no ionisation.

If 1 mol of an electrolyte is capable of ionising into v ions, the total number of particles will be $i = 1 - \alpha + v\alpha$. Hence

$$\alpha = (i - 1)/(v - 1) \qquad \text{. . . (XII, 15)}$$

At low concentrations i approaches 1 for non-electrolytes, while for strong electrolytes such as KCl, $CaCl_2$ and $K_3Fe(CN)_6$ i approaches 2, 3 and 4 respectively. In more concentrated strong electrolyte solutions i first decreases and then may increase (*see* 22 below). The values of i obtained for dilute solutions of weak electrolytes give values of α which agree well with those obtained from conductivity measurements, using (XII, 13). The correlation of the i factor with the number of ions expected from various electrolytes provides strong confirmation of the Arrhenius theory.

15. Strong electrolytes. Application of (XII, 14) to strong electrolytes yields unsatisfactory results, K_c varying markedly with concentration. It is now known that salts which behave as strong electrolytes are completely ionised in the solid state, and the process of solution merely allows the ions to separate from one another. It is apparent that a new approach is required, and this is supplied by the Debye–Hückel theory of interionic attraction. Before going on to discuss this theory, it is necessary to consider the role of the solvent in the formation of an electrolyte solution.

16. Solvation. The solvation of ions in water is discussed in **XX, 13**, and an ion in aqueous solution is surrounded by a sheath of bound water molecules. The dissociation of NaCl(s) into gaseous ions requires 766 kJ mol^{-1} (*see* XVI, 7–9), while the solution reaction

$$NaCl(s) + aq = Na^+(aq) + Cl^-(aq)$$

requires the expenditure of only about 6 kJ mol^{-1}. This is because of the energy released on solvation, and enables many processes which are energetically unfavourable in the gas phase to occur readily in solution.

17. The dielectric constant effect. The electrostatic force of attraction between two ions of charge z_+e and z_-e, a distance d apart in a medium of dielectric constant ϵ is given by

$$\text{Force} = z_+z_-e^2/4\pi\epsilon_0\epsilon d^2$$

NOTE: ϵ represents the effect of the medium in decreasing the force between the ions.

In water ($\epsilon \approx 80$) the force necessary to separate ions is much less than it is in air ($\epsilon \approx 1$), and the solution of electrolytes in water and other polar solvents is favoured.

THE DEBYE–HÜCKEL THEORY

The effects of solvation and interionic attraction, which were neglected in the Arrhenius theory, were successfully considered by P. Debye and E. Hückel (1923) to explain the variation of molar conductivity with concentration for strong electrolytes assuming complete dissociation. These effects can be ignored for weak electrolytes because of the low ionic concentration.

18. The ion atmosphere. Because of interionic attractions, each ion in solution is surrounded by an *atmosphere* of other ions whose net charge is on average opposite to that of the central ion (*see* Fig. 46 (*a*)). The ions in this atmosphere are not permanently attached to the central ion but are continuously being replaced by other similarly charged ions in solution.

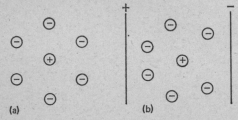

Fig. 46.—(*a*) *Symmetric ion atmosphere in the absence of an applied field.* (*b*) *Asymmetric ion atmosphere in the presence of an applied field*

Since the central ion and its ion atmosphere are oppositely charged, an applied field causes the ion to move in the direction opposite to that of its ion atmosphere. The result is that the ion atmosphere is distorted from its initial symmetrical configuration (*see* Fig. 46 (*b*)). Thus instead of experiencing a symmetrical field of force due to the ion atmosphere, the central ion now experiences a retarding force which slows down the forward motion of the ion.

19. The electrophoretic effect. This effect arises because an ion moving through the solution is not travelling through a stationary medium but through one which is moving in a direction opposite to that of the ion. The effect is magnified because of the solvated nature of the ions, *i.e.* the ion must effectively thread its way through a counter-current of solvent molecules.

20. Solvation effects. These effects were first recognised by Onsager, and reduce the mobility of the ion because:

(*a*) Solvation increases the effective radius of the ion.
(*b*) Water molecules in the solvation shell of the ion are linked via hydrogen bonds to other water molecules, and this will cause a frictional drag.

21. The Onsager equation. The above effects were first treated mathematically by Debye and Hückel, whose treatment was later modified by Onsager to yield the following equation for the dependence of the molar conductivity on concentration:

$$\Lambda = \Lambda^\infty - (A\Lambda^\infty + B)c^{\frac{1}{2}} \quad . \quad . \quad . \quad (XII, 16)$$

where A and B are constants for a given ion and solvent.

NOTE: Once again the original formulation was in terms of equivalent conductances.

Since the terms in parentheses are constant, (XII, 16) can be rewritten

$$\Lambda = \Lambda^\infty - Kc^{\frac{1}{2}}$$

and has the correct form for the concentration dependence of the molar conductivity (*see* (XII, 14)).

Equation (XII, 16) is valid for most electrolytes up to concentrations of the order of 10^{-2} M, and this equation has been used to draw in the linear broken lines in Fig. 45.

22. Ion-pair and ion-triplet formation. At higher concentrations (≈ 1 M) the Debye–Hückel–Onsager theory breaks down because of ion association. In this concentration range there is specific attraction between oppositely charged ions, resulting in the formation of ion-pairs ($+\ -$). Such ion-pairs having a net zero charge will cause a decrease in conductivity with increasing concentration. For a given concentration the extent of ion-pair formation is greater (*a*) the smaller the size of the solvated ions, (*b*) the higher their valency and (*c*) the lower the dielectric constant of the medium.

At still higher concentrations, ion-triplets will be formed ($-\ +\ -$) and ($+\ -\ +$) which, being charged, will cause the conductivity to rise. These associated ions should not be regarded as permanent but rather as transient species which are continuously associating and dissociating

$$(+) + (-) \rightleftharpoons (+\ -)$$

23. Supporting evidence for the concept of an ion atmosphere.

(*a*) *Debye–Falkenhagen effect.* The effect is observed when conductivities are studied at high frequencies ($\approx 3 \times 10^6$ Hz).

As the frequency of the field is increased a point is reached when the ion atmosphere can no longer follow the rapidly changing field. Under such conditions the ion moves as if it had no ion atmosphere, and the expected increase in conductivity is observed.

(b) *Wien effect*. The conductivity is also observed to increase at very high field strengths ($\approx 10^3$ V m^{-1}), where the ion moves so quickly that its ion atmosphere is left behind.

TRANSPORT NUMBERS

24. Definition. Although cations and anions are discharged in equivalent amounts at the electrodes during electrolysis, these ions do not necessarily carry equal fractions of the total current passing through the cell. The fraction of the current carried by a given ionic species in solution is called its *transport number*, t. From Kohlrausch's law of independent migration of ions (**10** above), the transport number is also the fraction of the total conductivity contributed by a given ionic species. For an infinitely dilute solution of an electrolyte $A_{v_+}B_{v_-}$, the transport numbers of the cation and anion are given by:

$$t_+^\infty = v_+\lambda_+^\infty/\Lambda^\infty \text{ and } t_-^\infty = v_-\lambda_-^\infty/\Lambda^\infty \ . \ . \ . \text{ (XII, 17)}$$

By virtue of the definition, the sum of the transport numbers of all the ions in a given electrolyte solution is unity, *i.e.* $t_+^\infty + t_-^\infty = 1$.

Transport numbers may be measured in three different ways: (a) the Hittorf method, (b) the moving boundary method, (c) from electrochemical cells. The first two methods are discussed below.

25. The principle of the Hittorf method. The method utilises the changes in the electrolyte concentration around each electrode which result not only from the movement of ions but also from the electrode reactions. The simplest case is where the electrodes and the cation of the electrolyte are of the same metal. In this case there is no apparent chemical change in the cell, but rather the apparent transport of electrolyte from one electrode region to the other.

To illustrate the technique we shall consider the electrolysis of silver nitrate solution between silver electrodes. Fig. 47 is a diagrammatic representation of the experimental cell divided into three compartments at A and C. The electrode reactions are:

$$\text{Anode:} \qquad Ag(s) = Ag^+ + e^-$$
$$\text{Cathode: } Ag^+ + e^- = Ag(s)$$

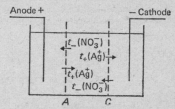

Fig. 47.—*Schematic representation of the Hittorf transport number apparatus containing silver nitrate solution*

The passage of $9 \cdot 649 \times 10^4$ C through the cell results in:

(*a*) 1 mol of chemical change at either electrode.
(*b*) The movement of t_+ mol of Ag^+ and t_- mol of NO_3^- in opposite directions across any imaginary boundary in the cell.

Effect (*b*) follows from the definition of transport number. The changes in the three cell compartments resulting from the passage of $9 \cdot 649 \times 10^4$ C are given below:

ANODE	CENTRAL	CATHODE
Gain by migration across $A : t_-$ mol of NO_3^-	t_- mol of NO_3^- *lost* by migration across A, but *gained* by migration across C	*Loss* by migration across $C : t_-$ mol of NO_3^-
Loss by migration across $A : t_+$ mol of Ag^+	t_+ mol of Ag^+ *gained* by migration across A, but *lost* by migration across C	*Gain* by migration across $C : t_+$ mol of Ag^+
Gain by electrode reaction: 1 mol of Ag^+		*Loss* by electrode reaction: 1 mol of Ag^+
Net change *Gain* $(1 - t_+) = t_-$ mol of Ag^+ and t_- mol of NO_3^-	No net change	*Loss* $(1 - t_+) = t_-$ mol of Ag^+ and t_- mol of NO_3^-

The overall effect of the passage of $9 \cdot 649 \times 10^4$ C through the cell is the apparent transport of t_- mol of $AgNO_3$ from the cathode to the anode compartment.

26. The experimental procedure. The amount of electrolyte lost from the cathode compartment or gained by the anode compartment after the passage of a measured quantity of electricity is determined. In the above example the amount of Ag^+ gained by the anode compartment after the passage of Q C is given by

$$n_{Ag^+} = Qt_-/9 \cdot 649 \times 10^4$$

Hence t_- may be found and t_+ obtained from the relation $t_+ = 1 - t_-$.

In the actual apparatus shown in Fig. 48 the boundaries A and C are replaced by stopcocks so that the three portions of the solution can be drawn off separately. Since the ions are solvated, their movement during electrolysis results in the transfer of solvent from one compartment to another. It is therefore

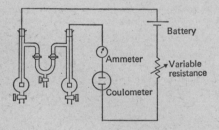

FIG. 48.—*Hittorf transport number apparatus*

necessary to determine the changes in the concentration of the electrolyte associated with a given amount of solvent. The composition of the central compartment should be unchanged, indicating that there has been no transport of electrolyte by diffusion or convection. The quantity of electricity passed through the cell is determined by a chemical coulometer (*see* **6** above) placed in series with the Hittorf apparatus.

27. The moving boundary method. This is a more direct method which utilises the movement of the boundary between two electrolyte solutions under an applied field. Fig. 49 represents a vertical tube containing two electrolytes, AX and BX, with a common anion, separated by a sharp boundary ab. This boundary may be observed because of differences in colour or in refractive index between the two electrolytes. Since the anions are the same, the boundary is due solely to

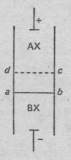

FIG. 49.—*The moving boundary method*

the cation. Under an applied field the cations will move upwards, resulting in the movement of the boundary to the new position cd. When the boundary reaches cd, all the A^{z+} ions originally present in the volume (V dm^3, say) will have been transported across dc, *i.e.* Vc mol, where c is the concentration of AX in mol dm^{-3}. If QC have passed during this time the amount of A^{z+} transported across the boundary will also be given by $Qt_+/z_+ \times 9\cdot649 \times 10^4$.

$$\therefore Vc = Qt_+/z_+ \times 9\cdot649 \times 10^4$$

t_+ may be found from measurements of V, c and Q.

In order to maintain a sharp boundary during the experiment the following conditions should be fulfilled:

(a) The tube must be vertical, and BX must be denser than AX.

(b) The speed of the leading ion A^{z+} must be greater than that of the following ion.

(c) Convection currents must be minimised by keeping the electric current low.

28. Transport numbers of ions. The following observations can be drawn from the numerical values of transport numbers:

(a) Transport numbers vary slightly with concentration, as might be expected from ionic interactions.

(b) H_3O^+ and OH^- ions are exceptional in carrying about 80% of the current in HCl and NaOH solutions respectively, suggesting a special mechanism for the movement of these ions (see **32** below).

(c) Li^+ has the smallest radius of any alkali metal ion in the gas phase and yet has the smallest transport number. The reason is that the small size of the bare ion gives rise to a strong electrostatic field, thus causing heavier solvation than with the other alkali metal ions. The size of the solvated alkali metal ions in water decreases in the order $Li^+ < Na^+ < K^+ < Rb^+$, and both the amount of solvation and the transport number decrease in the reverse order.

Ionic conductivities at infinite dilution may be obtained from transport numbers by application of (XII, 17). The abnormal ions, H_3O^+ and OH^-, have Λ^∞ values of 349·8 and 198·6Ω^{-1} cm^2 mol^{-1} respectively in aqueous solution at 25°C, while the values for other ions are usually in the range 40–90 Ω^{-1} cm^2 mol^{-1}.

IONIC MOBILITIES

29. Definition. The average velocity with which an ion moves towards an electrode depends on:

(a) The potential gradient between the electrodes.

(b) The viscosity of the solvent.

(c) The concentration of the solution, which influences the asymmetry and electrophoretic effects (see **18, 19** above).

The average velocity of an ion under a potential gradient of 1 V cm^{-1} is called its *mobility*, u, and is usually quoted for conditions of infinite dilution (u^∞).

30. Calculation of ionic mobilities. Ionic mobilities may be calculated from ionic conductivities as follows. Consider the electrolytic cell shown in Fig. 50, where the two ends of the cell form electrodes which are 1 cm^2 in area and 1 cm apart. The solution therefore has a volume of 1 cm^3 and contains 1 mol of an electrolyte $A_{v_+}B_{v_-}$.

Under these conditions $\Lambda = \kappa$ and (XII, 9) becomes $G = \Lambda$. Equation (XII, 8) may then be written

$$I = \Lambda V \qquad \text{. . . (XII, 18)}$$

At infinite dilution and putting $V = 1$ cm³, (XII, 18) becomes

$$I = \Lambda^\infty = v_+\Lambda_+^\infty + v_-\Lambda_-^\infty \quad \text{. . . (XII, 19)}$$

Since the cell contains 1 mol of $A_{v_+}B_{v_-}$ there will be v_+L cations and v_-L anions. If u_+ and u_- are the ionic mobilities, then for an ion to cross the imaginary boundary B (*see* Fig. 50)

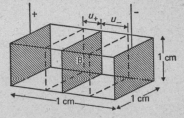

FIG. 50.—*Electrolyte cell to illustrate the calculation of ionic mobility (cell elongated for clarity)*

in one second, it must start at or less than a distance u_+ or u_- from the boundary. Thus in one second u_+v_+L cations and u_-v_-L anions cross the boundary, with associated charges $u_+ \mid z_+ \mid ev_+L$ and $u_- \mid z_- \mid ev_-L$. Since the quantity of charge transferred per second is the current

$$I_+ = u_+ \mid z_+ \mid ev_+L = u_+ \mid z_+ \mid v_+F \quad \text{. . . (XII, 20)}$$

and $\quad I_- = u_- \mid z_- \mid ev_-L = u_- \mid z_- \mid v_-F \text{. . . (XII, 21)}$

At infinite dilution u is replaced by u^∞.
The total current, I, is given by

$$I = I_+ + I_- = v_+\lambda_+^\infty + v_-\lambda_-^\infty \quad \text{. . . (XII, 22)}$$

Comparing (XII, 20), (XII, 21) and (XII, 22)

$$v_+\lambda_+^\infty = u_+^\infty \mid z_+ \mid v_+F \quad \text{and} \quad v_-\lambda_-^\infty = u_-^\infty \mid z_- \mid v_-F$$
$$\therefore u_+^\infty = \lambda_+^\infty \mid z_+ \mid F \quad \text{and} \quad u_-^\infty = \lambda_-^\infty / \mid z_- \mid F$$
$$\text{. . . (XII, 23)}$$

31. Ionic mobilities and transport numbers. From (XII, 23)

$$\lambda_\infty^\pm = u_+^\infty \mid z_+ \mid F \quad \text{and} \quad \lambda_-^\infty = u_-^\infty \mid z_- \mid F$$

Now $\quad \Lambda^\infty = v_+\lambda_+^\infty + v_-\lambda_-^\infty$
$$= v_+u_+^\infty \mid z_+ \mid F + v_-u_-^\infty \mid z_- \mid F$$

Hence $\quad t_+ = v_+ \lambda_+^\infty / \Lambda^\infty = \dfrac{v_+ u_+^\infty \mid z_+ \mid F}{v_+ u_+^\infty \mid z_+ \mid F + v_- u_-^\infty \mid z_- \mid F}$

For electrical neutrality $v_+ \mid z_+ \mid = v_+ \mid z_- \mid$

$$\therefore t_+ = u_+^\infty / (u_+^\infty + u_-^\infty) \quad \ldots \text{(XII, 24)}$$

and $\qquad\qquad\qquad t_- = u_-^\infty / (u_+^\infty + u_-^\infty) \quad \ldots \text{(XII, 25)}$

Thus the trends observed for transport numbers are also apparent for ionic mobilities. Apart from H_3O^+ and OH^- most ions have mobilities around $6 \times 10^{-4} \text{ cm}^2 \text{ s}^{-1} \text{ V}^{-1}$ in water at 25°C. These values are very low under typical potential gradients of less than 100 V cm^{-1}, and an ion follows a slow, winding path through a solution. It is more realistic to think of the solution between the electrodes as being displaced by the electric field, rather than of the ions travelling distances comparable with the electrode separation.

The high mobilities of H_3O^+ and OH^- are due to proton transfer through the water. Because of the hydrogen-bonded nature of water, the mechanism illustrated below is thought to apply.

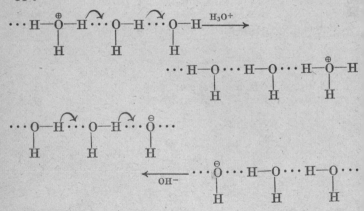

This mechanism results in a much more rapid transfer than would be possible if H_3O^+ and OH^- were physically transported through the solution, as with other ions. The mobilities of these ions are limited by the fact that, in both cases, after one proton has been transferred the orientation is wrong for the next transfer and the water molecule must relax back to its original orientation before a second proton transfer can take place. Thus the mobility is determined by the relaxation time of the water molecule. In solvents other than water, where the above mechanism cannot operate, H_3O^+ and OH^- show mobilities close to those of other ions.

APPLICATIONS OF CONDUCTANCE MEASUREMENTS

32. Conductimetric titrations. In principle, any titration in which ions are produced, removed or replaced by others of a different kind may be followed by observing the changes in conductance as the reagent is added. For example, in the titration of HCl with NaOH, the highly mobile H_3O^+ ions are replaced by the less mobile Na^+ ions up to the equivalence point. The conductance will therefore decrease until the equivalence point is reached, when the solution effectively contains only Na^+ and Cl^- ions. On further addition of NaOH, the conductance increases due to the addition of Na^+ and OH^- ions. Thus the plot of conductance $(1/R)$ against volume of NaOH added consists of two branches whose point of intersection represents the equivalence point (*see* Fig. 51).

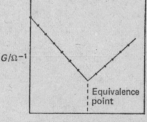

G/Ω^{-1}

Equivalence point

VOLUME OF NaOH ADDED

Fig. 51.—*Conductimetric titration curve for HCl with NaOH*

By making the alkali about ten times more concentrated than the acid, conductance changes due to dilution are minimised, and the two branches of the conductivity plot are virtually linear. The point of intersection may then be found by extrapolation. This method has advantages in titrations involving very dilute solutions, very weak acids or bases, mixtures of strong and weak acids, coloured solutions, precipitation reactions and reactions in non-aqueous media, for which classical indicator methods are often unavailable.

33. The ionic product for water. The conductivity of the purest water so far obtained is $0.58 \times 10^{-7} \, \Omega^{-1} \, cm^{-1}$ at 25°C, and is due to self-ionisation

$$2H_2O \rightleftharpoons H_3O^+ + OH^-$$

The molar conductivity $\Lambda = \kappa/c$,
Since κ is given in $\Omega^{-1} \, cm^{-1}$, c must be found in mol cm^{-3}
The density of pure water at 25°C is $0.997 \, g \, cm^{-3}$
Hence $c = 1 \times 0.997/18.02 = 5.53 \times 10^{-2} \, mol \, cm^{-3}$

$$\therefore \Lambda = 0.58 \times 10^{-7}/5.53 \times 10^{-2}$$
$$= 1.05 \times 10^{-6} \, \Omega^{-1} \, cm^2 \, mol^{-1}$$

If water were completely ionised, its molar conductivity would be

$$\Lambda^\infty(H_2O) = \lambda^\infty(H_3O^+) + \lambda^\infty(OH^-)$$
$$= 349.8 + 198.0 = 547.8 \, \Omega^{-1} \, cm^2 \, mol^{-1}$$

For a weak electrolyte such as water, α, the degree of ionisation, is given by

$$\alpha = \Lambda/\Lambda^\infty = 1.05 \times 10^{-6}/547.8$$
$$= 1.9 \times 10^{-9}$$

Thus $[H_3O^+] = [OH^-] = \alpha c = 1.9 \times 10^{-9} \times 5.53 \times 10^{-2}$
$$= 1.05 \times 10^{-10} \, mol \, cm^{-3}$$
$$= 1.05 \times 10^{-7} \, mol \, dm^{-3}$$

and the ionic product for water is

$$K_w = [H_3O^+][OH^-] = 1.1 \times 10^{-14} \, mol^2 \, dm^{-6}$$

34. The ionisation constant of a weak electrolyte. This application of conductivity measurements relies on a generalisation of the above argument. The ionisation constant of a weak electrolyte is given by (XII, 14)

$$K_a = \Lambda^2 c/\Lambda^\infty(\Lambda^\infty - \Lambda)$$

This equation may be rearranged to give

$$c\Lambda^2 = (K_a\Lambda^{\infty 2}/\Lambda) - K_a\Lambda^\infty$$

Hence a plot of $c\Lambda^2$ against $1/\Lambda$ should give a straight line of slope $K_a\Lambda^{\infty 2}$ and intercept $-K_a\Lambda^\infty$. Since $K_a = (\text{intercept})^2/\text{slope}$, the value of Λ^∞ need not be known.

35. The solubility of sparingly soluble salts.
Conductivity measurements offer a simple means of determining the solubilities of sparingly soluble salts such as $BaSO_4$, $AgCl$, etc. The conductivity of a saturated solution of the salt is measured, and the conductivity of water is subtracted to give the conductivity of the salt alone.

$$\kappa(\text{salt}) = \kappa(\text{solution}) - \kappa(\text{water})$$

$\Lambda^\infty(\text{salt})$ may be found from the tabulated λ^∞ values of the individual ions. Since the solution is so dilute, Λ^∞ may to a good approximation be put equal to Λ.

Hence $\Lambda^\infty = \Lambda = \kappa/c$, and the solubility may be calculated from the value of c, the ionic concentration. The limitations of the method are:

(a) The saturated solution must be sufficiently dilute to put $\Lambda = \Lambda^\infty$.

(b) The electrolyte must ionise simply and completely.

PROGRESS TEST 12

1. Calculate the volume of oxygen liberated from an aqueous solution of silver nitrate using platinum electrodes by a current of 1 A flowing for two hours. The temperature is 25°C and the pressure 101·3 kN m^{-2}. (2, 5)

2. Calculate the degree of ionisation and the ionisation constant of benzoic acid at 25°C, given the following information: Λ^∞ for sodium benzoate, sodium chloride and hydrochloric acid at 25°C are 82·5, 126·5 and 426·2 Ω^{-1} cm^2 mol^{-1} respectively and κ for benzoic acid at a concentration of $1\cdot5 \times 10^{-3}$ mol dm^{-3} is $1\cdot14 \times 10^{-4}$ Ω^{-1} cm^{-1} at 25°C. (9, 11–13)

3. A 0·02 molal solution of silver nitrate was electrolysed between silver electrodes. After passing 0·02 A through the cell for twenty minutes 26·312 g of the anode solution was found to contain 0·0371 g of silver. Calculate the transport numbers of the silver and nitrate ions. (25–26)

4. The boundary between a 0·01 M solution of potassium chloride and a 0·02 M solution of lithium chloride moves 14·7 cm in a tube 1 cm in diameter in sixty minutes when the current is

5×10^{-3} A. What are the transport numbers of the potassium and chloride ions ? (27)

5. Given that $\Lambda^{\infty}(BaCl_2) = 280 \cdot 0 \ \Omega^{-1} \ cm^2 \ mol^{-1}$ and $\lambda^{\infty}(Cl^-)$ $= 76 \cdot 4 \ \Omega^{-1} \ cm^2 \ mol^{-1}$, calculate the transport numbers and the mobilities of Ba^{2+} and Cl^- in infinitely dilute solution. (11, 30–31)

6. A saturated solution of silver chloride has a conductivity of $3 \cdot 24 \times 10^{-6} \ \Omega^{-1} \ cm^{-1}$ at 25°C, while the water used to make up the solution has a conductivity of $1 \cdot 44 \times 10^{-6} \ \Omega^{-1} \ cm^{-1}$. Given that the ionic conductivities at infinite dilution of Ag^+ and Cl^- are $61 \cdot 9$ and $76 \cdot 3 \ \Omega^{-1} \ cm^2 \ mol^{-1}$ respectively at 25°C, calculate the solubility of silver chloride. (35)

7. Draw and explain the conductimetric titration curves obtained on titration of (a) acetic acid with sodium hydroxide, (b) a mixture of hydrochloric acid and acetic acid with sodium hydroxide, (c) sodium chloride with silver nitrate. (32)

8. Describe the electrode processes which take place during the electrolysis of a silver nitrate solution using (a) silver electrodes, (b) platinum electrodes. (2)

9. Define, by means of equations, (a) conductance (7), (b) conductivity (7), (c) molar conductivity (9).

10. Describe, graphically, the dependence of molar conductivity on concentration for strong and weak electrolytes. (10)

11. State Kohlrausch's law of independent migration of ions, and explain its significance. (11)

12. Use the Arrhenius theory of ionisation to derive an expression for the ionisation constant of a weak electrolyte. (12–13)

13. Outline the principles behind the Debye–Hückel–Onsager theory of interionic attraction. (15–21)

14. What is meant by a transport number ? Describe two methods for its determination. (24–27)

15. Explain some of the more important conclusions to be drawn from transport number values. (28)

16. Derive the relationships between ionic mobilities and (a) molar conductances, (b) transport numbers. (29–31)

17. Describe some applications of conductance measurements. (32–35)

CHAPTER XIII

IONIC EQUILIBRIA

In electrolyte solutions equilibrium exists between ions and neutral molecules. These neutral molecules may be true covalent molecules, as in the case of weak electrolytes, or ion pairs held together by electrostatic forces, as in the case of strong electrolytes (*see* XII, 22).

THEORIES OF ACIDS AND BASES

1. The Arrhenius classification. This states that acidic and basic properties are exhibited by substances which ionise in solution to give H^+ and OH^- ions respectively, *i.e.*

Acids: $HA \rightleftharpoons H^+ + A^-$
Bases: $BOH \rightleftharpoons B^+ + OH^-$

All neutralisation reactions must therefore be described by equilibria such as

$$H^+ + A^- + B^+ + OH^- \rightleftharpoons B^+ + A^- + H_2O$$
$$\ldots \text{(XIII, 1)}$$

The results of early measurements of the relative strengths of acids and bases, based on the ability of H^+ and OH^- ions to catalyse the hydrolysis of esters, were explained as being due to differences in the degree of ionisation.

The main objections to the Arrhenius classification are that:

(*a*) Certain substances (*e.g.* ammonia, sodium carbonate) neutralise acids even though their ionisation cannot directly produce OH^- ions.

(*b*) Unionised acids and bases and certain substances not containing H^+ or OH^- ions also catalyse the hydrolysis of esters.

(*c*) Strong electrolytes such as sodium hydroxide are completely ionic in the crystal and cannot therefore be pictured as ionising in the same way as acids, which are covalent in the pure state.

(*d*) The radius of a free proton is about 10^{-4} nm, whereas all other ions have radii of the order of 10 nm. Thus the proton can become very intimately associated with the electronic system of a solvent molecule, *i.e.* the proton should be very strongly and specifically solvated. In water H_3O^+ ions are formed and were first detected as separate entities in an X-ray diffraction study of crystalline perchloric acid monohydrate, *i.e.* $HClO_4 \cdot H_2O = H_3O^+ ClO_4^-$.

The enthalpy change for the addition of a proton to a water molecule to form H_3O^+ is approximately 1200 kJ mol^{-1}. Using the Boltzmann expression (*see* IV, 13), the proportion of free H^+ ions to H_3O^+ ions can thus be calculated as approximately $10^{-200}:1$.

NOTE: The representation of the proton as H_3O^+ in aqueous solution is an over-simplification and the species is undoubtedly further hydrated. All ions in aqueous solution are hydrated (*see* XX, 13).

2. The Brønsted–Lowry classification.

This more general and powerful classification satisfies the above objections by defining acids and bases as substances which tend to donate and to accept *protons* respectively. Some typical acid–base reactions may therefore be formulated as:

$$HCl + H_2O \rightleftharpoons H_3O^+ + Cl^- \qquad \ldots \text{(XIII, 2)}$$
$$H_2SO_4 + C_6H_5NH_2 \rightleftharpoons C_6H_5NH_3^+ + HSO_4^- \qquad \ldots \text{(XIII, 3)}$$
$$H_2O + CO_3^{2-} \rightleftharpoons HCO_3^- + OH^- \qquad \ldots \text{(XIII, 4)}$$

acid 1   base 2   acid 2   base 1

Taking reaction (XIII, 2) as typical, the following important points about this classification should be noted:

(*a*) Since an acid can donate a proton only if a base is present to accept (and vice versa), it is obvious that the exhibition of acidic or basic character will be dependent on the solvent. Water can function either as an acid (XIII, 4) or as a base (XIII, 2).

(*b*) HCl (acid) donates a proton to H_2O (base) forming H_3O^+ and Cl^-. However, since the reaction is reversible, H_3O^+ (acid) may also transfer a proton to Cl^- (base), reforming HCl and H_2O. Since HCl and Cl^- differ only by a single proton they are called a *conjugate acid–base pair*, and likewise H_3O^+ and H_2O.

(*c*) If the acid of a conjugate acid–base pair is strong (indicating a pronounced tendency to donate protons) then the conjugate base will be weak (indicating a small tendency to accept protons) and vice versa (*see* 7 below).

3. The Lewis classification. The Brønsted–Lowry classification still suffers from a certain lack of generality, *e.g.* the reaction of MgO(s) with SO_3(g) to form $MgSO_4$(s) involves no protons but is obviously a neutralisation. The Lewis classification of acids and bases as substances which can accept and donate *a pair of electrons* respectively thus extends the range of acid–base character still further.

Since Brønsted–Lowry bases transfer electrons to protons, any Brønsted–Lowry base is also a Lewis base. However, a Lewis acid need not necessarily have a proton available for donation to a base.

ACID–BASE EQUILIBRIA

The Brønsted–Lowry definition will be used in the succeeding sections, since we shall be principally concerned with aqueous systems.

4. Ionisation of weak acids and bases. The behaviour of weak acids in aqueous solution can be represented by the reaction

$$HA + H_2O \rightleftharpoons H_3O^+ + A^- \quad \ldots \text{(XIII, 5)}$$
$$\text{acid} \quad \text{base} \quad \text{acid} \quad \text{base}$$

and the thermodynamic acid ionisation constant K_A is given by

$$K_A = (a_{H_3O^+} \times a_{A^-})/(a_{HA} \times a_{H_2O})$$

Since the activity of the pure solvent may be taken as unity (*see* VIII, 28) the above equation becomes

$$K_A = (a_{H_3O^+} \times a_{A^-})/(a_{HA})$$

If the solution is sufficiently dilute, activities may be replaced by concentrations, and the acid ionisation constant is given by

$$K_a = [H_3O^+][A^-]/[HA] \quad \ldots \text{(XIII, 6)}$$

The omission of the concentration of water from (XIII, 6) is justified since in dilute solution the concentration of water may be taken as constant and included in the equilibrium constant (see XI, 10). K_A is a true constant, while K_a varies slightly with concentration, but this is insignificant for dilute solutions. We shall use concentrations instead of activities except where specifically mentioned. K_a is identical with the constant of Ostwald's dilution law (*see* XII, **13**). Similarly, the ionisation of a base B in aqueous solution may be represented by

$$B + H_2O \rightleftharpoons BH^+ + OH^-$$
$$\text{base} \quad \text{acid} \qquad \text{acid} \qquad \text{base}$$

and the base ionisation constant is

$$K_b = [BH^+][OH^-]/[B] \quad . \quad . \quad . \quad (XIII, 7)$$

the concentration of water again being taken as constant. The greater the value of K_a (K_b), the greater the strength of the acid (base) and the greater the degree of ionisation.

5. Strengths of acids and bases—the pK scale. K_a (K_b) gives a measure of the strength of the acid (base), *i.e.* of its tendency to donate (accept) protons. Since values of K_a and K_b for weak acids and bases tend to be small, it has become customary to compare their relative strengths using a logarithmic scale, *i.e.* values of pK_a and pK_b are used rather than K_a and K_b, where pK is defined by the equation

$$pK = -\lg K \quad . \quad . \quad . \quad (XIII, 8)$$

For strong acids in water, equilibrium (XIII, 5) is so far over to the right that it is not possible to obtain an order for their relative strengths (water is said to exert a levelling effect). However, in a solvent with less affinity for protons (*e.g.* glacial acetic acid) the equilibrium is less far over to the right and the following order of relative strengths has been obtained: $HClO_4 > HBr > H_2SO_4 > HCl > HNO_3$.

In a similar manner ammonia, being a stronger base than water, exerts a levelling effect on acids which are weak in water, *e.g.* HCl, HNO_3, HCOOH and CH_3COOH are apparently equally strong in liquid ammonia. Likewise carbonyl compounds, which are non-conductors in water, behave as quite strong bases in HF owing to the reaction

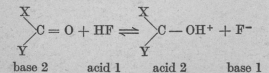

base 2   acid 1   acid 2   base 1

6. The self-ionisation of water (*see also* XII, 33). Since water is capable of acting either as an acid or as a base, it is not surprising to discover that pure water always contains a small concentration of H_3O^+ and OH^- ions produced by self-ionisation (autoprotolysis)

$$H_2O + H_2O \rightleftharpoons H_3O^+ + OH^-$$
$$\text{acid 1 base 2} \qquad \text{acid 2 base 1}$$

The equilibrium constant for this reaction (treating the concentration of water as a constant) is given by

$$K_w = [H_3O^+][OH^-] \quad \ldots \text{(XIII, 9)}$$

K_w, the ion product of water, has a value of 10^{-14} mol³ dm⁻⁶ at 24·85°C and we shall use this value in all succeeding calculations.

In neutral solutions the concentrations of H_3O^+ and OH^- are equal,

$$[H_3O^+] = [OH^-] = 10^{-7} \text{ M} \ldots \text{(XIII, 10)}$$

It is often convenient to use pK values in calculations, and pK_w is simply defined by the equation

$$\text{p}K_w = -\lg K_w \quad \ldots \text{(XIII, 11)}$$
or, at about 25°C $\qquad \text{p}K_w = 14 \qquad \ldots \text{(XIII, 12)}$

7. The relationship between K_a and K_b for conjugate acids and bases. The following equilibrium is established in an aqueous solution of a weak acid, HA,

$$HA + H_2O \rightleftharpoons H_3O^+ + A^-$$

A^-, the conjugate base of the acid, may also react with the solvent

$$A^- + H_2O \rightleftharpoons HA + OH^-$$

The respective equilibrium constants are:

$$K_a = [H_3O^+][A^-]/[HA] \qquad K_b = [HA][OH^-]/[A^-]$$

Therefore $K_a \times K_b = [H_3O^+][OH^-] = K_w \ldots \text{(XIII, 13)}$

Hence the greater the strength of the acid (*i.e.* the larger K_a), the weaker the conjugate base, and vice versa.

8. The pH scale. The concentrations of H_3O^+ and OH^- ions obtained from the self-ionisation of water or from the ionisation of weak acids or bases are often very small and may be expressed on a logarithmic scale, using the definitions

$$pH = -lg[H_3O^+], \quad pOH = -lg[OH^-]$$

where concentrations are in mol dm^{-3}.

NOTE: As before, strictly, $pH = -lg\ a_{H_3O^+}$.

Thus, for any solution, $pH + pOH = pK_w = 14$.

The pH scale is taken to extend from 0 to 14. However, in principle, pH's can be outside this range, *i.e.* negative or greater than 14. In practice, such pH's are very rarely encountered and in any case cannot be measured with great accuracy.

EXAMPLE 1: Calculate the pH and pOH of a solution containing 5×10^{-4} M of H_3O^+ ions.

$$pH = -lg(5 \times 10^{-4}) = 4 - lg\ 5$$
$$= 4 - 0.699 = 3.301$$

The pOH is found as follows:

$$pOH = pK_w - pH = 14 - 3.301$$
$$= 10.699$$

9. Calculations of pH and degree of ionisation. Such calculations for weak acids and bases are illustrated in the following examples.

EXAMPLE 2: Calculate the degree of ionisation and pH of a 0.01M solution of acetic acid. $K_a = 1.85 \times 10^{-5}$ mol dm^{-3}.

We shall first derive a general formula for the degree of ionisation, α, of a c M solution of acetic acid (abbreviated to to HAc). Since HAc is a stronger acid than water, Le Chatelier's principle predicts that the self-ionisation of water will be suppressed. Thus to a first approximation we may neglect the concentration of H_3O^+ ions resulting from self-ionisation, so that at equilibrium the concentrations of H_3O^+ and Ac^- ions will be equal. The concentration of water is treated as constant.

$$HAc + H_2O \rightleftharpoons H_3O^+ + Ac^-$$

At equilibrium: $c(1 - \alpha)$M αc M αc M

$$\therefore K_a = c\alpha^2/(1 - \alpha)$$

and, rearranging, $c\alpha^2 + K_a\alpha - K_a = 0$

This quadratic equation may be solved for α:

$$\alpha = \frac{-K_a + (K_a{}^2 + 4cK_a)^{\frac{1}{2}}}{2c} \quad \ldots \text{(XIII, 16)}$$

or, since

$$[H_3O^+] = \alpha c; \; [H_3O^+] = -\frac{K_a}{2} + \left(\frac{K_a{}^2}{4} + cK_a\right)^{\frac{1}{2}}$$
$$\ldots \text{(XIII, 17)}$$

Substituting the given values of K_a and c:

From (XIII, 16), $\qquad \alpha = 4 \cdot 21 \times 10^{-2}$
and from (XIII, 17), $[H_3O^+] = 4 \cdot 21 \times 10^{-4} \, \text{mol dm}^{-3}$
$$\therefore \; \text{pH} = 3 \cdot 38$$

A further approximation could have been made, since, if K_a is small, α must also be small and $(1 - \alpha)$ is approximately unity

$$\therefore \; \alpha \approx (K_a/c)^{\frac{1}{2}} \quad \text{and} \quad [H_3O^+] = (cK_a)^{\frac{1}{2}} \quad \ldots \text{(XIII, 18)}$$

Substituting the given values of K_a and c gives:

$$\alpha = 4 \cdot 30 \times 10^{-2}, \quad \text{pH} = 3 \cdot 37$$

which are close to the values previously obtained. α and pOH may be calculated in an analogous manner for the ionisation of a weak base.

EXAMPLE 3: Calculate the degree of ionisation and pH of a solution containing $0 \cdot 01 \, \text{mol dm}^{-3}$ of HAc and $0 \cdot 1 \, \text{mol dm}^{-3}$ of sodium acetate (NaAc).

Again, we first derive a general equation for a solution containing $c \, \text{mol dm}^{-3}$ of HAc and $c' \, \text{mol dm}^{-3}$ of NaAc. Ac^- ions are obtained from the ionisation of HAc and from the NaAc (which may be considered to be completely ionised). If we ignore the concentration of H_3O^+ ions from self-ionisation and treat the concentration of water as constant:

$$HAc + H_2O \rightleftharpoons H_3O^+ + Ac^-$$

At equilibrium: $c(1 - \alpha) \, \text{M} \qquad c\alpha \, \text{M} \qquad c\alpha + c' \, \text{M}$

$$\therefore \; K_a = \alpha(c' + c\alpha)/(1 - \alpha) \quad \ldots \text{(XIII, 19)}$$

Rearranging and solving for α,

$$\alpha = \frac{-(c' + K_a) + \{(c' + K_a)^2 + 4cK_a\}^{\frac{1}{2}}}{2c}$$

Substituting the given values of c, c' and K_a:

$$\alpha = 1\cdot74 \times 10^{-4}$$

and

$$[H_3O^+] = c\alpha = 1\cdot74 \times 10^{-4} \times 10^{-2} = 1\cdot74 \times 10^{-6}\ \text{mol dm}^{-3}$$
$$\therefore pH = 5\cdot76$$

The treatment can be further simplified as follows. Since NaAc is completely ionised, a large excess of Ac^- ions is present in solution and Le Chatelier's principle predicts that this will suppress the ionisation of the HAc. Since the degree of ionisation was small even in the absence of the added salt, we may assume that all the Ac^- ions are derived from the salt, *i.e.* $c' \gg c\alpha$ and also $1 - \alpha \approx 1$. Equation (XIII, 19) reduces to

$$K_a = \alpha c' \quad \text{or} \quad \alpha = K_a/c' \quad \text{. . . (XIII, 20)}$$

and substituting the given values of c' and K_a:

$$\alpha = 1\cdot85 \times 10^{-4} \quad \text{and} \quad pH = 5\cdot74.$$

10. Hydrolysis. If the salt of a weak acid and a strong base, *e.g.* NaAc, is dissolved in water, the Ac^- ion, being a fairly strong base (conjugate base of a weak acid—*see* **4** above), reacts with water to produce OH^- ions

$$Ac^- + H_2O \rightleftharpoons HAc + OH^- \quad \text{. . . (XIII, 21)}$$

This process is called *hydrolysis* and corresponds to the partial reversal of the neutralisation reaction by which the salt was formed. The amount of hydrolysis is measured by the hydrolysis constant K_h, where

$$K_h = [HAc][OH^-]/[Ac^-] \quad \text{. . . (XIII, 22)}$$

Noting that for the ionisation of acetic acid, $K_a = [H_3O^+[Ac^-]/[HAc]$ and for the self-ionisation of water, $K_w = [H_3O^+][OH^-]$, (XIII, 22) can be rearranged to

$$K_h = K_w/K_a$$

i.e. for the salts of a series of weak acids with the same strong base, the amount of hydrolysis increases as the acid becomes weaker.

Aqueous solutions of salts of strong acids and weak bases, *e.g.* NH_4Cl, are acidic, owing to the hydrolytic equilibria

$$NH_4^+ + H_2O \rightleftharpoons NH_3 + H_3O^+$$

The hydrolysis constant is

$$K_h = [NH_3][H_3O^+]/[NH_4^+] = K_w/K_b \quad \ldots \quad (XIII, 23)$$

where K_b is the ionisation constant of ammonia, given by

$$K_b = [NH_4^+][OH^-]/[NH_3]$$

Both the cation and the anion in aqueous solutions of salts of weak bases and weak acids, *e.g.* NH_4Ac, undergo hydrolysis

$$NH_4^+ + Ac^- + H_2O \rightleftharpoons NH_4OH + HAc$$

and $K_h = [NH_4OH][HAc]/[NH_4^+][Ac^-] = K_w/K_aK_b$
$$\ldots \quad (XIII, 24)$$

NOTE: The salts of strong bases and strong acids, *e.g.* NaCl, do not undergo hydrolysis.

11. pH of aqueous salt solutions. Taking as an example a c M solution of NaAc (degree of hydrolysis α), a general formula for the pH may be derived as follows. The hydrolysis of Ac^- is represented by

$$Ac^- \quad + H_2O \rightleftharpoons HAc + OH^-$$

At equilibrium $c(1 - \alpha)$ M $\qquad c\alpha$ M $\quad c\alpha$ M

$$\therefore K_h = K_w/K_a = \alpha^2 c/(1 - \alpha)$$

Assuming that α is small, we may write $(1 - \alpha) \approx 1$

$$\therefore \alpha = (K_w/K_a c)^{\frac{1}{2}}$$

But $\qquad [OH^-] = \alpha c$
$$\therefore [OH^-] = c(K_w/K_a c)^{\frac{1}{2}} = (K_w c/K_a)^{\frac{1}{2}}$$

and $[H_3O^+] = K_w/[OH^-]$ from (XIII, 10)

$$\therefore [H_3O^+] = (K_w K_a/c)^{\frac{1}{2}}$$

Taking negative logarithms

$$pH = \tfrac{1}{2}pK_w + \tfrac{1}{2}pK_a + \tfrac{1}{2}\lg c \quad \ldots \quad (XIII, 25)$$

The corresponding equations for the salts of a strong acid–weak base and weak acid–weak base are respectively:

$$pH = \tfrac{1}{2}pK_w - \tfrac{1}{2}pK_a - \tfrac{1}{2}\lg c \quad \ldots \quad (XIII, 26)$$
$$pH = \tfrac{1}{2}pK_w + \tfrac{1}{2}pK_a - \tfrac{1}{2}pK_b \quad \ldots \quad (XIII, 27)$$

It is important to note that since (XIII, 25) and (XIII, 26) involve the salt concentration, hydrolysis will increase as the solution is diluted. However, the salt concentration does not appear in (XIII, 27), indicating that the hydrolysis of a salt of a weak acid–weak base is independent of concentration.

12. Buffer solutions. Such solutions consist of a mixture of a weak acid and its salt with a strong base (or a weak base and its salt with a strong acid), and have the very useful property that their pH changes only marginally on dilution or addition of small quantities of acids or bases.

The hydrogen ion concentration is determined by the equilibrium

$$HA + H_2O \rightleftharpoons A^- + H_3O^+ \quad . . . \text{(XIII, 28)}$$

which, because of the relatively large concentration of A^- ions from the salt, lies well over to the left. If excess hydrogen ions are added to this mixture the equilibrium is pushed further to the left, and the total hydrogen ion concentration remains approximately constant. The addition of hydroxyl ions, on the other hand, causes the equilibrium

$$H_3O^+ + OH^- \rightleftharpoons 2H_2O$$

to be shifted to the right. This reduces the hydrogen ion concentration, causing more HA molecules to ionise. Again, this tends to restore the hydrogen ion concentration to its original value.

The pH of a buffer solution may be calculated using the *Henderson–Hasselbalch* equation (*see* 13 below).

13. The Henderson–Hasselbalch equation. From (XIII, 28) the ionisation constant of the weak acid is given by

$$K_a = [H_3O^+][A^-]/[HA]$$

Since this equilibrium lies well over to the left, [HA], the concentration of the unionised acid, may be replaced by the total concentration of acid in the buffer solution, [acid], and [A^-] may be replaced by [salt]. With these approximations, the expression for K_a may be written

$$K_a = [H_3O^+][\text{salt}]/[\text{acid}]$$

Taking negative logarithms

$$pK_a = pH - \lg([\text{salt}]/[\text{acid}])$$

or $\qquad pH = pK_a + \lg([\text{salt}]/[\text{acid}]) \quad . . . \text{(XIII, 29)}$

This is called the *Henderson–Hasselbalch* equation.

Similarly, for a mixture of a weak base and its salt with a strong acid

$$pOH = pK_b + \lg([salt]/[base])$$

The use of (XIII, 29) in calculating the pH of a buffer solution is illustrated in the following example.

EXAMPLE 4: Compare the changes in pH occurring when 1 cm^3 of 1M HCl is added to 1 dm^3 of (a) pure water, (b) a buffer solution containing 0·5M NaAc and 0·5M HAc. K_a for acetic acid is $1·85 \times 10^{-5}$ mol dm^{-3}.

(a) Initial pH = 7·0.

Addition of 1 cm^3 of 1M HCl to 1000 cm^3 of water gives a solution containing 10^{-3} mol of H_3O^+ ions in 1001 cm^3 of water. The concentration of H_3O^+ ions formed by self-ionisation may be ignored, since this must be less than 10^{-7}M.

$$\therefore [H_3O^+] = 10^{-3}M \qquad pH = 3.$$

i.e. a change of four pH units (corresponding to a 10 000-fold increase in $[H_3O^+]$).

(b) The initial pH is found from (XIII, 29)

$$pH = 4·73 + \lg (0·499/0·501) = 4·74$$

The added H_3O^+ ions can all be regarded as reacting with Ac^- ions to produce unionised HAc (HAc is a weak acid and Ac^- a strong conjugate base)

$$new [HAc] = 0·50 + 0·001 = 0·501 M$$
$$new [Ac^-] = 0·50 - 0·001 = 0·499 M$$

and using these values in (XIII, 29)

$$pH = 4·73 + \lg (0·501/0·499) = 4·74$$

i.e. a change of only 0·01 pH units.

14. Buffering capacity. The pH of the buffer solution depends on the ratio [salt]/[acid], and the buffering capacity is a maximum (*i.e.* the change in pH is a minimum) when the change in this ratio is as small as possible for the addition of any given amount of acid or base. This is realised when [salt] = [acid] and both concentrations are large relative to the hydrogen ion concentration. When [salt] = [acid], $\lg$[salt]/[acid] = 0 and pH = pK_a. For maximum buffering capacity, a particular buffer should be used to control pH's close to the pK_a value of its weak acid constituent. Buffers may be used to prepare a solution of any given pH by first choosing a weak acid with a suitable pK_a value and adjusting the ratio [salt]/[acid] to give the desired pH.

15. Acid–base titrations. The calculation of the concentration of H_3O^+ ions present at various stages in the titration of 100 cm³ of 0·1 M HCl with 0·1 M NaOH is illustrated in Table 14.

TABLE 14   pH AT VARIOUS STAGES IN THE TITRATION OF 100 cm³ OF 0·1 M HCl WITH 0·1 M NaOH

Volume of NaOH added/cm³	Total volume/cm³	Concentration of un-neutralised acid or base/mol dm⁻³	pH
0	100	10^{-1}	1·00
50	150	$3·33 \times 10^{-2}$	1·48
99	199	$5·025 \times 10^{-4}$	3·30
99·9	199·9	$5·025 \times 10^{-5}$	4·30
100	200	0·00	7·00
100·1	200·1	$5·025 \times 10^{-5}([OH^-])$	9·70

The clearest representation of the variation of pH during the titration is given by a titration curve (*i.e.* a plot of pH against volume of added base), as in Fig. 52. A noticeable feature is

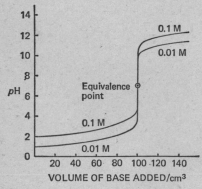

FIG. 52.—*Titration curves for 0·1M HCl with 0·1M NaOH and for 0·01M HCl with 0·01M NaOH*

the very rapid change of pH in the region of the equivalence point—5·4 units between 99·9 and 100·1 cm³ of added base. Also shown in this figure is the titration curve for 0·01 M HCl with 0·01 M NaOH; here the change in pH around the equivalence point is not so rapid and an accurate titration is consequently more difficult.

A similar calculation for the titration of 100 cm³ of 0·1 M HAc with 0·1 M NaOH is illustrated in Table 15. Remembering that at any stage before the equivalence point we have effectively a buffer solution of NaAc and HAc, we may use (XIII, 29) to calculate the pH, where [salt] is taken as the concentra-

TABLE 15 pH AT VARIOUS STAGES DURING THE TITRATION OF 100 cm³ OF 0·1 M HAc WITH 0·1 M NaOH

Volume of NaOH added/cm³	Total volume/cm³	Concentration of acid/mol dm⁻³	Concentration of salt/mol dm⁻³	pH
0	100	10^{-1}	0·00	2·87
50	150	$3·33 \times 10^{-2}$	$3·33 \times 10^{-2}$	4·73
90	190	$5·26 \times 10^{-3}$	$4·74 \times 10^{-2}$	5·68
95	195	$2·56 \times 10^{-3}$	$4·87 \times 10^{-2}$	6·01
99	199	$5·03 \times 10^{-4}$	$4·97 \times 10^{-2}$	6·73
99·9	199·9	$5·00 \times 10^{-5}$	$5·00 \times 10^{-2}$	7·73
100	200	–	$5·00 \times 10^{-2}$	8·43

tion of added base and [acid] as the concentration of un-neutralised acid. At, or very close to, the equivalence point the solution contains a large concentration of Ac^- ions and few H_3O^+ ions, hydrolysis is appreciable and the pH is calculated using (XIII, 25). Beyond the equivalence point the pH may be simply determined from the concentration of un-neutralised base.

The titration curve is shown in Fig. 53, and it is noticeable

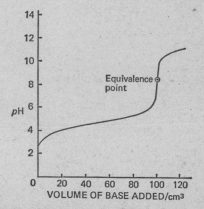

FIG. 53.—*Titration curve for 0·1M HAc with 0·1M NaOH*

that the pH does not change very rapidly in the region of the equivalence point, making accurate titration more difficult, and also that the equivalence point occurs in alkaline rather than neutral solution.

At the half-neutralisation point [salt] = [acid] and hence the pH at this point is equal to pK_a. This provides a general method for measuring the ionisation constant of a weak acid. Figs. 52 and 53 may also be obtained experimentally (*see* XIV, 35).

16. Acid–base indicators. Acid–base indicators are weak acids or bases whose conjugate forms are of different colours. For an indicator which is a weak acid and may be represented by HIn, the following equilibrium occurs in solution

$$HIn + H_2O \rightleftharpoons In^- + H_3O^+$$
$$\textit{Colour A} \qquad \textit{Colour B}$$

and $[H_3O^+] = K_{In}[HIn]/[In^-]$ . . . (XIII, 30)

where K_{In} is the indicator ionisation constant.

This equation is formally analogous to (XIII, 29) for buffer solutions. However, buffers are present in large concentrations and effectively control the pH of the solution, whereas indicators are present in such small amounts that the pH of the solution entirely controls the indicator equilibrium. The $[H_3O^+]$ from the indicator is so small as not to affect the main acid–base equilibrium of the solution.

From (XIII, 30) it can be seen that the $[H_3O^+]$ determines the ratio $[HIn]/[In^-]$, since K_{In} is a constant. If the ratio $[HIn]/[In^-]$ is greater than about 10 only colour A is visible, whereas if the ratio is less than about 0·1 only colour B is visible. Thus the indicator changes colour in a range of $[H_3O^+]$ of about 100 around the value of K_{In}, *i.e.* in a pH range of two units around pK_{In}. For example, in a titration having an equivalence point at pH 7, an indicator would be chosen having $pK_{In} = 7$ so that complete colour change would be observed between pH's 6 and 8.

The pK values of several common indicators are shown in Table 16. In the strong acid–strong base titration considered in the previous section a change of six pH units occurred around the equivalence point, so that almost any indicator could have been used. In other cases the indicator must be chosen carefully.

TABLE 16 pK VALUES OF SOME COMMON INDICATORS

Indicator	pK	Colour	
		Acidic	Basic
Bromophenol blue	3·98	Yellow	Blue
Methyl orange	3·7	Red	Yellow–Orange
Methyl red	5·0	Red	Yellow
Bromothymol blue	7·1	Yellow	Blue
Phenol red	7·7	Yellow	Red
Phenolphthalein	9·7	Colourless	Red

17. Multi-stage equilibria. If a weak acid can ionise in two or more ways, its solution will contain more than one weak acid. The following equilibria are present in a carbonic acid solution:

$$H_2CO_3 + H_2O \rightleftharpoons HCO_3^- + H_3O^+$$
$$K_1 = 4·2 \times 10^{-7} \text{ mol dm}^{-3}$$
$$HCO_3^- + H_2O \rightleftharpoons CO_3^{2-} + H_3O^+$$
$$K_2 = 4·8 \times 10^{-11} \text{ mol dm}^{-3}$$

$$K_1 = [H_3O^+][HCO_3^-]/[H_2CO_3],$$
$$K_2 = [H_3O^+][CO_3^{2-}]/[HCO_3^-]$$
$$\therefore K = K_1 K_2 = [H_3O^+]^2[CO_3^{2-}]/[H_2CO_3]$$

The method of calculation is illustrated in the following example.

EXAMPLE 5: Calculate the concentrations of all the ions present in a 0·1 M solution of H_2CO_3.

Since $K_2 \ll K_1$, the equilibrium concentration of H_3O^+ ions is determined primarily by K_1, i.e. $[H_3O^+]$ can be calculated as if the second equilibrium did not exist. Using (XIII, 17)

$$[H_3O^+] = -\frac{K_1}{2} + \left(\frac{K_1^2}{4} + cK_1\right)^{\frac{1}{2}} \quad \text{. . . (XIII, 17)}$$

Substituting the given values:

$$[H_3O^+] \approx [HCO_3^-] = 6·46 \times 10^{-5} \text{ M}$$
$$[CO_3^{2-}] = K_2[HCO_3^-]/[H_3O^+] \approx K_2 = 4·8 \times 10^{-11} \text{ M}$$
$$[OH^-] = K_w/[H_3O^+] = 1·54 \times 10^{-10} \text{ M}$$

HETEROGENEOUS IONIC EQUILIBRIA

18. The solubility product. In a saturated solution equilibrium exists between the solid salt and its ions in solution, e.g.

$$AB(s) \rightleftharpoons A^+ + B^-$$

Since AB(s) will have a constant concentration (unit activity):

$$K_s = [A^+][B^-] \quad \text{. . . (XIII, 31)}$$

K_s is called the *solubility product* of the salt, and has the same characteristics as other equilibrium constants.

NOTE: The solubility product should strictly be defined in terms of activities rather than concentrations. For sparingly soluble salts this approximation is not too serious (however, *see* 20 below).

EXAMPLE 6: The solubility product of $PbCl_2$ is $1 \cdot 7 \times 10^{-5}$ mol² dm⁻⁶. Calculate its solubility in mol dm⁻³.

$$PbCl_2(s) \rightleftharpoons Pb^{2+} + 2Cl^- \text{ . . . (XIII, 32)}$$
$$K_s = [Pb^{2+}][Cl^-]^2$$

But from (XIII, 32): $2[Pb^{2+}] = [Cl^-]$

$$\therefore K_s = [Pb^{2+}](2[Pb^{2+}])^2$$
$$4[Pb^{2+}]^3 = 1 \cdot 7 \times 10^{-5}$$
$$\therefore [Pb^{2+}] = 2 \cdot 6 \times 10^{-2} \text{ M}$$

Since one mole of Pb^{2+} dissolves per mole of salt, the solubility of $PbCl_2$ is $2 \cdot 6 \times 10^{-2}$ mol dm⁻³.

19. The common-ion effect.

Most salts are more soluble in pure water than in solutions containing electrolytes with which they have an ion in common.

NOTE: This is only true provided no complexes are formed—*see* 20 below.

The reason for this is qualitatively obvious from Le Chatelier's principle; the quantitative calculation is illustrated in the following example.

EXAMPLE 7: The solubility product of AgCl is $2 \cdot 8 \times 10^{-10}$ mol² dm⁻⁶. Calculate its solubility in (*a*) pure water, (*b*) a 0·1 M solution of $AgNO_3$.

(*a*) $$AgCl(s) \rightleftharpoons Ag^+ + Cl^-$$

$$K_s = [Ag^+][Cl^-] = [Cl^-]^2$$
$$\therefore [Cl^-]^2 = 2 \cdot 8 \times 10^{-10}$$
$$[Cl^-] = 1 \cdot 7 \times 10^{-5} \text{ M}$$
$$\therefore \text{Solubility} = 1 \cdot 7 \times 10^{-5} \text{ M}$$

(b) The concentration of Ag^+ from $AgNO_3$ is much greater than that from $AgCl$ (which must be less than 1.7×10^{-5} M), so that to a first approximation we may write $[Ag^+] = 0.1$ M

$$\therefore [Cl^-] = K_s/[Ag^+] = 2.8 \times 10^{-10}/0.1$$
$$= 2.8 \times 10^{-9} \text{ M}$$
$$\therefore \text{Solubility} = 2.8 \times 10^{-9} \text{ M}$$

Whenever the product of the concentrations of the ions exceeds the solubility product, solid will be precipitated until the equality given in (XIII, 31) is re-established. This effect is extremely important in both qualitative and quantitative analysis.

20. The uncommon-ion effect.

In contrast to the above effect, the presence of salts without a common ion often markedly increases the solubility. This may be due to one of two causes:

(a) A complex may be formed, e.g. silver chloride is much more soluble in solutions containing cyanides than in water because of the equilibrium

$$Ag^+ + 2CN^- \rightleftharpoons [Ag(CN)_2]^-$$

which effectively removes Ag^+ ions from the solution, thus causing more $AgCl(s)$ to dissolve to maintain the value of the solubility product.

(b) Consider the approximation of using concentrations rather than activities to define the solubility product, i.e.

$$K_s = a_{M+}\, a_{A-} = \gamma_+[M^+]\gamma_-[A^-] \approx [M^+][A^-]$$

where γ is the activity coefficient (see VIII, 26–27). In dilute solutions, such as those formed by sparingly soluble salts, γ is close to 1. However, in solutions containing appreciable concentrations of other ions γ is considerably less than 1, and in order to preserve the true equality

$$K_s = \gamma_+[M^+]\gamma_-[A^-]$$

$[M^+]$ and $[A^-]$ must increase.

PROGRESS TEST 13

1. Calculate the degree of ionisation and the pH of a solution prepared by dissolving 0.1 mol of $CH_2ClCOOH$ in sufficient water to make 1 dm^3 of solution. $K_a = 1.5 \times 10^{-3}$ mol dm^{-3}. (9)

2. Calculate the concentrations of NH_3, OH^- and NH_4^+ present in equilibrium in a solution prepared by dissolving 0.1 mol of

NH_4Cl and 0·2 mol of NaOH in 1 dm^3 of water. Assume that the NaOH is completely dissociated and that K_b for NH_3 is 1·8 × 10^{-5} mol dm^{-3}. (9)

3. Calculate the hydrolysis constant, degree of hydrolysis and pH of a solution prepared by dissolving 0·1 mol of NH_4Cl in 1 dm^3 of water. K_b for NH_3 is 1·8 × 10^{-5} mol dm^{-3}. (10–11)

4. What masses of $Na_2HPO_4.2H_2O$ (relative molecular mass = 178) and $NaH_2PO_4.H_2O$ (relative molecular mass = 138) are required to prepare 100 cm^3 of a 0·1 M buffer solution of pH 7·5? The pK_a values for phosphoric acid are $pK_1 = 2·1$, $pK_2 = 7·2$, $pK_3 = 12·3$. (13)

5. A saturated solution of H_2S in water contains 0·1 M H_2S. The concentration of H_3O^+ ions is fixed at 0·1 M by addition of a strong acid. Find the concentration of S^{2-} ions, given that the ionisation constants of H_2S are $K_1 = 1·0 × 10^{-7}$ mol dm^{-3}, $K_2 = 1·3 × 10^{-13}$ mol dm^{-3}. (17)

6. Using the sulphide ion concentration obtained in the previous question, find which of the following cations will be precipitated by bubbling hydrogen sulphide through a 0·01 M solution of the cation in 0·1 M acid: Mn^{2+}, Fe^{2+}, Pb^{2+}, Zn^{2+}. The relevant solubility products (in mol^3 dm^{-6}) are:

(a) $MnS = 2·5 × 10^{-10}$ (b) $FeS = 1·0 × 10^{-19}$
(c) $PbS = 1·0 × 10^{-29}$ (d) $ZnS = 4·5 × 10^{-24}$ (18–19)

7. Describe three methods for classifying acids and bases. (1–3)

8. What is meant by (a) a conjugate acid–base pair (2), (b) pK (5), (c) pK_w (6), (d) pH (8)?

9. Derive an expression for (a) the ionisation constant of a weak acid (4), (b) the ionisation constant of a weak base (4), (c) the relationship between (a) and (b) (7).

10. Derive an expression for the degree of ionisation and the pH of a solution of a weak acid. Point out what approximations can be made. (9)

11. Explain why the solution of a salt of a weak acid and a strong base is alkaline, and derive an expression for the pH of such a solution. (10–11)

12. Explain how a buffer solution can be used (a) to control pH (12), (b) to produce any desired pH (13–14).

13. Outline the theory of acid–base indicators. (16)

14. Define the solubility product and explain the common-ion effect. (18–19)

CHAPTER XIV

ELECTRODE POTENTIALS

1. Electron transfer reactions. All electron transfer reactions involve the oxidation of one species and the reduction of another. Thus when zinc is added to copper sulphate solution, copper is deposited, the reaction being

$$Zn(s) + Cu^{2+} = Zn^{2+} + Cu(s) \quad . \quad . \quad (XIV, 1)$$

In effect metallic zinc transfers two electrons to the copper ion, *i.e.* metallic zinc is oxidised to Zn^{2+} ions while Cu^{2+} ions are reduced to metallic copper. The reaction may be considered to be made up of two *half-reactions*:

$$Zn(s) = Zn^{2+} + 2e^- \quad . \quad . \quad (XIV, 2)$$
$$Cu^{2+} + 2e^- = Cu(s) \quad . \quad . \quad (XIV, 3)$$

Addition of (XIV, 2) and (XIV, 3) gives (XIV, 1).

The reaction proceeds in the direction indicated because metallic zinc is a stronger reducing agent than metallic copper. If the reverse were true, (XIV, 1) would be reversed.

Another well-known example of an electron transfer reaction is the reduction of iodine by thiosulphate ions:

$$2S_2O_3{}^{2-} + I_2 = S_4O_6{}^{2-} + 2I^- \quad . \quad . \quad (XIV, 4)$$

This reaction may again be divided into half-reactions:

$$2S_2O_3{}^{2-} = S_4O_6{}^{2-} + 2e^- \quad . \quad . \quad (XIV, 5)$$
$$I_2 + 2e^- = 2I^- \quad . \quad . \quad (XIV, 6)$$

$S_2O_3{}^{2-}$ is a stronger reducing agent than I^-, and the reaction proceeds in the direction indicated.

2. Electrochemical cells. The direction in which an electron transfer reaction takes place depends on the ability of the reductant to give up electrons and of the oxidant to receive electrons. Electron transfer reactions such as (XIV, 1) and (XIV, 4) take place at the electrodes in *electrochemical cells,*

237

where each half-reaction occurs at one of the electrodes. By making measurements on electrochemical cells it is possible to examine the feasibility of any electron transfer reaction.

TYPES OF ELECTRODE

The electrodes discussed below are reversible in a thermodynamic sense (see **9** below), *i.e.* a slight change in the external conditions completely reverses the electrode process.

3. Reversible cation electrodes. When a metal electrode is placed in a solution of its own ions, there will be two opposing tendencies, which may be represented by the equation

$$M^{n+} + ne^- \rightleftharpoons M$$

The position of the equilibrium depends on the metal. If the equilibrium lies over to the left, the electrode will tend to become negatively charged with respect to the solution, since electrons are left behind on the electrode. If the equilibrium lies over to the right, the electrode will tend to become positively charged with respect to the solution, owing to the removal of electrons. For example, a zinc electrode in a zinc sulphate solution tends to accumulate negative charge, while a copper electrode in a copper sulphate solution tends to accumulate positive charge.

The potential set up between the electrode and the solution in its immediate vicinity is known as the *electrode potential*, and its magnitude depends on the nature of the metal, the concentration of the solution and the temperature.

Hydrogen gas may also give rise to a reversible cation electrode. If hydrogen is bubbled through a solution containing hydrogen ions, the following equilibrium is set up.

$$H^+ + e^- \rightleftharpoons \tfrac{1}{2}H_2(g)$$

NOTE: In this chapter it will be convenient to represent the hydrogen ion in aqueous solution by H^+ rather than by the more correct H_3O^+.

An inert platinum electrode covered with finely divided platinum (known as a platinised platinum electrode) must be present which plays no part in the reaction but:

    (a) adsorbs molecular hydrogen as hydrogen atoms and facilitates the transfer of electrons,

    (b) provides a path by which electrons may enter or leave the system.

4. Reversible anion electrodes. This type of electrode often involves a metal and a sparingly soluble salt of the metal in contact with a soluble salt containing a common anion. In such electrodes loss or gain of electrons is accompanied by a reversible change involving oxidation or reduction of the anion,

$$MA + ne^- \rightleftharpoons M + A^{n-}$$

Examples of this type of electrode are:

(a) A silver wire coated with silver chloride dipping into a solution containing chloride ions (silver–silver chloride electrode). The electrode reaction is

$$AgCl(s) + e^- \rightleftharpoons Ag(s) + Cl^-$$

(b) A pool of mercury in contact with a suspension of solid mercurous chloride and a solution of chloride ions (calomel electrode).

$$Hg_2Cl_2(s) + 2e^- \rightleftharpoons 2Hg(l) + 2Cl^-$$

(c) Reversible anion gas electrodes are also possible, e.g. chlorine gas bubbled through a solution of chloride ions in the presence of a platinised platinum electrode gives rise to the electrode reaction

$$\tfrac{1}{2}Cl_2(g) + e^- \rightleftharpoons Cl^-$$

5. Reversible oxidation–reduction electrodes. These consist of an inert electrode such as platinum or gold dipping into a solution containing a chemical species in two different oxidation states. The equilibrium involved in a system containing a platinum electrode immersed in a solution of ferrous and ferric ions is

$$Fe^{3+} + e^- \rightleftharpoons Fe^{2+}$$

The platinum electrode supplies and removes electrons, as in the case of the gas electrodes. The position of the equilibrium depends on the relative concentrations of the oxidised and reduced species. These species need not be ions but may be molecular species, e.g. quinone and hydroquinone

$$C_6H_4O_2 + 2H^+ + 2e^- \rightleftharpoons C_6H_4(OH)_2$$
quinone (Q)   hydroquinone (QH$_2$)

All the electrodes mentioned in 3–5 above involve an oxidised and a reduced state, and a general electrode reaction may be written

$$\text{Oxidised form} + ne^- \rightleftharpoons \text{Reduced form.}$$

GALVANIC CELLS

6. Galvanic cells. A *galvanic cell* is a device which utilises chemical changes which take place at two electrodes to establish a potential difference between them, thus causing a flow of current through an external circuit. There are two types of galvanic cell:

(a) *Chemical cells*, in which the flow of current results in an overall chemical reaction.

(b) *Concentration cells*, in which the flow of current results in the effective transfer of electrolyte from one electrode solution (at a higher concentration) to another (at a lower concentration), *i.e.* transfer of material from one electrode to the other, but no overall chemical reaction.

When the electrodes are not externally connected, no current flows and no reaction occurs at the electrodes. We shall be almost exclusively concerned with chemical cells.

7. Combination of electrodes. In the Daniell cell the following chemical changes take place at the electrodes and drive current round the external circuit:

$$Zn(s) = Zn^{2+} + 2e^-$$
$$Cu^{2+} + 2e^- = Cu(s)$$

The overall reaction is

$$Zn(s) + Cu^{2+} = Zn^{2+} + Cu(s)$$

Thus 2 mol of electrons, *i.e.* $2 \times 9.649 \times 10^4$ C of charge, are transferred around the external circuit from the zinc to the copper electrode for each mole of zinc dissolved or copper deposited. Notice that the same reaction takes place when zinc is added to copper sulphate solution or when current is taken from a Daniell cell.

In order to determine whether a given metal can reduce the ions of another metal it is only necessary to set up a cell containing the two metal–metal ion electrodes and to measure its potential difference.

8. Measurement of potential difference. The potential difference generated by a cell cannot be determined with a voltmeter, since this instrument draws current from the cell. This causes electrode reactions to occur, thereby changing the concentration of the solution and altering the potential difference. Instead, the potential difference must be deter-

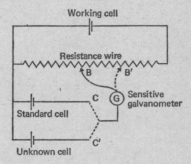

Fig. 54.—*Potentiometer circuit for the measurement of the e.m.f. of a cell*

mined under conditions of zero current flow (*i.e.* the *electromotive force*, or e.m.f., of the cell must be determined). A potentiometer circuit (*see* Fig. 54) may be used to measure the e.m.f. The working cell (a lead—acid accumulator) provides a potential difference across the resistance wire, and a variable potential difference may be tapped off by the movable contact *B*. *B* is attached to a sensitive galvanometer *G* and to either a standard cell of accurately known e.m.f. (by making contact at *C*) or to the unknown cell (by making contact at *C'*). The standard cell is usually a Weston cell with an e.m.f. of 1·01859 V at 20°C.

9. Reversible cells. In order that the electrical energy produced by a galvanic cell may be related to the processes occurring in the cell, it is essential that the cell should behave reversibly in a thermodynamic sense (*see* VI, 12). The concept of reversibility as applied to galvanic cells can be readily understood from Fig. 54. If the e.m.f. of the unknown cell is exactly balanced by the externally applied potential (at *B'*), then no current flows and no chemical change takes place. If the movable contact is moved by an infinitesimal amount either side of B', then on one side a minute current will flow through the cell in one direction, and on the other side in the opposite direction. For the cell to behave reversibly, the chemical changes which take place must be exactly reversed when the direction of the current is reversed.

In the Daniell cell the chemical change which drives current around the external circuit is

$$Zn(s) + Cu^{2+} = Zn^{2+} + Cu(s)$$

while if an external potential difference is applied which is slightly greater than that of the cell, the reverse reaction occurs.

Just as the complete cell reaction is reversed, so are the individual electrode reactions, and electrodes which have this property are known as reversible electrodes (see 3–5).

10. Cell diagrams. The Daniell cell may be represented diagrammatically by

$$Zn \mid Zn^{2+} \mid Cu^{2+} \mid Cu$$

where $\mid$ indicates a phase boundary. The two solutions are separated by a semi-permeable membrane, which is usually in the form of a porous pot.

By convention, the right-hand electrode in the cell diagram is taken as positive and the left-hand electrode as negative. Hence reduction occurs at the former and oxidation at the latter. In the case of the Daniell cell as written above, the copper electrode is the positive electrode and the individual electrode reactions and the complete cell reaction have been given in 7 above.

In the cell

$$Pt, \overset{-}{H_2} \mid H^+ \mid Cl^- \mid Ag\overset{+}{Cl}, Ag$$

the electrode reactions will be:

Left hand electrode: $\frac{1}{2}H_2(g) = H^+ + e^-$
Right hand electrode: $AgCl(s) + e^- = Ag(s) + Cl^-$
Net reaction: $\frac{1}{2}H_2(g) + AgCl(s)$
$$= H^+ + Cl^- + Ag(s)$$

If the cell was represented in the reverse manner the reaction would be reversed, and the sign of the e.m.f. would change.

11. Liquid junction potentials. Some diffusion will occur at the interface between the two solutions in the Daniell cell, *i.e.* Zn^{2+} ions diffuse into the $CuSO_4$ solution and Cu^{2+} ions into the $ZnSO_4$ solution (neglecting the movement of the SO_4^{2-} ions). Since the speeds of migration of the cations will be different, there will be a separation of charge across the interface and a potential difference will be set up which is known as the *liquid junction potential*. This potential acts in such a way as to retard the more rapidly moving ions and to accelerate

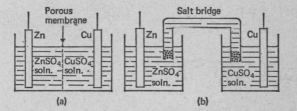

Fɪɢ. 55.—(a) *Normal Daniell cell.*   (b) *Daniell cell with salt bridge*

the more slowly moving ones. Liquid junction potentials do not normally exceed 0·1 V, but are an experimental nuisance since they are difficult to reproduce. Such potentials may be reduced to a negligible value by using a *salt bridge*. This consists of a tube containing either saturated potassium chloride or saturated ammonium nitrate solution (*see* Fig. 55). To prevent excessive diffusion the ends of the salt bridge may be plugged with cotton wool or the salt dissolved in gelatine. The salt bridge replaces one liquid junction with two, and at each junction the vast majority of the current is carried by the electrolyte of the salt bridge, this being much more concentrated than the electrolyte solutions. Thus, at the boundaries, the ions of unequal speeds (Zn^{2+}, Cu^{2+}) are swamped with an enormously greater concentration of ions of approximately equal speeds (K^+, Cl^- or NH_4^+, NO_3^-).

Such an arrangement is conventionally represented by a double vertical line, *e.g.*

$$\bar{Zn} \mid Zn^{2+} \parallel Cu^{2+} \mid \overset{+}{Cu}$$

THERMODYNAMIC ASPECTS OF CELL REACTIONS

12. The relationship between cell e.m.f. and free energy.
The e.m.f. of a cell is the potential difference between the
electrodes when no current flows through the cell. Under these
reversible conditions the amount of chemical energy made
available is equal to the decrease in free energy (*see* VIII,
18–21). If E represents the cell e.m.f. and the cell reaction
involves the passage of n mol of electrons per mole of chemical
change, then the decrease in free energy accompanying the
reaction is given by

$$-\Delta G = nFE \qquad \ldots \text{ (XIV, 8)}$$

where F is the Faraday constant (*see* XII, 5).

If the constituents of the cell are in their standard states

$$-\Delta G^\ominus = nFE^\ominus \qquad \ldots \text{ (XIV, 9)}$$

$E^\ominus$ is a measure of the work the cell can do when operating
under standard conditions and, like $\Delta G^\ominus$, is a measure of the
tendency of reactants in their standard states to yield products
in their standard states. A positive value of $E^\ominus$ indicates that
the cell reaction proceeds spontaneously in the direction
written, and a negative value that the reaction proceeds
spontaneously in the reverse direction. The entropy change
accompanying the reaction may be found by substituting
(XIV, 8) into (VIII, 31)

$$\Delta S = -\left(\frac{\partial \Delta G}{\partial T}\right)_p = nF(\partial E/\partial T)_p \ldots \text{ (XIV, 10)}$$

13. The Gibbs–Helmholtz equation as applied to cells. Sub-
stituting (XIV, 9) and (XIV, 10) into $\Delta G^\ominus = \Delta H^\ominus - T\Delta S^\ominus$
gives

$$-nFE^\ominus = \Delta H^\ominus - nFT(\partial E^\ominus/\partial T)_p$$
$$\therefore \Delta H^\ominus = nF[T(\partial E^\ominus/\partial T)_p - E^\ominus]$$

The standard enthalpy change for a cell reaction may be
found from measurements of $E^\ominus$ and its variation with temperature at constant pressure. The results obtained are always
in excellent agreement with those found calorimetrically for
the same reaction (now occurring outside the cell), thus
indicating the validity of (XIV, 8).

14. The effect of concentration on cell e.m.f.s. The free energy change accompanying a chemical reaction depends on the activities of the reactants and the products (*see* VIII, 30). For the cell reaction

$$aA + bB = cC + dD$$

$$\Delta G = \Delta G^{\ominus} + RT \ln\left[\frac{(a_C)^c(a_D)^d}{(a_A)^a(a_B)^b}\right] \quad \cdots \quad (VIII, 34)$$

a_C, a_D, etc., do not represent equilibrium activities and the quantity in the square brackets is *not* an equilibrium constant. Substituting (XIV, 8) and (XIV, 9) into (VIII, 34) gives

$$E = E^{\ominus} - \frac{RT}{nF} \ln\left[\frac{(a_C)^c(a_D)^d}{(a_A)^a(a_B)^b}\right] \quad \cdots \quad (XIV, 10)$$

It is usual to write (XIV, 10), the *Nernst equation*, in the form

$$E = E^{\ominus} + \frac{RT}{nF} \ln\left[\frac{(a_A)^a(a_B)^b}{(a_C)^c(a_D)^d}\right] \quad \cdots \quad (XIV, 11)$$

As previously, activities may be replaced by concentrations if the solutions are sufficiently dilute.

The e.m.f. of the Daniell cell, when the cell reaction is given by (XIV, 7), is

$$E = E^{\ominus} + \frac{RT}{2F} \ln\left[\frac{(a_{Zn})(a_{Cu^{2+}})}{(a_{Zn^{2+}})(a_{Cu})}\right]$$

Now $a_{Zn} = a_{Cu} = 1$, and replacing the remaining activities by concentrations

$$E = E^{\ominus} + \frac{RT}{2F} \ln\frac{[Cu^{2+}]}{[Zn^{2+}]}$$

The standard state of concentration is 1 mol dm³, and under these conditions $E = E^{\ominus}$.

15. The dependence of the sign of E on concentration. For the cell

$$\overset{-}{Fe} \mid Fe^{2+} \parallel Cd^{2+} \mid \overset{+}{Cd}$$

$E^{\ominus}$ is $+0.04$ V at 25°C and the cell reaction is

$$Fe(s) + Cd^{2+} \rightleftharpoons Fe^{2+} + Cd(s)$$

Under non-standard conditions the cell e.m.f. is given by

$$E = E^{\ominus} + \frac{2.303\ RT}{2F} \lg \frac{[Cd^{2+}]}{[Fe^{2+}]}$$

Table 17 shows the effect on the value of E of changing the ratio $[Cd^{2+}]/[Fe^{2+}]$. $2.303\ RT/F$ has the value 0.059 V at 25°C.

TABLE 17   THE DEPENDENCE OF CELL E.M.F. ON CONCENTRATION

$[Cd^{2+}]$/mol dm^{-3}	$[Fe^{2+}]$/mol dm^{-3}	$E^{\ominus}$/V	E/V		
1	1	0.04	0.04 + 0	= 0.04	
1	0.1	0.04	0.04 + 0.03	= 0.06	
0.01	1	0.04	0.04 − 0.06	= −0.02	

The effect of changing the ratio $[Cd^{2+}]/[Fe^{2+}]$ from 10 to 10^{-2} is to change E from $+0.06$ V to -0.02 V. The value of -0.02 V indicates that the direction of the cell reaction has been reversed.

16. Concentration cells. The Nernst equation indicates that a cell potential can be set up simply by a difference in concentration. Consider the cell

$$\overset{-}{Zn} \mid Zn^{2+}\ (10^{-3}\ M) \parallel Zn^{2+}\ (10^{-1}\ M) \mid \overset{+}{Zn}$$

The complete cell reaction is

$$Zn^{2+}\ (10^{-1}\ M) = Zn^{2+}\ (10^{-3}\ M)$$

At 25°C the cell e.m.f. is

$$E = E^{\ominus} + \frac{0.059}{1} \lg \frac{10^{-1}}{10^{-3}} = 0 + 0.118 = +0.118\ V$$

It is well known that if a concentrated solution is brought into contact with a more dilute solution of the same substance, the two mix spontaneously to give a uniform solution of intermediate composition. This natural tendency is measured by the e.m.f. of the concentration cell.

17. The relationship between cell e.m.f. and equilibrium constants. If the cell reaction is at equilibrium, (XIV, 10) may be written

$$E = 0 = E^{\ominus} - \frac{RT}{nF} \ln K_A$$

where

$$K_A = \left[\frac{(a_C)^c (a_D)^d}{(a_A)^a (a_B)^b} \right]_{\text{equil.}}$$

$$\therefore \lg K_A = nFE^{\ominus}/2 \cdot 303 \, RT$$

If $E^{\ominus}$ is positive, K_A is greater than one, and the more positive $E^{\ominus}$, the larger K_A. If $E^{\ominus}$ is negative, K_A is less than one, and the more negative $E^{\ominus}$, the smaller K_A. As usual a large value for K_A indicates that the reaction goes virtually to completion.

SINGLE ELECTRODE POTENTIALS

18. The need for single electrode potentials. In the case of the Daniell cell the value of $E^{\ominus} = +1 \cdot 1$ V at 25°C is a measure of the tendency of zinc metal to lose electrons and of copper ions to accept electrons, *i.e.* $E^{\ominus}$ is a simultaneous measure of the strength of zinc metal as a reductant and of copper ion as an oxidant. To compare the strengths of oxidising and reducing agents we need a measure of the tendency of the separate electrode reactions to occur, *i.e.* we need to know single electrode potentials.

19. The standard hydrogen electrode. Since only differences between electrode potentials rather than individual values can be measured, it is convenient to refer all electrodes to a single reference electrode and arbitrarily to assign the value zero to the potential of this electrode at all temperatures. The reference electrode used is the standard hydrogen electrode (see Fig. 56), consisting of a platinised platinum electrode

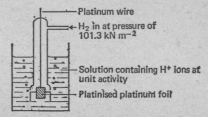

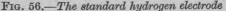

Fig. 56.—*The standard hydrogen electrode*

surrounded by hydrogen gas at a pressure of $101 \cdot 325$ kN m^{-2} and immersed in a solution containing hydrogen ions at unit activity (in practice a 1 mol dm^{-3} solution is used). The electrode reaction is

$$H^+ + e^- = \tfrac{1}{2} H_2(g)$$

and the electrode diagram

$$H^+ \mid H_2, Pt$$

20. Single electrode potentials. If a cell is made up in which a given electrode is combined with the standard hydrogen electrode, the e.m.f. of the cell is equal to the potential of the given electrode. *By convention, all electrode potentials are tabulated as reduction potentials and the associated electrode reactions are reduction reactions, e.g.*

$$Zn^{2+} + 2e^- = Zn(s) \qquad \tfrac{1}{2}Cl_2(g) + e^- = Cl^-$$

This means that the given electrode must form the right-hand electrode of the cell, and the standard hydrogen electrode the left-hand electrode, *e.g.*

$$Pt, \overset{-}{H_2} \mid H^+ \parallel Zn^{2+} \mid \overset{+}{Zn} \qquad Pt, \overset{-}{H_2} \mid H^+ \parallel Cl^- \mid \overset{+}{Cl_2}, Pt$$

The e.m.f. of the cell is equal to the potential of the right-hand electrode (V_R) *minus the potential of the left-hand electrode* (V_L) *under conditions of zero current flow, i.e.* $E = V_R - V_L$.

For the hydrogen–zinc combination under standard conditions

$$E^\ominus = V^\ominus(Zn^{2+}/Zn) - V^\ominus(H^+/H_2) = -0 \cdot 763 \text{ V}$$

where $V^\ominus(Zn^{2+}/Zn)$ is the shorthand notation for the zinc electrode potential and implies the reaction

$$Zn^{2+} + 2e^- = Zn(s)$$

Since $V^\ominus(H^+/H_2) = 0$, $V^\ominus(Zn^{2+}/Zn) = -0 \cdot 763 \text{ V}$

21. Tabulated standard electrode potentials. Electrode potentials provide a quantitative comparison of the strengths of oxidising and reducing agents. Several standard electrode potentials are shown in Table 18.

Oxidants appear on the left-hand side of the electrode reaction, and any oxidant in the table is stronger than all those above it, *i.e.* F_2 is the strongest and Li$^+$ the weakest. Conversely, reductants, which appear on the right-hand side of the electrode reaction, increase in strength in ascending the table, *i.e.* Li is the strongest and F^- the weakest.

TABLE 18. STANDARD ELECTRODE POTENTIALS AT 25°C

Electrode reaction	$V^\ominus$/V
$Li^+ + e^- = Li$	$-3\cdot045$
$Zn^{2+} + 2e^- = Zn$	$-0\cdot763$
$Fe^{2+} + 2e^- = Fe$	$-0\cdot44$
$Cr^{3+} + e^- = Cr^{2+}$	$-0\cdot41$
$Cd^2 + 2e^- = Cd$	$-0\cdot402$
$2H^+ + 2e^- = H_2$	0 (by definition)
$Cu^{2+} + e^- = Cu^+$	$+0\cdot153$
$AgCl + e^- = Ag \mid Cl^-$	$+0\cdot222$
$Hg_2Cl_2 + 2e^- = 2Hg + Cl^-$	$+0\cdot2676$
$Cu^{2+} + 2e^- = Cu$	$+0\cdot34$
$I_2 + 2e^- = 2I^-$	$+0\cdot536$
$Fe^{3+} + e^- = Fe^{2+}$	$+0\cdot771$
$Ag^+ + e^- = Ag$	$+0\cdot799$
$Cl_2 + 2e^- = 2Cl^-$	$+1\cdot36$
$F_2 + 2e^- = 2F^-$	$+2\cdot65$

The electrode reactions are frequently called "half-cell reactions" and the electrode potentials "half-cell potentials."

If the electrode reactions were written in the reverse direction, *i.e.* as oxidations, the signs of the electrode potentials would be reversed, *i.e.* $V^\ominus(Li^+/Li) = -3\cdot045$ V but $V^\ominus(Li/Li^+) = +3\cdot045$ V and is an oxidation potential.

22. Cell e.m.f.s from electrode potentials. The method may be seen with the aid of two examples.

EXAMPLE 1: Find $E^\ominus$ for the cell

$$Zn \mid Zn^{2+} \parallel Cu^{2+} \mid Cu$$

and write down the cell reaction

$$E^\ominus = V^\ominus(Cu^{2+}/Cu) - V^\ominus(Zn^{2+}/Zn)$$
$$= 0\cdot34 - (-0\cdot763) = +1\cdot103 \text{ V}$$

In a similar manner, the overall cell reaction is found by subtracting the right-hand electrode reaction from the left-hand electrode reaction (*both written as reductions*)

Right-hand electrode	$Cu^{2+} + 2e^- = Cu(s)$
Left-hand electrode	$Zn^{2+} + 2e^- = Zn(s)$

Overall reaction	$Cu^{2+} - Zn^{2+} = Cu(s) - Zn(s)$
Rearranging	$Cu^{2+} + Zn(s) = Cu(s) + Zn^{2+}$

Since $E^{\ominus}$ is positive, the cell reaction proceeds simultaneously as written. *It must be emphasised that, since the left hand electrode is the negative electrode, the actual electrode reaction is*

$$Zn(s) = Zn^{2+} + 2e^-$$

and the above convention of writing both electrode reactions as reductions is merely one of convenience for finding the cell e.m.f. and the cell reaction.

EXAMPLE 2: Find $E^{\ominus}$ for the cell

$$Ag \mid Ag^+ \parallel Cd^{2+} \mid Cd$$

and write down the cell reaction.

The electrode reactions (written as reductions) are

$$Cd^{2+} + 2e^- = Cd(s) \quad V^{\ominus}(Cd^{2+}/Cd) = -0{\cdot}402 \text{ V}$$
$$\text{. . . (XIV, 13)}$$
$$Ag^+ + e^- = Ag(s) \quad V^{\ominus}(Ag^+/Ag) = 0{\cdot}799 \text{ V . . . (XIV, 14)}$$

In order to balance the number of electrons transferred from one electrode to the other, (XIV, 14) must be multiplied by 2

$$2Ag^+ + 2e^- = 2Ag(s) \quad V^{\ominus}(Ag^+/Ag) = 0{\cdot}799 \text{ V}$$
$$\text{. . . (XIV, 15)}$$

It is important to note that $V^{\ominus}(Ag^+/Ag)$ is *not* multiplied by 2, since $V^{\ominus}$ is an intensive property and as such does not depend on the quantity of material present. Subtraction of (XIV, 15) from (XIV, 13) gives the complete cell reaction

$$Cd^{2+} + 2Ag(s) = Cd(s) + 2Ag^+$$
$$E^{\ominus} = -0{\cdot}402 - 0{\cdot}799 = -1{\cdot}201 \text{ V}$$

The negative value of $E^{\ominus}$ indicates that the cell reaction as written is forbidden and instead proceeds spontaneously in the reverse direction.

23. Summary of rules regarding electrode potentials

(*a*) The standard hydrogen electrode is arbitrarily assigned a potential of zero at all temperatures.

(*b*) All electrode reactions are written as reductions

$$\text{oxidant} + ne^- = \text{reductant}$$

(*c*) All electrode reactions that proceed to the right more readily than $H^+ + e^- = \frac{1}{2}H_2(g)$ are assigned a positive value and those that proceed less readily are given a negative value.

(d) The magnitude of the electrode potential is a quantitative measure of the tendency of the electrode reaction to proceed from left to right.

(e) If the direction of an electrode reaction is reversed, the sign of its electrode potential is reversed. However, when an electrode reaction is multiplied by a positive number, the value of its potential is unchanged.

24. The effect of concentration on electrode potentials.

For a single electrode reaction such as that involving a metal electrode and a solution of its own ions, i.e.

$$M^{n+} + ne^- = M$$

(XIV, 11) becomes $V = V^{\ominus} + \dfrac{RT}{nF} \ln \dfrac{a_{M^{n+}}}{a_M}$ . . . (XIV, 16)

$$= V^{\ominus} + \frac{RT}{nF} \ln \frac{a(\text{oxidised form})}{a(\text{reduced form})}$$

Since $a_M = 1$ and replacing activities by concentrations, (XIV, 16) may be written

$$V = V^{\ominus} + \frac{RT}{nF} \ln [M^{n+}]$$

For the ferrous–ferric system, where both the oxidised and reduced forms are in solution, the electrode reaction is

$$Fe^{3+} + e^- = Fe^{2+}$$

and the electrode potential is given by

$$V = V^{\ominus} + \frac{RT}{F} \ln \frac{a(Fe^{3+})}{a(Fe^{2+})}$$

In dilute solutions

$$V = V^{\ominus} + \frac{RT}{F} \ln \frac{[Fe^{3+}]}{[Fe^{2+}]}$$

25. The calomel reference electrode.

The hydrogen electrode is not a convenient practical standard because:

(a) The platinised platinum is very susceptible to poisoning both from the solution and from the gas, and this may cause the electrode to become irreversible.

(b) The potential of the electrode is affected by oxidants and reductants in the solution.

(c) The potential of the electrode is altered by changes in barometric pressure.

A subsidiary reference electrode is usually employed, whose potential has been accurately determined with respect to the hydrogen electrode (the primary standard). The most commonly used subsidiary reference electrode is the mercury-

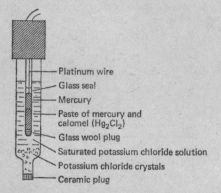

Platinum wire
Glass seal
Mercury
Paste of mercury and calomel (Hg_2Cl_2)
Glass wool plug
Saturated potassium chloride solution
Potassium chloride crystals
Ceramic plug

Fig. 57.—*Saturated calomel electrode*

mercurous chloride–potassium chloride electrode, which is known as the *calomel electrode*. Two types of calomel electrode are in general use and differ in the concentration of potassium chloride. For normal work the saturated calomel electrode containing saturated potassium chloride is used, while for accurate work an electrode containing 1 M potassium chloride solution is preferred, since its potential is less sensitive to changes in temperature. One form of the calomel electrode is shown in Fig. 57.

The electrode reaction is the sum of two reactions:

$$\tfrac{1}{2} Hg_2Cl_2(s) = \tfrac{1}{2}Hg_2^{2+} + Cl^-$$
$$\tfrac{1}{2} Hg_2^{2+} + e^- = Hg(l)$$

Net reaction: $\tfrac{1}{2} Hg_2Cl_2(s) + e^- = Hg(l) + Cl^-$

The electrode potential of the calomel electrode is the e.m.f. of the cell

$$Pt, \bar{H}_2 \mid H^+ \parallel KCl \mid Hg_2\overset{+}{Cl}_2, Hg$$

At 25°C the potential of the saturated calomel electrode is +0·244 V, and this value must be added to e.m.f.s measured using the saturated calomel electrode in order to bring them on to the standard hydrogen scale.

POTENTIOMETRIC TITRATIONS

26. Oxidation–reduction titrations. Oxidation–reduction (or redox) titrations may be followed by the measurement of cell potentials. As an example we will consider the titration of ferrous ions with ceric ions. The standard potential of the Fe^{3+}/Fe^{2+} electrode is +0·771 V, while that of the Ce^{4+}/Ce^{3+} electrode is +1·61 V. Thus when ceric ions are mixed with ferrous ions the latter will be oxidised according to the equation

$$Fe^{2+} + Ce^{4+} = Fe^{3+} + Ce^{3+} \quad . \quad . \quad . \quad (XIV, 17)$$

Consider the cell

$$\text{calomel } \overset{-}{\text{electrode}} \parallel Fe^{2+}, Fe^{3+} \mid \overset{+}{Pt}$$

The potential of the right-hand electrode is given by

$$V(Fe^{3+}/Fe^{2+}) = V^{\ominus}(Fe^{3+}/Fe^{2+}) + \frac{2·303\,RT}{F} \lg \frac{[Fe^{3+}]}{[Fe^{2+}]}$$

$$. \quad . \quad . \quad (XIV, 18)$$

and depends on the ratio $[Fe^{3+}]/[Fe^{2+}]$. The potential of the cell, $E = V(Fe^{3+}/Fe^{2+}) - V(\text{calomel})$, also depends on this ratio. Thus the concentration ratio may be determined by e.m.f. measurements.

27. The shape of the redox titration curve. The titration cell consists of a calomel electrode and a platinum electrode, both dipping into a solution of Fe^{2+} ions. Initially the ratio $[Fe^{3+}]/[Fe^{2+}]$ is very small (ideally zero). As Ce^{4+} ions are added, Fe^{2+} ions will be oxidised, the ratio $[Fe^{3+}]/[Fe^{2+}]$ will increase and the cell potential will change as reaction (XIV, 17) takes place. The ratio $[Fe^{3+}]/[Fe^{2+}]$ reaches a maximum value at the equivalence point of the titration and thereafter does not change appreciably. Up to the equivalence point, the potential of the right-hand electrode is given by (XIV, 18).

Beyond the equivalence point, the cell e.m.f. is determined by the Ce^{4+}/Ce^{3+} electrode, and the new cell diagram is

$$\text{calomel } \overset{-}{\text{electrode}} \parallel Ce^{3+}, Ce^{4+} \mid \overset{+}{Pt}$$

The ratio $[Ce^{4+}]/[Ce^{3+}]$ now controls the potential of the right-hand electrode and therefore that of the cell as a whole.

$$V(Ce^{4+}/Ce^{3+}) = V^{\ominus}(Ce^{4+}/Ce^{3+}) + \frac{2 \cdot 303\, RT}{F} \lg \frac{[Ce^{4+}]}{[Ce^{3+}]}$$

As excess Ce^{4+} ions are added, the ratio $[Ce^{4+}]/[Ce^{3+}]$ increases, causing an increase in the cell potential. The resulting titration curve is S-shaped (*see* Fig. 58), the abrupt change in cell potential being due to the change of the right-hand electrode from a Fe^{3+}/Fe^{2+} electrode to a Ce^4/Ce^{3+} electrode.

28. Feasibility of a redox titration. It is apparent from Fig. 58 that the two electrode potentials should be as far apart as possible. The minimum separation can be calculated as follows.

When equivalent amounts of reactants have been added, the reaction should be at least 99·9% complete. In the above reaction this means that, at equilibrium, the ratios $[Fe^{3+}]/$

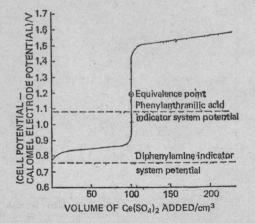

FIG. 58.—*Potentiometric titration curve for 1M FeSO₄ with 1M Ce(SO₄)₂*

$[Fe^{2+}]$ and $[Ce^{3+}]/[Ce^{4+}]$ will be $99·9/0·1 \approx 10^3$. The equilibrium constant for (XIV, 17) is given by

$$K_c = [Fe^{3+}][Ce^{3+}]/[Fe^{2+}][Ce^{4+}] = 10^6$$

This corresponds to a cell potential of 0·355 V. Since $V^\ominus(Ce^{4+}/Ce^{3+}) - V^\ominus(Fe^{3+}/Fe^{2+}) = 0·84$ V, it is apparent that the titration of Fe^{2+} by Ce^{4+} is quantitative and may be followed potentiometrically.

29. Redox indicators. A redox indicator is a substance which undergoes reversible oxidation and reduction and exhibits different colours in its oxidised and reduced forms. For example, diphenylamine exhibits the following equilibrium:

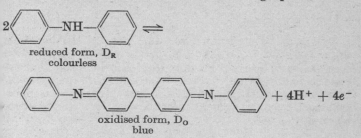

reduced form, D_R
colourless

oxidised form, D_O
blue

and the system has an electrode potential, $V^\ominus(D_O/D_R)$, of $+0·76$ V. For an indicator to be suitable, its electrode potential should lie on the vertical portion of the potentiometric titration curve (just like the pK values of indicators in normal acid–base titrations). Only a small amount of the redox indicator is added, so that the major redox reaction is barely disturbed. Under these conditions the indicator will adopt the potential of the system to which it is added and hence will follow the titration curve. If the potential of the system is less than 0·76 V the reduced form of the indicator is present, while if the potential is greater than 0·76 V the oxidised form will be present.

Diphenylamine cannot be used in the straightforward titration of Fe^{2+} with Ce^{4+}, since $V^\ominus(D_O/D_R)$ is less than $V^\ominus(Fe^{3+}/Fe^{2+})$ (*see* Fig. 58). However, the whole titration curve may be lowered by the addition of phosphoric acid, which forms phosphate complexes with Fe^{3+}. This reduces the concentration of free Fe^{3+}, thereby lowering the cell potential for the first part of the curve. Under these conditions the indicator system electrode potential lies on the vertical portion of the curve.

The phenylanthranilic acid indicator system has a redox potential of $+1.08$ V and can be used directly for the above titration (*see* Fig. 55).

30. The range of colour change of redox indicators. If the indicator equilibrium is represented by

$$I_{ox} + e^- \rightleftharpoons I_{red}$$

then the redox potential is

$$V(I_{ox}/I_{red}) = V^{\ominus}(I_{ox}/I_{red}) + \frac{2 \cdot 303\, RT}{F} \lg \frac{[I_{ox}]}{[I_{red}]}$$

$$\ldots \text{(XIV, 19)}$$

As with acid–base indicators, (*see* XIII, **16**), the visible colour change covers the range $[I_{ox}]/[I_{red}] = 10$ to $[I_{ox}]/[I_{red}] = 0 \cdot 1$. Substitution of these ratios into (XIV, 19) indicates that at 25°C for a one-electron transfer the potential will change from approximately $V^{\ominus}(I_{ox}/I_{red}) + 0 \cdot 06$ V to $V^{\ominus}(I_{ox}/I_{red}) - 0 \cdot 06$ V, *i.e.* the total change in potential for the observed colour change will be $0 \cdot 12$ V.

31. Precipitation titrations. In Progress test 12, 7 we saw how the precipitation reaction between sodium chloride and silver nitrate could be followed using conductance measurements. The same titration could be followed potentiometrically. The cell consists of a calomel electrode and a silver electrode. Any change in the cell potential will be due to changes in the concentration of silver ions around the silver electrode. Initially this concentration will be zero, but as silver nitrate is added, silver chloride will be precipitated and the solution will contain a small concentration of silver ions formed by the slight dissociation of silver chloride. This concentration will increase slightly as chloride ions are removed in order to maintain the solubility product $K_s = [Ag^+] [Cl^-]$. After the equivalence point the concentration of silver ions, and therefore the silver electrode potential, will rise very sharply owing to the presence of excess silver nitrate (*see* Fig. 59). The use of this technique for precipitation titrations is limited by the availability of suitable indicator electrodes.

Although all metal electrodes should in theory indicate the concentration of their own ions, only silver and mercury are suitable in practice.

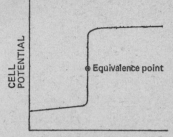

FIG. 59.—*Potentiometric titration curve for KCl with AgNO₃*

THE MEASUREMENT OF pH

One of the most important uses of cells is in the determination of the pH of solutions. The standard for all pH measurements is the hydrogen electrode, although from a practical point of view it is most unsatisfactory because of the care required in its operation. For most routine work it has been superseded by the glass electrode.

32. The hydrogen electrode. (*See also* **19** above.) The electrode reaction is

$$H^+ + e^- \rightleftharpoons \tfrac{1}{2} H_2(g)$$

$$V(H^+/H_2) = V^{\ominus}(H^+/H_2) + \frac{2\cdot303\,RT}{F} \lg \frac{a_{H^+}}{a_{H_2}^{\frac{1}{2}}}$$

For the hydrogen electrode $V^{\ominus}(H^+/H_2) = 0$

$$\therefore V(H^+/H_2) = \frac{2\cdot303\,RT}{F} \lg a_{H^+} - \frac{2\cdot303\,RT}{F} \lg a_{H_2}^{\frac{1}{2}}$$

$$\qquad\qquad\qquad\qquad \text{. . . (XIV, 20)}$$

The activity of hydrogen gas can be taken as unity if its partial pressure is $101\cdot325\,kN\ m^{-2}$ (*see* VIII, **26**), and (XIV, 20) simplifies to

$$V(H^+/H_2) = \frac{2\cdot303\,RT}{F} \lg a_{H^+}$$

By definition $pH = -\lg a_{H^+}$

$$\therefore V(H^+/H_2) = -\frac{2 \cdot 303\, RT}{F}\, pH$$

The hydrogen electrode is invariably combined with a calomel electrode to form the cell. The main difficulties with this electrode have been given in 25 above.

33. The quinhydrone electrode. (*See also* 4 above.) The electrode equilibrium is

$$Q + 2H^+ + 2e^- \rightleftharpoons QH_2$$

$$V(Q/QH_2) = V^{\ominus}(Q/QH_2) + \frac{2 \cdot 303\, RT}{2F} \lg \frac{a_Q\, a_{H^+}{}^2}{a_{QH_2}}$$

$$\qquad\qquad\qquad\qquad \ldots \text{(XIV, 21)}$$

In practice, quinone and hydroquinone are added in the form of the compound $Q . QH_2$ (quinhydrone), which is only sparingly soluble in water. Under these conditions in the saturated solution the activities of quinone and hydroquinone are equal and (XIV, 21) becomes

$$V(Q/QH_2) = V^{\ominus}(Q/QH_2) + \frac{2 \cdot 303\, RT}{2F} \lg a_{H^+}{}^2$$

$$\qquad\quad = V^{\ominus}(Q/QH_2) - \frac{2 \cdot 303\, RT}{F}\, pH$$

The quinhydrone electrode consists of a gold or platinum wire immersed in a solution of unknown pH saturated with quinhydrone. A calomel electrode also dips into the solution, and the cell is calibrated using a buffer solution of known pH. The quinhydrone electrode is unsatisfactory above pH 8 because of atmospheric oxidation of hydroquinone which alters the ratio a_Q/a_{QH_2}.

34. The glass electrode. This electrode depends on the fact that a potential difference is set up across a thin glass membrane separating two solutions of different pH. If one of these solutions has a constant pH, then the potential difference varies linearly with the pH of the other solution. The most convenient arrangement is to seal the solution of constant pH inside the membrane. Fig. 60 shows a typical glass electrode

consisting of a silver–silver chloride electrode dipping into a
0·1 M hydrochloric acid solution contained in a very thin
glass bulb. The electrode is immersed in the solution of un-

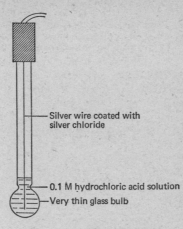

Silver wire coated with
silver chloride

0.1 M hydrochloric acid solution
Very thin glass bulb

FIG. 60.—*Glass electrode*

known pH and a calomel electrode completes the cell, which
may therefore be represented by

$$Ag, \overline{Ag}Cl \mid 0·1 \text{ M HCl} \mid \text{glass membrane} \mid \text{solution}$$
$$\parallel \overset{+}{\text{calomel}} \text{ electrode}$$

The potential of the glass electrode varies with the pH of the
unknown solution according to the equation

$$V(G) = V^{\ominus}(G) + \frac{2·303\,RT}{F}\,pH$$

$V^{\ominus}(G)$ is not a true constant, since it depends on the particular
glass electrode and may also vary with time. It is therefore
essential to calibrate the system, using a buffer solution of
known pH. Hence the glass electrode, like the quinhydrone
electrode, only *compares* pH values, while the hydrogen
electrode measures pH *absolutely*.

The potentials of cells which incorporate glass electrodes
cannot be measured using an ordinary potentiometer circuit

because of the very high resistance of the glass membrane ($\approx 100 \text{ M}\Omega$). A vacuum tube voltmeter is required, and the scale of this instrument is calibrated directly in pH units.

The advantages of the glass electrode are that it is simple to operate, is not easily poisoned, and it is not affected by strong oxidising or reducing agents. In highly alkaline solutions (pH > 13) special glass membranes must be used.

35. Acid–base titrations.

Titration curves such as those calculated in XIII, **15** and drawn in Figs. 52 and 53 can be determined experimentally using the glass electrode. The pH of the solution is given directly by the pH meter and the experimental curves are found to follow the calculated ones very closely. Potentiometric acid–base titrations enjoy the same advantages as conductimetric acid–base titrations (*see* XIII, **32**).

PROGRESS TEST 14

1. Draw the cell diagram and write out the individual electrode reactions for the cell reaction

$$Hg_2Cl_2(s) + H_2(g) = 2Hg(l) + 2H^+ + 2Cl^-$$

Given that under standard conditions the e.m.f. of the cell is $+0 \cdot 2676$ V and its variation with temperature is $-3 \cdot 2 \times 10^{-4}$ V K^{-1} calculate $\Delta G^\ominus$, $\Delta S^\ominus$ and $\Delta H^\ominus$ for the reaction. (**10, 12**)

2. Calculate the e.m.f. of the cell

$$Pb \mid Pb^{2+}(10^{-3} \text{ M}) \parallel Cu^{2+}(10^{-2} \text{ M}) \mid Cu$$

given $V^\ominus(Pb/Pb^{2+}) = -0 \cdot 126$ V and $V^\ominus(Cu/Cu^{2+}) = +0 \cdot 337$ V. (**14**)

3. In the cell, Ag $\mid$ Ag$^+$ $\parallel$ Cl$^-$ $\mid$ AgCl(s), Ag the standard electrode potentials of the right- and left-hand electrodes are $+0 \cdot 222$ V and $-0 \cdot 799$ V. Write down the cell reaction and hence calculate the solubility product of AgCl. (**17**)

4. The equilibrium constant for the reaction

$$MnO_4^- + 8H^+ + 5Fe^{2+} \rightleftharpoons Mn^{2+} + 5Fe^{3+} + 4H_2O$$

is 10^{63} at 25°C. Given that $V^\ominus(Fe^{3+}/Fe^{2+}) = +0 \cdot 771$ V, calculate $V^\ominus(MnO_4^-/Mn^{2+})$. (**17**)

5. Calculate the equilibrium constant for the reaction

$$Fe^{2+} + Ce^{4+} \rightleftharpoons Fe^{3+} + Ce^{3+}$$

at 25°C given $V^{\ominus}(Fe^{3+}/Fe^{2+}) = 0 \cdot 771$ V and $V^{\ominus}(Ce^{4+}/Ce^{3+}) = 1 \cdot 61$ V. Also, calculate the right-hand electrode potential at the equivalence point in the titration of Fe^{2+} with Ce^{4+} when the cell diagram is initially

calomel electrode || Fe^{2+}, Fe^{3+} | Pt   **(17, 26–27)**

6. Calculate the pH of solution x in the following cell

$$Pt,H_2 \mid \text{solution } x \mid\mid \text{saturated calomel electrode}$$

given that at 25°C, the partial pressure of hydrogen $= 101 \cdot 3$ kN m^{-2}, the cell potential $= -0 \cdot 073$ V and the saturated calomel electrode potential is $+0 \cdot 244$ V. **(32)**

7. Why are electrode potentials important? **(1)**

8. Describe the various types of reversible electrode and write down the electrode reaction in each case. **(2–4)**

9. How may cell e.m.f.s be measured? What is a reversible cell? **(8–9)**

10. Explain the convention used in writing down cell diagrams. **(10)**

11. Explain the purpose of a salt bridge. **(11)**

12. What is the relationship between $E^{\ominus}$ and (a) $\Delta G^{\ominus}$ **(12)**, (b) $\Delta S^{\ominus}$ **(12)**, (c) $\Delta H^{\ominus}$ **(13)**, (d) K **(17)**?

13. Derive the Nernst equation. **(14)**

14. What is meant by a standard electrode potential and why are they so important? What convention is employed in tabulating standard electrode potentials? **(18–22)**

15. Explain the principle involved in potentiometric oxidation–reduction titrations and sketch a typical titration curve. **(26–27)**

16. Explain the theory of redox indicators. **(29–30)**

17. Describe the principles behind three methods for measuring pH. **(32–34)**

PART SIX

ATOMIC AND MOLECULAR STRUCTURE

Chapter XV begins with a brief account of the atomic nucleus. The developments which have led to the presently accepted wave mechanical picture of the atom are then considered, and the structure of the periodic table is explained in terms of the arrangement of the extranuclear electrons. The chapter closes with an account of ionisation energies and electron affinities.

The elementary theory of molecular structure is outlined in the first part of Chapter XVI, while the latter half of the chapter deals with the ionic bond.

A wave mechanical approach is used in Chapter XVII to describe covalent bonding, and the hydrogen molecule is treated in some detail. The bonding and structures of some simple molecules are considered.

No chemical bond is completely ionic or completely covalent and intermediate types of bonding are discussed in Chapter XVIII. Intermolecular forces are also considered and the chapter closes with an account of metallic bonding.

CHAPTER XV

ATOMIC STRUCTURE

THE ATOMIC NUCLEUS

1. The Thomson model of the atom. Up to the end of the nineteenth century, atoms were considered to be rigid spheres, strictly indivisible and devoid of any internal structure. However, the discovery of the electron by J. J. Thomson in 1897 led him to the proposal of a structure consisting of a sphere of positive charge occupying the total volume of the atom with sufficient electrons to maintain electrical neutrality residing in discrete holes in this sphere. This model soon fell into disrepute and was replaced by Rutherford's nuclear model.

2. The Rutherford model of the atom. Rutherford proposed that the atom consisted of a heavy positively charged mass, called the *nucleus*, surrounded by sufficient electrons to make the whole unit electrically neutral. On this model the electrons revolve around the nucleus in orbits in much the same way as the planets revolve around the sun. The diameters of the nucleus and the atom are about 10^{-15} m and 10^{-10} m respectively and although the nucleus occupies only about $10^{-13}\%$ of the total volume of the atom, it accounts for some $99 \cdot 95\%$ of the mass. Later investigations have shown that the nucleus is essentially built up of two kinds of particle of roughly equal size, the positively charged proton of mass $m_p = 1 \cdot 672\ 52 \times 10^{-27}$ kg and the uncharged neutron of mass $m_n = 1 \cdot 674\ 84 \times 10^{-27}$ kg; the electron has a mass $m_e = 9 \cdot 109\ 1 \times 10^{-31}$ kg. The charges carried by the electron and proton are equal and opposite and of magnitude $1 \cdot 602\ 10 \times 10^{-19}$ C.

NOTE: Although the existence of the neutron was first postulated by Rutherford around 1910, it was not until 1932 that this particle was observed by Chadwick during the bombardment of beryllium by energetic particles.

3. Definitions

(a) *The atomic number.* The atomic number, Z, of an atom is defined as the number of protons in the nucleus. For electrical neutrality Z must also represent the number of extranuclear electrons, and since the chemical properties of an element depend almost exclusively on such electrons, Z fixes the position of the element in the periodic table.

(b) *The mass number.* All nuclei apart from hydrogen also contain a number of neutrons, N. The total number of protons and neutrons in a nucleus (collectively called nucleons) is the mass number A ($= Z + N$) and determines the mass of the nucleus.

The atomic numbers of the known elements range from 1 for hydrogen to 103 for the heaviest known "man-made" element, lawrencium, while neutron numbers range from 0 to 156.

4. Radioactive isotopes. (For a definition of isotopes see III, **10**.) Since the isotopes of a given element have the same atomic number, their properties are virtually identical except for the light elements, where the isotopes exhibit their greatest fractional difference in mass, and some differences are apparent. Certain combinations of protons and neutrons are unstable and undergo changes leading to the spontaneous disintegration of the nucleus, with the emission of particles and electromagnetic radiation. This phenomenon is known as *radioactivity* and such isotopes are said to be *radioactive*. Some radioactive isotopes occur naturally, but the vast majority are man-made.

5. Stable isotopes. The majority of naturally occurring isotopes are stable and persist indefinitely. Thus lead has four stable isotopes with mass numbers of 204 (1·48%), 206 (23·6%), 207 (22·6%) and 208 (52·3%), the figures in brackets indicating the percentage abundances. In general the stable isotopes of a given element occur together in constant proportions, and hence the relative atomic masses of such elements will be constant. However, some elements, such as boron, potassium and lead, show very slight variations in their relative atomic masses, depending on the source, due to differences in the percentage abundances of the component isotopes.

THE NATURE OF RADIATION

The physical and chemical properties of an element are almost exclusively determined by the number and arrangement of the extranuclear electrons. Before going on to consider these electrons we must first look at the *electromagnetic theory of radiation*, which was the basis of the physical sciences in the nineteenth century.

6. The electromagnetic theory of radiation. In 1864 Maxwell concluded that visible light is propagated through space by means of oscillating electric and magnetic fields. The oscillations of the two fields are at right angles to one another and to the direction of propagation of the light and may be represented pictorially by sinusoidal waves (Fig. 61(a)).

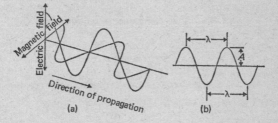

Fig. 61.—(a) *An electromagnetic wave.* (b) *The electric or magnetic component of an electromagnetic wave where* λ *is the wavelength and* A *the amplitude*

Visible light is only one form of electromagnetic radiation and other forms include X-rays, ultra-violet rays, infra-red rays and radio waves. All forms of electromagnetic radiation travel through vacuum with the same speed ($c = 2\cdot997\ 925 \times 10^8$ m s^{-1}) and are characterised either by their *wavelength*, λ, or their *frequency* ν. The wavelength may be defined as the distance between two successive maxima (Fig. 61 (b)), and the relationship between λ and ν is given by $\nu = c/\lambda$.

The complete electromagnetic spectrum is shown in Fig. 62.

7. The emission and absorption of radiation. The emission of radiation by a body is brought about by the oscillation of electrically charged particles within the body which give rise to the oscillating electric and magnetic fields by which the radiation is propagated. The frequency of the emitted radiation is determined by the frequency of the electric oscillators. Similarly, the absorption of radiation by a body induces oscillations of the electrically charged particles. All bodies

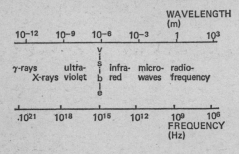

FIG. 62.—*The electromagnetic spectrum*

which are not at the absolute zero of temperature are continually emitting and absorbing radiation from their surroundings. According to the electromagnetic theory, this emitted or absorbed radiation is *continuous*, *i.e.* it is not divided into small units.

8. Black body radiation. The electromagnetic theory of light enables diffraction and interference phenomena to be satisfactorily interpreted and also explains why the speed of light is smaller the denser the medium through which it is travelling. However, in one area, that of black body radiation, the electromagnetic theory fails to explain the experimental observations.

When the relative amounts of energy radiated from different coloured surfaces are measured at a given temperature, it is found that a black surface radiates most energy. A black surface is also a better absorber of radiation than a coloured surface, and in general a good absorber of radiation is a good emitter of radiation and vice versa. This has led to the concept of a perfect absorber and emitter of radiation, which is usually called a *black body*. A black body would absorb all the radiation falling on it and, at a given temperature, emit the maximum amount of radiation that can be emitted. In practice, a small hole in the wall of a hollow sphere at constant temperature behaves as a black body.

The spectral distribution of the energy emitted by a black body at two different temperatures is shown in Fig. 63. The dashed line represents the spectrum predicted using the electromagnetic theory of radiation and is obviously not in accord with the experimental observations.

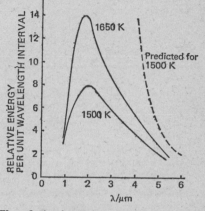

FIG. 63.—*The relative intensity of the radiation emitted from a black body at two temperatures*

9. The quantum theory. In 1900 Planck advanced the theory that energy can be taken up or given out only in discrete units, or *quanta*. The magnitude of the quantum of energy E_q is related to its frequency by the equation

$$E_q = h\nu \qquad . . . (XV, 1)$$

where h is Planck's constant ($6 \cdot 625\ 6 \times 10^{-34}$ J s). The quantum hypothesis was successful in explaining black body radiation, and an extension of the hypothesis enabled Einstein to interpret the photoelectric effect (the release of electrons when light is incident upon a clean metallic surface under vacuum).

10. The dual nature of light. Although the above effects provide decisive evidence in support of the quantum theory, diffraction and interference phenomena indicate that light also possesses wave properties. Both views are correct, and light has the dual property of wave and particle. In one experiment the wave nature of light may manifest itself, while in another a particulate view may provide the best representation. Equation (XV, 1) provides the link between the two theories, where the energy of the quantum of radiation is related to the wavelength of the radiation.

The dual nature of light can be extended to include matter such as neutrons and electrons (*see* **19, 20** below).

ATOMIC SPECTRA

11. The anomaly of the Rutherford nuclear atom. In the Rutherford nuclear atom the coulombic attraction between the orbiting electrons and the nucleus is balanced by the force due to centrifugal acceleration. According to classical theory, an electron rotating in an electric field such as that due to the nucleus must radiate energy continuously and eventually spiral into the nucleus. In practice, however, this does not occur.

12. Atomic spectra. Atoms will emit radiation if they are excited in some way, *i.e.* given more energy than they normally possess. One common method of producing excited atoms is to pass an electrical discharge through a gas such as hydrogen at low pressure. Radiation is emitted from the excited atoms produced by dissociation of the gaseous hydrogen molecules, and using a spectroscope it is possible to split up the radiation into its component wavelengths, producing a line spectrum which can be photographed (Fig. 64).

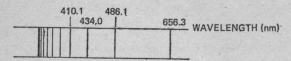

FIG. 64.—*The atomic spectrum of hydrogen in the visible region (Balmer Series)*

13. Atomic spectral series. The hydrogen atom spectrum consists of several series of lines, and the lines in any given series can be represented by the simple formula

$$\frac{1}{\lambda} = \sigma = R_H\left(\frac{1}{n_1{}^2} - \frac{1}{n_2{}^2}\right) \quad \ldots \text{(XV, 2)}$$

where σ is the wave number and R_H is the Rydberg constant for hydrogen ($1\cdot096\ 775\ 8 \times 10^7\ \mathrm{m}^{-1}$). n_1 can take only integral values and is constant for a given series; n_2 can take any integral value greater than n_1. The names of the various series (after their discoverers), the region of the spectrum where they appear and the values of n_1 and n_2 are shown in Table 19. Spectral series similar to, but more complex than, those obtained for hydrogen have been found in the atomic spectra of other elements.

TABLE 19   ATOMIC SPECTRAL SERIES OF HYDROGEN

Name of series	Region of spectrum	n_1	n_2
Lyman	Ultra-violet	1	2, 3, 4, etc.
Balmer	Visible	2	3, 4, 5, etc.
Paschen	Near infra-red	3	4, 5, 6, etc.
Brackett	Infra-red	4	5, 6, 7, etc.
Pfund	Far infra-red	5	6, 7, 8, etc.

The study of atomic spectra shows that radiation is emitted in a discrete manner from excited atoms, contrary to the predictions of the electromagnetic theory. Since every element produces a characteristic atomic spectrum determined by its atomic structure, any workable theory of the atom must explain atomic spectra. In 1913 Bohr put forward a revolutionary theory based on the quantum hypothesis, which was immediately successful in explaining the experimentally observed hydrogen atom spectrum.

THE BOHR THEORY OF THE HYDROGEN ATOM

14. The postulates. The Bohr model is based on the following postulates:

(a) An electron can rotate in a closed orbit in the potential field of the nucleus without radiating energy. Bohr called such orbits *stationary* states, since each has its own particular energy, and we now refer to them as energy levels.

(b) An electron can move from one orbit to another and in so doing either emit or absorb a quantum of radiation E_q, equal to the difference in energy ΔE, between the two orbits, *i.e.*
$$E_q = \Delta E = h\nu$$

(c) The permitted energy levels are determined by the condition that the angular momentum of the electron is an integral multiple of $h/2\pi$, *i.e.*
$$m_e v r_n = nh/2\pi$$

where v is the velocity of the orbiting electron, r_n the radius of the orbit corresponding to a given value of n, and n, called a *quantum number*, can only take integral values greater than 0.

It should be noted at this point that Bohr had no theoretical basis for his postulates, their main justification being that they gave the right answers (up to a point).

15. The energy of the electron. For simplicity Bohr regarded the electrons as moving in circular orbits, and derived the following expression for the energy of an electron in a given orbit of the hydrogen atom

$$E_n = -\frac{2\pi^2 m_e e^4}{(4\pi\epsilon_0)^2 h^2}\left(\frac{1}{n^2}\right) = -\frac{m_e e^4}{8\epsilon_0^2 h^2}\left(\frac{1}{n^2}\right) \; . \; . \; . \; \text{(XV, 3)}$$

where ϵ_0 is the permittivity of a vacuum ($8\cdot854\,185 \times 10^{-12}$ kg^{-1} m^{-3} s^4 A^2). For one electron species other than the hydrogen atom, the term Z^2 appears in the numerator.

Only certain energy levels are possible (the energy is said to be *quantised*) and these are given by the condition $n = 1, 2,$ $3,$ etc. The energies given by (XV, 3) are negative because the energy of the electron in the atom is less than the energy of the free electron, which is taken to be zero. Thus the smaller the value of n the more negative (*i.e.* the lower) the energy and the more stable the hydrogen atom. As n increases, E_n becomes less negative and in the limit when $n = \infty$, $E_\infty = 0$, which corresponds to complete separation of the electron and the nucleus.

16. Comparison with experiment. The lines in the atomic spectrum of hydrogen arise from electronic transitions between energy levels, and the theoretical frequencies corresponding to these transitions, say from E_2 to E_1, may be found by substitution into (XV, 3)

$$E_2 - E_1 = \Delta E = h\nu = \frac{m_e e^4}{8\epsilon_0^2 h^2}\left(\frac{1}{n_1^2} - \frac{1}{n_2^2}\right) \text{where } n_2 > n_1$$

Hence $\quad \sigma = \dfrac{\nu}{c} = \dfrac{m_e e^4}{8\epsilon_0^2 h^3 c}\left(\dfrac{1}{n_1^2} - \dfrac{1}{n_2^2}\right)$

This expression is analogous to that formulated from the experimentally observed spectra (XV, 2), and the constant term outside the bracket may be evaluated to $1\cdot097\,373$

$\times 10^7$ m^{-1}, which is in remarkably good agreement with the Rydberg constant for the hydrogen atom. This brilliantly confirms the original Bohr postulates.

NOTE: For an exact comparison, the motion of the nucleus must be taken into account and m_e should be replaced by the reduced mass, $\mu = m_p m_e / (m_p + m_e)$. When this is done, the calculated and observed values agree exactly.

17. The radius of an electron orbit. The radius of an electron orbit is given by

$$r_n = n^2 (\epsilon_0 h^2 / \pi m_e e^2) \qquad . \quad . \quad . \quad (XV, 5)$$

The term inside the brackets is constant and hence the radii of possible electron orbits are proportional to n^2. By combining (XV, 3) and (XV, 5) it can be seen that the energy of the electron (a negative quantity) is inversely proportional to the radius of its orbit. The larger the orbit the less negative (*i.e.* the greater) the energy of the electron. Some of the orbits are illustrated in Fig. 65, which also shows how the different

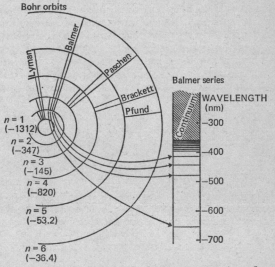

FIG. 65.—*The Bohr orbits of the hydrogen atom (not drawn to scale), indicating the transitions which give rise to the various spectral series. The energies of the orbits are shown beneath the n values*

spectral series originate. The electron is excited to a high-energy orbit, and this excitation energy is subsequently lost as the electron falls back to an orbit of lower energy. In the case of the Balmer series this is to the $n_1 = 2$ level, and the line spectrum produced is also shown in Fig. 65. The separation between the lines in the various series decreases with increasing values of n_2 until the continuum is reached and discrete lines are no longer observed.

The Bohr model has been applied with similar success to other one-electron systems such as He^+ and Li^{2+}.

WAVE MECHANICS

18. The inadequacies of the Bohr theory. Closer examination of atomic spectral lines show that they consist not of single lines, but rather of two or more closely spaced lines, *i.e.* they possess a fine structure. The simple Bohr theory was found to be inadequate to explain this observation, and although it was subsequently refined by introducing quantum numbers in addition to n, some spectral observations still could not be explained even in the relatively simple case of helium, with only two extranuclear electrons. It became clear that a new approach to atomic theory was necessary. Two important principles laid the foundations of this new approach, the *dual nature of matter* and the *Heisenberg uncertainty principle*.

19. The dual nature of matter. In 1924 De Broglie advanced the hypothesis that particles possess wave properties and that momentum is related to wavelength by the expression

$$mv = h/\lambda$$

The greater the mass and velocity of a particle the shorter its wavelength. The wave properties of macroscopic objects such as cricket balls can never be observed, but diffraction effects have been observed with electrons, protons, neutrons and even xenon atoms. The wavelengths of electrons and protons at room temperature are about 6 and 0·07 nm respectively. It should be noted that electrons, protons, etc., behave *either* as particles *or* as waves, depending on the property being observed.

20. The Heisenberg uncertainty principle (1925). This states that it is impossible to measure accurately both the position and the momentum of a moving particle.

NOTE: The very use of the term "particle" implies knowledge of momentum and position.

If Δx is the uncertainty in the measurement of position and Δp is the uncertainty in the measurement of momentum, then the product of the uncertainties is approximately equal to Planck's constant, i.e. $\Delta x \times \Delta p \approx h$. The importance of this equation can be illustrated by a rough calculation.

Suppose the position of an electron is known quite accurately, say to within 10^{-12} m. Then $\Delta p \approx h/\Delta x = 6 \cdot 6 \times 10^{-34}/10^{-12} = 6 \cdot 6 \times 10^{-22}$ kg m s^{-1}. Since the mass of the electron is $9 \cdot 1 \times 10^{-31}$ kg, the uncertainty in the velocity of the electron is given by

$$\Delta v = \Delta p/m_e = 6 \cdot 6 \times 10^{-22}/9 \cdot 1 \times 10^{-31} = 7 \cdot 2 \times 10^8 \text{ m s}^{-1}$$

which is of the same order as the velocity of light. The large uncertainty in the velocity arises because of the very small mass of the electron and would be quite negligible for macroscopic objects.

The consequences of the uncertainty principle are far-reaching, and *probability* takes the place of *exactness* in atomic theory. Thus statements regarding the *exact position and velocity* (which is related to kinetic energy) of an electron must be replaced by statements concerning the probability that the electron has a given position and velocity. The Bohr concept of the atom, which regards the electrons as rotating in definite orbits around the nucleus, must be abandoned and replaced by a theory which considers the probability of finding the electrons in a particular region of space.

21. The Schrödinger wave equation. Schrödinger applied the De Broglie expression, which is concerned with freely moving particles, to bound electrons in atoms, and derived an equation which is analogous to the equations which describe ordinary wave motion. This equation, which was formulated largely as a result of mathematical intuition, predicts the allowed energies of the bound electrons in atoms. For the motion of an electron in one direction, this equation is

$$\frac{d^2\Psi}{dx^2} + \frac{8\pi^2 m_e}{h^2}(E_T - E_p)\Psi = 0$$

where m_e and E_p, the mass and potential energy of the electron, are known, while E_T, the total energy of the electron, and Ψ (psi), the *wave function*, are unknown and are found by solution of the equation.

NOTE: In three dimensions the wave equation is

$$\frac{d^2\Psi}{dx^2} + \frac{d^2\Psi}{dy^2} + \frac{d^2\Psi}{dz^2} + \frac{8\pi^2 m_e}{h^2}(E_T - E_p)\Psi = 0$$

22. The significance of Ψ and Ψ^2. There are many different solutions of the wave equation Ψ_1, Ψ_2, Ψ_3, etc., each giving rise to a particlar value of E_T. The function Ψ is known as an *atomic orbital* and is related to the probability of finding the electron at a particular point. The physical meaning of Ψ can be seen if it is realised that Ψ is equivalent to the amplitude of a wave whose intensity, which is proportional to Ψ^2, gives the probability of finding the electron at various points in space. Thus $\Psi^2 dv$ may be taken as a measure of the probability of finding the electron in a small element of volume dv. An alternative, but completely equivalent approach, is to regard the electron as being spread out into a cloud of varying charge density. $\Psi^2 dv$ is then a measure of the electron charge density in dv. In both approaches it is possible to draw a boundary surface enclosing the regions of highest probability or density. Such a region is often loosely called an atomic orbital, but this term should strictly be reserved for the function Ψ rather than Ψ^2.

THE SCHRÖDINGER WAVE EQUATION APPLIED TO THE HYDROGEN ATOM

23. Agreement with the Bohr theory. The Schrödinger wave equation for the hydrogen atom gives satisfactory solutions for Ψ only when certain values of E_T are allowed, and moreover these are precisely the values predicted by the Bohr theory. At the same time, the additional quantum numbers which were proposed in the extension of Bohr's original theory to account for the fine structure of atomic spectral lines now appear as a necessity of the mathematics.

While agreeing with the Bohr theory for simple spectra, the wave mechanical approach is a tremendous advance. It is able to explain spectral observations which the older theory cannot, and in contrast to the older theory all the results can be derived from the one fundamental equation.

The four quantum numbers which emerge from the Schrödinger wave equation completely determine the behaviour of electrons in atoms.

24. The principle quantum number, n.

This number, which was first introduced by Bohr, completely determines the energy of the electron in a one-electron system such as the hydrogen atom, and is important in determining electron energies in multi-electron atoms. It can take only positive values from 1 upwards. The larger the value of n, the greater (*i.e.* the less negative) the energy.

25. The angular momentum quantum number, l.

This number determines the angular momentum of the electron, with the higher values of l corresponding to greater angular momentum. l also determines the shape of the orbital of the electron (*see* **28–32** below). For a particular value of n, l may take integral values from 0 to $n - 1$ inclusive. Orbitals with l values of 0, 1, 2, 3 are called s, p, d and f orbitals respectively. They are further classified by writing the value of the principle quantum number in front of the letter. Thus $2p$ refers to an orbital for which $n = 2$, $l = 1$, and an electron in such an orbital is a $2p$ electron.

26. The magnetic quantum number, m.

Because of its orbital motion, an electron generates an electric current and hence produces a magnetic field. The magnetic moment associated with this magnetic field is a vector quantity (since it possesses direction as well as magnitude), and can orientate itself in an externally applied magnetic field. Only certain orientations are possible, each giving rise to a definite energy level, and these are determined by the value of m. m retains its significance in the absence of an externally applied field, but in this case the electron energy levels are identical, and are said to be *degenerate*. m is restricted to integral values between

—1 and +1, including 0. Fig. 66 is a diagrammatic representation of the allowed orientations for the case where $l = 2$.

> NOTE: The above argument is somewhat simplified, since the orbital magnetic moment vector is not static but precesses about the axis of the externally applied magnetic field. The situation is analogous to the precession of a gyroscope in the earth's gravitational field.

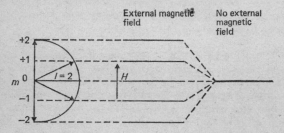

FIG. 66.—*The representation of angular momentum, indicating the allowed orientations of the magnetic moment vector. The degeneracy is removed by an external magnetic field and five definite energy levels are produced for $l = 2$*

27. The spin quantum number, s.

The spin of the electron about its own axis also generates a magnetic moment vector, which may be orientated in two ways with respect to any fixed axis. These two orientations are allowed for by specifying a fourth quantum number, s, which may take the values $+\frac{1}{2}$ and $-\frac{1}{2}$ and may be pictured as corresponding to clockwise and anti-clockwise spin respectively. The existence of electron spin was first postulated by Uhlenbeck and Goudsmit in 1925 to explain the appearance of closely spaced lines (doublets) in the spectra of alkali metal atoms. The original Schrödinger wave equation gave no indication of the existence of s, but subsequent modification has shown that s does indeed appear in the solution for the hydrogen atom.

> NOTE: The spin magnetic moment vector also precesses about the axis of an applied magnetic field.

THE SHAPES OF ATOMIC ORBITALS

28. Polar co-ordinates.

The three-dimensional solution of the Schrödinger wave equation involves three co-ordinates.

It is more convenient if these are spherical polar co-ordinates (r, θ, ϕ) instead of the more usual cartesian co-ordinates (x, y, z). The relationship between the two co-ordinate systems is shown in Fig. 67.

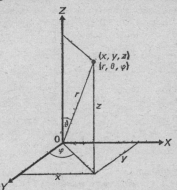

FIG. 67.—*The relationship between cartesian co-ordinates and spherical polar co-ordinates*

The solution of the Schrödinger wave equation can be expressed as the product of three functions

$$\Psi = R(r)\Theta(\theta)\Phi(\phi)$$

where $R(r)$ is a function of r alone, $\Theta(\theta)$ a function of θ alone and $\Phi(\phi)$ a function of ϕ alone. It is convenient to divide the wave function into two parts: the radial function $R(r)$ and the angular function $\Theta(\theta)\Phi(\phi)$. It is physically more significant to discuss the square of these functions, *i.e.* $R^2(r)$ and $\Theta^2(\theta)\Phi^2(\phi)$, since these represent the radial and angular probability distributions of electron density.

29. The radial probability distribution. Since the probability of finding an electron in a volume element dv is $\Psi^2 dv$ or $R^2(r)dv$ if only the radial part of the wave function is considered, the probability of its being at a distance between r and $r + dr$ from the nucleus, *i.e.* in a spherical shell of volume $4\pi r^2 dr$, is $4\pi r^2 R^2(r)dr$. The quantity $4\pi r^2 R^2(r)$ is known as the *radial distribution function*, and the variation of this function with distance from the nucleus for the $1s$, $2s$, $2p$, $3s$, $3p$ and $3d$

electrons of the hydrogen atom is shown in Fig. 68. The maxima in the radial distribution curves occur at distances from the nucleus where there are high probabilities of finding the electron in the spherical shell. For the $1s$ orbital of the

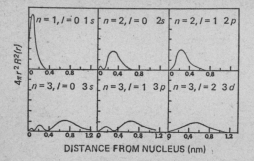

FIG. 68.—*The radial distribution functions of the hydrogen atom*

hydrogen atom, the distance of maximum probability occurs at 0·0591 nm, which corresponds to the radius of the first Bohr orbit (with $n = 1$). However, the electron has a finite probability of being closer to the nucleus or further away from it.

NOTE: The volume of a spherical shell close to the nucleus is very small and this leads to a small value for the radial distribution function. In contrast, the actual probability of finding the electron in a given volume is greatest close to the nucleus.

These "smeared out" pictures of electron charge distribution are in complete contrast to the Bohr theory.

30. The angular probability distribution. The angular probability distribution of electron density may be found by plotting $\Theta^2(\theta)\Phi(\phi)$ against both θ and ϕ simultaneously. Since there are two independent variables, the result is a three-dimensional closed surface. The resulting closed surfaces for all s electrons ($l = 0$) are spherically symmetrical and that for the $1s$ electron is shown in Fig. 69. The surfaces for electrons with l values greater than 0 do not possess spherical symmetry but have directional characteristics. There are three such surfaces for p electrons ($l = 1$), one for each allowed value of

m, and each has a dumb-bell shape, with the electron density being concentrated about an axis passing through the nucleus. The three equivalent surfaces are mutually perpendicular to one another and for convenience are considered to be directed along the x, y and z axes (Fig. 69). Fig. 69 also shows the

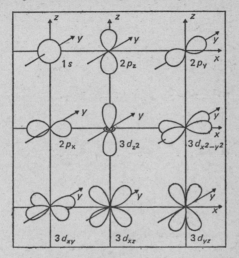

FIG. 69.—$\Theta^2(\theta)\Phi^2(\phi)$ *surfaces for the 1s, 2p and 3d electrons in the hydrogen atom*

diagrams of the angular probability distributions for the $3d$ electrons ($n = 3$, $l = 2$). There are five surfaces, one for each allowed value of m (0, ±1, ±2). Although it is not apparent from the diagram, these five surfaces are all equivalent and differ only in their orientation in space.

31. The total probability distribution. The *total* probability of finding an electron in a given volume dv, Ψ^2dv is obtained by multiplying $R^2(r)$ by $\Theta^2(\theta)\Phi^2(\phi)$. For an s electron, the total probability distribution has the same spherical symmetry as for the angular probability distribution. This is not the case for the p and d electrons, whose angular probability distributions have directional characteristics. Thus the angular probability distribution gives no indication of the change in electron density as the distance from the nucleus increases.

However, a reasonable, though approximate, physical picture of the total probability distributions for p and d electrons is obtained by assuming that they have the same shapes as the respective angular probability distributions. The enclosed surface then represents the volume where there is, say, a 95% chance of finding the electron.

32. The relationship between Ψ and Ψ^2. The total probability distribution (and its radial and angular components), which involves the square of the wave function, Ψ^2, gives a convenient physical picture of electron density. However, the actual orbitals involve the function Ψ and hence the surfaces drawn in Fig. 69 may not be identical with those for Ψ. In the case of s orbitals, the two boundary surfaces (Ψ and Ψ^2) have the same symmetrical shape, but this is not true for p and d orbitals, although the directional characteristics are the same. Fig. 70 compares the shape of the two surfaces for the $2p_x$ electrons. The angular component of Ψ will be used to represent diagrammatically covalent bonding between atoms in molecules (see XVII).

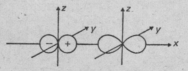

FIG. 70.—(a) $\Theta(\theta)\Phi(\phi)$ and (b) $\Theta^2(\theta)\Phi^2(\phi)$ surfaces for the $2p_x$ electrons

The two lobes of a p orbital are separated by a *nodal plane* (the yz plane in the case of the p_x orbital), for which $\Psi = 0$. The five degenerate d orbitals each have two nodal planes. Whenever a nodal plane is crossed, the sign of Ψ changes (see Fig. 70). These signs are important in the consideration of covalent bonding in molecules (see XVII). The sign of Ψ^2 is always positive.

MULTI-ELECTRON ATOMS

33. Comparison with the hydrogen atom. In the normal or ground state of the hydrogen atom, the single electron resides in the orbital of lowest energy ($1s$) and can occupy higher energy orbitals only if it is given excess energy. For multi-electron atoms, the higher energy orbitals are occupied, and we must consider the arrangement of electrons in these orbitals and the influence they exert on one another. Because of the additional nuclear charge and the interactions between electrons, the energy levels of the orbitals are changed, although their types and shapes still correspond to those previously described for the hydrogen atom.

34. The Pauli exclusion principle. An exact solution of the Schrödinger wave equation has been obtained for only one electron species. When more than one electron is present the situation is very complicated, although each electron may still be described by a set of four quantum numbers. The values of the quantum numbers which are allowed for a particular electron in an atom are determined by the *Pauli exclusion principle*, which states that *no two electrons can have the same set of values of all four quantum numbers*. There are therefore limits to the number of electrons which can be accommodated in the various orbitals, and the way these arise is shown in Table 20.

TABLE 20.   THE ACCOMMODATION OF ELECTRONS IN ORBITALS

Quantum numbers				Type of orbital	Total number of electrons	Shell letter
n	l	m	s			
1	0	0	$-\frac{1}{2}, +\frac{1}{2}$	$1s$	2	K
2	0	0	$-\frac{1}{2}, +\frac{1}{2}$	$2s$	2 } 8	L
	1	$-1, 0, +1$	$-\frac{1}{2}, +\frac{1}{2}$	$2p$	6	
3	0	0	$-\frac{1}{2}, +\frac{1}{2}$	$3s$	2	
	1	$-1, 0, +1$	$-\frac{1}{2}, +\frac{1}{2}$	$3p$	6 } 18	M
	2	$-2, -1, 0, +1, +2$	$-\frac{1}{2}, +\frac{1}{2}$	$3d$	10	
4	0	0	$-\frac{1}{2}, +\frac{1}{2}$	$4s$	2	
	1	$-1, 0, +1$	$-\frac{1}{2}, +\frac{1}{2}$	$4p$	6	
	2	$-2, -1, 0, +1, +2$	$-\frac{1}{2}, +\frac{1}{2}$	$4d$	10 } 32	N
	3	$-3, -2, -1, 0, +1, +2, +3$	$-\frac{1}{2}, +\frac{1}{2}$	$4f$	14	

35. The maximum number of electrons per value of n. When $n = 1$, only a single s orbital is available which can accommodate two electrons of opposite spin. When $n = 2$, both s and p orbitals are available. The former can accommodate two electrons as before, while the $2p_x$, $2p_y$ and $2p_z$ orbitals can each accommodate two electrons, making a total of eight electrons for $n = 2$. Using a similar procedure, the maximum possible numbers of electrons for $n = 3$, and $n = 4$, are seen to be 18 and 32 respectively. For any given value of n, the maximum possible number of electrons is $2n^2$. Electrons with the same value of n are said to belong to the same *shell*, and these shells are designated by capital letters (Table 20). The s, p, d and f levels within a shell are termed *sub-shells*.

THE PERIODIC TABLE

It is now possible to understand the structure of the periodic table in terms of the electronic configurations of the elements.

36. The *aufbau* principle. The elements beyond hydrogen may be thought of as being formed by the addition of (a) protons and neutrons to the nucleus and (b) extranuclear electrons. This is the so called *aufbau* or *building up principle*. The addition of electrons occurs in a definite way, governed by the following rules:

(a) The Pauli exclusion principle must be obeyed, so that no more than two electrons with opposite spins may occupy any given orbital.

(b) The allowed orbitals are occupied in order of increasing energy, *i.e.* $1s$, $2s$, $2p$, $3s$, $3p$, $4s$, $3d$, $4p$, $5s$, $4d$, $5p$, $6s$, $4f$, $5d$, $6p$, $7s$, $6d \approx 5f$. This order is governed by the penetration of the outer orbitals into regions of space already occupied by filled inner orbitals. After the $1s$ orbital is filled, two types of orbital are available for $n = 2$ ($2s$ and $2p$). The $2s$ orbital has a much greater degree of penetration into the $1s$ core than has the $2p$ orbital (*see* Fig. 68), hence a $2s$ electron is less well shielded from the positive charge on the nucleus than a $2p$ electron. This lowers the potential energy of a $2s$ electron relative to a $2p$ electron, and hence the former is more stable. Similar considerations account for the complete order of stability given above.

(c) For the p, d and f orbitals, the electron enters that orbital which is not already occupied by another electron. Obviously, from an electrostatic viewpoint the most stable arrangement is with the electrons as far apart as possible. In addition, the spins of any unpaired electrons are parallel, an arrangement which is energetically favoured. The foregoing statements constitute *Hund's first rule*.

37. Filling the s and p orbitals. The single electron in the ground state of the hydrogen atom must occupy the $1s$ orbital, giving rise to the electron configuration $1s^1$, where the superscript indicates the number of electrons. Helium, the next element in the periodic table, has two electrons both of which go into the $1s$ orbital to give the $1s^2$ configuration. These electrons must have opposite spins and are said to be *paired*. This completes the filling of the K shell.

TABLE 21. THE ELECTRONIC CONFIGURATION OF THE ATOMS OF THE ELEMENTS IN THEIR GROUND STATES

Element	Atomic number	K	L		M			N				O				
		$1s$	$2s$	$2p$	$3s$	$3p$	$3d$	$4s$	$4p$	$4d$	$4f$	$5s$	$5p$	$5d$	$5f$	
H	1	1														
He	2	2														
Li	3	2	1													
Be	4	2	2													
B	5	2	2	1												
C	6	2	2	2												
N	7	2	2	3												
O	8	2	2	4												
F	9	2	2	5												
Ne	10	2	2	6												
Na	11				1											
Mg	12				2											
Al	13	NEON			2	1										
Si	14				2	2										
P	15				2	3										
S	16	CORE			2	4										
Cl	17				2	5										
A	18				2	6										
K	19							1								
Ca	20	ARGON CORE						2								
Sc	21						1	2								

The filling of the L shell begins with lithium, $1s^2\ 2s^1$ and the $2s$ orbital is completed at beryllium $1s^2\ 2s^2$. Boron, with five electrons, has the configuration $1s^2 2s^2 2p^1$. For carbon, with six electrons, and nitrogen, with seven, the additional electrons go into the remaining unoccupied $2p$ orbitals, and this may be stressed by writing the configurations as $1s^2 2s^2 2p_x^1\ 2p_y^1$ and $1s^2 2s^2 2p_x^1 2p_y^1\ 2p_z^1$ respectively. Thus boron has one unpaired electron, carbon two and nitrogen three. Oxygen, with eight electrons, has the configuration $1s^2 2s^2 2p_x{}^2 2p_y{}^1 2p_z{}^1$. The filling of the $2p$ orbitals continues until neon, $1s^2 2s^2 2p^6$, where the L shell is complete. The electronic configurations of the first twenty-one elements in the periodic table are shown in Table 21. The $4s$ orbital is more stable than the $3d$ orbital, since it has greater penetration into the neon core, and hence for the nineteenth element, potassium, the electron goes into the $4s$ rather than the $3d$ orbital.

38. Filling the d and f orbitals. The filling of the $3d$ orbital begins once the $4s$ orbital is complete, starting with scandium, [Neon core] $3d^1 4s^2$, and ending nine elements later with zinc [Neon core] $3d^{10} 4s^2$. These ten elements constitute the first transition metal series, and the second and third transition metal series are formed by the filling of the $4d$ and $5d$ orbitals respectively. Two other series are important, arising from the filling of the $4f$ and $5f$ orbitals, and are called the lanthanide and actinide series respectively. Since the f orbitals can accommodate fourteen electrons, each series comprises fourteen members. The modern periodic table (Fig. 71) ends with the completion of the $5f$ orbitals at lawrencium.

39. Relationships in the periodic table. The similarities in the chemical properties of some groups of elements may be understood from a knowledge of their atomic structures. Thus the noble gases (group 0) owe their exceptional chemical stability to the fact that their electrons are in completely filled shells or sub-shells. The arrangement $ns^2 np^6$ occurs in all the noble gases except helium.

The alkali metals (group IA except hydrogen) all have a single electron outside a noble gas core. Since the size of the alkali metal atoms increases in descending the periodic table, their properties change in a fairly regular manner with increase in size.

In addition to such vertical relationships, horizontal relationships are also apparent. Thus the first transition metal series, formed by the filling of the $3d$ orbitals, are all typical metals, forming alloys with other metals, forming coloured ions and exhibiting variable valency (for a definition of valency *see* XVII). The lanthanide series, formed by the filling of the $4f$ orbitals, is also a closely allied group, whose members have very similar chemical and physical properties.

Period	Orbitals being filled	IA	IIA	IIIA	IVA	VA	VIA	VIIA	VIII	VIII	VIII	IB	IIB	IIIB	IVB	VB	VIB	VIIB	0
1	1s	1 H																	2 He
2	2s, 2p	3 Li	4 Be											5 B	6 C	7 N	8 O	9 F	10 Ne
3	3s, 3p	11 Na	12 Mg											13 Al	14 Si	15 P	16 S	17 Cl	18 Ar
4	4s, 3d, 4p	19 K	20 Ca	21 Sc	22 Ti	23 V	24 Cr	25 Mn	26 Fe	27 Co	28 Ni	29 Cu	30 Zn	31 Ga	32 Ge	33 As	34 Se	35 Br	36 Kr
5	5s, 4d, 5p	37 Rb	38 Sr	39 Y	40 Zr	41 Nb	42 Mo	43 Tc	44 Ru	45 Rh	46 Pd	47 Ag	48 Cd	49 In	50 Sn	51 Sb	52 Te	53 I	54 Xe
6	6s, 5d, 4f, 6p	55 Cs	56 Ba	57 [1] La	72 Hf	73 Ta	74 W	75 Re	76 Os	77 Ir	78 Pt	79 Au	80 Hg	81 Tl	82 Pb	83 Bi	84 Po	85 At	86 Rn
7	7s, 5f, 6d	37 Fr	88 Ra	89 [2] Ac															

Transition elements

s — *d* — *f* — *p*

Lanthanide Series [1]	58 Ce	59 Pr	60 Nd	61 Pm	62 Sm	63 Eu	64 Gd	65 Tb	66 Dy	67 Ho	68 Er	69 Tm	70 Lu	71 Yb
Actinide Series [2]	90 Th	91 Pa	92 U	93 Np	94 Pu	95 Am	96 Om	97 Bk	98 Cf	99 Es	100 Fm	101 Md	102 No	103 Lw

FIG. 71.—The periodic table of the elements

IONISATION ENERGY AND ELECTRON AFFINITY

40. Definition of ionisation energy. *The ionisation energy*, ΔH_I, is the minimum energy required to remove completely an electron from an isolated atom, ion or molecule. The energy required to remove the first electron is called the first ionisation energy and a larger amount of energy is required to remove subsequent electrons, since the effective nuclear charge has increased. An atom, ion or molecule has as many ionisation energies as there are electrons. The first ionisation energy is most important, and its variation with atomic number for the isolated gaseous atoms of the elements is shown in Fig. 72. All ionisation energies are taken by convention to be

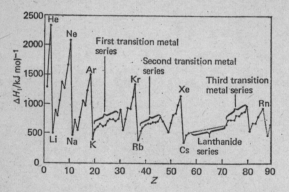

Fig. 72.—*The first ionisation energies of the elements as a function of atomic number*

positive, since energy must be supplied to remove the electron. For atoms the units of ionisation energy are eV atom^{-1} or J mol^{-1}. The *ionisation potential*, measured in volts, is numerically equal to the ionisation energy in eV atom^{-1}.

NOTE: For the relationship between the units see Progress Test 15, 2.

41. The periodicity exhibited by the ionisation energies. It is apparent from Figs. 71 and 72 that the periodicity exhibited by the ionisation energies closely parallels that shown in the periodic table—in fact the periodic table could be built up from a knowledge of ionisation energies. Among the many features that should be noted are:

(a) The key positions occupied by the alkali metals and the noble gases.

(b) The principal breaks in the ionisation energy curve fit in exactly with the periods in the periodic table.

(c) The vertical relationships observed between elements in the periodic table agree with their position on the ionisation energy curve.

(d) The similarities in the ionisation energies of the transition metal elements and the lanthanide elements suggest the horizontal relationships found in the periodic table.

42. The significance of ionisation energies.

The reason behind the periodicity shown by the ionisation energies lies in the arrangement of the electrons in the atoms. In particular, the inner shell electrons shield the full effect of the nuclear charge from the outer shell electrons, and hence the *effective* nuclear charge experienced by the outer electrons is much less than the *true* nuclear charge. The outer electrons are less well shielded if their orbitals can penetrate the inner shells.

The effect of shielding is most easily seen in the alkali metals, where the lone electron outside a closed shell is effectively shielded from the nuclear charge by the inner electrons and hence is easily removed. At the other end of the scale the high ionisation energies of the noble gases illustrate the fact that it is extremely difficult to remove an electron from a closed shell. An outer electron in a noble gas atom is shielded from the nucleus only by the inner shell electrons and to a lesser extent by the other electrons in the outer shell. The fine structure between the main maxima and minima in the ionisation energy curve may likewise be explained in terms of the nuclear charge and shielding. Thus, because the $2s$ electrons can penetrate the inner K shell better than the $2p$ electrons, (*i.e.* they have a higher probability of being close to the nucleus; *see* Fig. 68) the $2s$ electrons are more tightly bound and therefore more difficult to remove than the $2p$ electrons. This explains why boron, $1s^2 2s^2 2p^1$, has a lower ionisation energy than beryllium, $1s^2 2s^2$, even though the nuclear charge has increased by one.

43. Electron affinities. The electron affinity, ΔH_E, is the energy change when an isolated atom, ion or molecule and an electron, initially separated by an infinite distance, are brought together. The process involved is the reverse of that giving rise to ionisation energies. The sign of the electron affinity may be either positive or negative, the latter sign indicating that the species involved will readily accept the additional electron. As with ionisation energies, the magnitude (and now also the sign) of the electron affinity is governed by the arrangement of the electrons.

Halogen atoms readily accept an additional electron to yield a halide ion with the noble gas structure ($\Delta H_E \approx -350$ kJ mol^{-1}). The extra electron fills the vacancy in the p orbital and is only slightly shielded by the remaining electrons. The high ionisation energies of the noble gases indicate how tightly the p electrons are bound. The situation is very different for the alkali metal atoms, since the additional electron enters an s orbital outside the noble gas electronic configuration (and as such is not strongly bound) and low electron affinities result (≈ -60 kJ mol^{-1}). Similarly, the noble gases have essentially zero electron affinities, indicating that a lone electron outside the closed shell would be very weakly bound.

PROGRESS TEST 15

1. Write down the electronic configuration of the elements with the following atomic numbers: 3, 15, 26, 36, 51, 67, 92. (35–38)

2. Using the Bohr formula, calculate the ionisation energy for the hydrogen atom in the ground state in kJ mol^{-1} and in eV atom^{-1}. (16, 40)

3. Define (a) atomic number (3), (b) mass number (3), (c) ionisation energy (40), (d) electron affinity (43).

4. Outline (a) the Rutherford theory of the atom (2), (b) the electromagnetic theory of radiation (6), (c) the quantum theory (9).

5. What is meant by black body radiation? (8)

6. List the postulates of the Bohr theory of the hydrogen atom, and explain how this theory accounts for atomic line spectra. (14–17)

7. What two important principles were necessary to explain the inadequacies of the Bohr theory? (18–20)

8. Write down the Schrödinger wave equation and explain the meaning of the terms. (21–22)

9. What four quantum numbers arise from the solution of the Schrödinger wave equation and how are they related? (24–27)

10. Explain the difference between the radial probability distribution and the angular probability distribution and plot both distributions for the 1s and 2p orbitals of hydrogen. (29–32)

11. State (a) the Pauli exclusion principle (34), (b) the *aufbau* principle (36), (c) Hund's first rule (36)

12. Explain the significance of the periodicity exhibited by the ionisation energies of the atoms of the elements. (41–42)

CHAPTER XVI

INTRODUCTION TO MOLECULAR STRUCTURE; THE IONIC BOND

Except for the noble gases, atoms do not normally exist as separate entities but rather they combine to form molecules. Any theory of molecular structure must explain:

(a) How and why atoms combine.
(b) The distances between atoms.
(c) The strength of the forces between atoms.
(d) The spatial distribution of atoms.

Why is the distance between carbon atoms 0·154 nm in ethane, 0·135 nm in ethylene and 0·121 nm in acetylene? Why are the three atoms in CO_2 arranged linearly while those in H_2O are bent at an angle of about 105°? The theories which provide the answers to these questions are outlined in the following sections.

ELEMENTARY CONCEPTS

1. Molecular and structural formulae. The allocation of the correct molecular formula to a compound requires knowledge of the atomic composition and the relative molecular mass of the compound. Thus water may be represented by H_2O, sodium chloride by NaCl and acetic acid by $C_2H_4O_2$.

Two completely different compounds may have the same molecular formula, *e.g.* ethanol and dimethyl ether both have the molecular formula C_2H_6O. However, sodium reacts with ethanol to replace one of the hydrogen atoms and therefore ethanol is better represented by the formula C_2H_5OH, which stresses the fact that one of the hydrogen atoms is different from the remainder. Similar deductions resulted in the development of structural formulae, in which ethanol and dimethyl ether are represented by

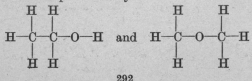

The lines in the structural formulae represent chemical bonds between the atoms concerned.

2. Valency. The term was initially introduced as a measure of the combining capacity of an atom, *i.e.* the number of atoms of hydrogen or chlorine with which an atom would combine. A more specific meaning of the term is to denote the number of bonds or lines radiating from a particular atom in a structural formula. Thus we speak of a carbon atom having a valency of four, an oxygen atom a valency of two, etc. This convention introduces the need for double and triple bonds for compounds such as ethylene, C_2H_4, and acetylene C_2H_2

$$\begin{array}{c} H \\ H \end{array}\!\!\!\!\diagdown C\!\!=\!\!C\!\!\diagup\begin{array}{c} H \\ H \end{array} \qquad H\!\!-\!\!C\!\!\equiv\!\!C\!\!-\!\!H$$

The presence of these multiple bonds is shown by the marked chemical reactivity of ethylene and acetylene when compared with ethane, which has only single bonds.

There is another, more modern meaning of valency, namely to describe the type of force existing between atoms in a molecule. It is unfortunate that the term "valency" has come to have two rather different meanings, but both are well established.

THE ELEMENTARY ELECTRONIC THEORY OF VALENCY

3. The noble gas electronic configuration. With the advent of the electronic theory of atomic structure, attempts were made to explain the concept of valency in terms of electrons. The foundations of the electronic theory of valency were laid independently by Kossel and by Lewis in 1916. The former was concerned with the transfer of electrons between atoms and the latter with the sharing of electrons by atoms, but both theories depend on the experimental observation that the noble gases are monatomic and chemically unreactive. As we have seen in XV, **39**, the reason for this exceptional stability lies in the presence of the octet grouping ns^2np^6 in the outer shells of these atoms (except for helium). This stable grouping must also be the reason behind the monatomic nature of these gases, and Kossel and Lewis reasoned that atoms would tend to attain this grouping by loss or gain of electrons. This may be done in two ways, giving rise to two types of bond.

(*a*) *The ionic or electrovalent bond,* in which an atom completely transfers one or more electrons to another atom.

(*b*) *The covalent bond,* in which two atoms share one or more pairs of electrons.

4. The ionic bond. The ionic bond may be illustrated by the alkali metal halides. The alkali metals and the halogens are positioned on either side of a noble gas in the periodic table, the halogens having one electron fewer than and the alkali metals one electron in excess of the noble gas octet of outer electrons. By the transfer of one electron from the metal atom to the halogen atom both atoms become charged ions, each with the noble gas electronic configuration. For example, the combination of potassium and chlorine atoms may conveniently be represented by the electron dot convention

$$K \cdot + \cdot \overset{\cdot\cdot}{\underset{\cdot\cdot}{Cl}} \colon = K^+ + \colon \overset{\cdot\cdot}{\underset{\cdot\cdot}{Cl}} \colon^-$$

where only the outermost or *valency* electrons are shown. The ions, each with the electronic configuration of an argon atom, are held together by electrostatic attraction. The presence of ions in fused potassium chloride is readily demonstrated by its high electrical conductivity.

5. The covalent bond. An example of covalent bonding is provided by the formation of molecular chlorine from its constituent atoms.

$$\colon \overset{\cdot\cdot}{\underset{\cdot\cdot}{Cl}} \cdot + \cdot \overset{\cdot\cdot}{\underset{\cdot\cdot}{Cl}} \colon = \colon \overset{\cdot\cdot}{\underset{\cdot\cdot}{Cl}} \colon \overset{\cdot\cdot}{\underset{\cdot\cdot}{Cl}} \colon$$

Each chlorine atom contributes one electron to the shared pair and in so doing attains the noble gas structure. The shared pair of electrons constitutes the chemical bond which we previously represented by a line in the structural formula. The electron dot picture for ethanol is

$$\begin{matrix} & H & H & & \\ H & \colon \overset{\cdot\cdot}{C} & \colon \overset{\cdot\cdot}{C} & \colon O & \colon H \\ & H & H & & \end{matrix}$$

6. The dative bond. This is a variation on normal covalency where the shared pair of electrons is contributed by one atom. An example is the molecule NH_3BF_3

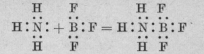

The nitrogen atom is called the *donor* and the boron atom the *acceptor*. This donation of a pair of electrons is often stressed by representing a dative bond by an arrow pointing in the direction of the donation, *e.g.* $H_3N \rightarrow BF_3$.

When a dative bond is formed between uncharged species some degree of charge separation occurs, since the donor atom effectively loses one electron to the acceptor. Hence an alternative method of indicating that dative bond is $\overset{+}{H_3N} - \overset{-}{BF_3}$. The unshared pair of electrons on the nitrogen atom of ammonia is known as the *lone pair*.

ENERGETICS OF IONIC BOND FORMATION

Since ions are formed by the complete transfer of electrons between atoms, the formation of ionic compounds is related to the ease with which ions can be produced from neutral atoms. The formation of an ionic bond is favoured when the energy required for cation formation (the ionisation energy) is low and the energy liberated in anion formation (the electron affinity) is high. The energy changes involved in ionic bond formation may be readily understood with the aid of an example. Let us consider first the formation of Na^+Cl^- ion pairs from sodium and chlorine atoms in the gas phase.

7. The formation of $Na^+(g)$ and $Cl^-(g)$. The energies involved in the formation of $Na^+(g)$ and $Cl^-(g)$ from their gaseous atoms are the ionisation energy and the electron affinity respectively.

$$Na(g) = Na^+(g) + e^-\ \Delta H_I = 494\ \text{kJ} \quad \text{. . . (XVI, 1)}$$
$$e^- + Cl(g) = Cl^-(g) \qquad \Delta H_E = -365\ \text{kJ . . . (XVI, 2)}$$

In the formation of these ions $494 - 365 = 129\ \text{kJ mol}^{-1}$ of energy has to be supplied to the system, *i.e.* $Na^+(g)$ and

$Cl^-(g)$ are unstable with respect to $Na(g)$ and $Cl(g)$. However, when $Na^+(g)$ and $Cl^-(g)$ combine to form an ion pair, an additional energy must be considered, namely the coulombic potential energy which stabilises the ion pair.

8. The coulombic potential energy. Let us consider the two gaseous ions, initially separated by an infinite distance and then allowed to approach one another. The potential energy of the system will vary as illustrated in Fig. 73. Zero energy

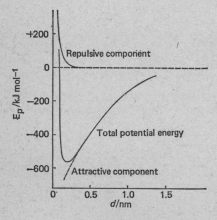

FIG. 73.—*The potential energy diagram for $Na^+ + Cl^-$*

in this diagram represents the energy of the system when the ions are separated by an infinite distance. As the two ions are brought together the potential energy decreases because of increasing coulombic attraction. At very small internuclear separations the potential energy rises very sharply as the two electron clouds repel one another. A stable situation exists when the attractive and repulsive forces are balanced, at the equilibrium bond distance, and the curve passes through a minimum (the so-called "potential well").

From Fig. 73 it may be seen that the potential energy of the system at the equilibrium bond distance is effectively the same as the coulombic attraction energy. This energy may be calculated for an isolated Na^+Cl^- ion pair on the assumption that $Na^+(g)$ and $Cl^-(g)$ are point charges and using the known bond distance in $Na^+Cl^-(g)$ of 0·238 nm.

Coulombic potential energy $= e^2/4\pi\epsilon_0 d$
$$= -(1\cdot602 \times 10^{-19})^2/(1\cdot113 \times 10^{10} \times 10^{-9})$$
$$= -9\cdot68 \times 10^{-19} \text{ J (ion pair)}^{-1}$$

For one mole of ion pairs this energy will be $-9\cdot68 \times 10^{-19} \times 6\cdot023 \times 10^{23} = -583$ kJ mol^{-1}. The negative sign indicates that the energy of $Na^+Cl^-(g)$ is less than that of the separate gaseous ions. Hence the potential energy gained by the system on bond formation must be added to the energy changes involved in (XVI, 1) and (XVI, 2). The energy released when one mole of $Na^+Cl^-(g)$ is formed will be $494 - 365 - 583 = -454$ kJ. Thus even though energy is required to form ions from the gaseous atoms, this is more than compensated by the gain in energy on formation of the ionic bond.

9. The lattice energy. The above example, although instructive, is artificial, since sodium chloride does not exist as ion pairs under normal conditions; nor do sodium and chlorine exist as gaseous atoms. The normal stable form of sodium chloride is an ionic solid consisting of an infinite array, or lattice (*see* XIX, **4**, **14–17**), of positive and negative ions, while sodium is a solid and chlorine a diatomic gas. In the case of an ionic solid, the simple coulombic potential energy is replaced by the *lattice energy*, which allows for all the interactions (attractive and repulsive) between the ions in the lattice. The lattice energy, ΔH_L, will be the sum of a very large number of terms (equal to the total number of ions in the crystal) and may be represented by

$$\Delta H_L = -MLe^2/4\pi\epsilon_0 d \quad . \quad . \quad . \quad \text{(XVI, 3)}$$

where M, the Madelung constant, depends on the geometrical arrangement in the crystal and for sodium chloride is 1·75. d, the ionic separation in the crystal, is 0·28 nm, a somewhat larger value than for the gaseous ion pair (0·238 nm). ΔH_L may be evaluated as -863 kJ mol^{-1} for $Na^+Cl^-(s)$.

Equation (XVI, 3) is an over-simplification of the situation and should be modified to take into account the fact that:

(*a*) Ions have a finite size and are not point charges.

(*b*) Short-range repulsive forces exist between neighbouring ions due to the interaction of electron clouds.

The modified value of ΔH_L is -766 kJ mol^{-1}. Fig. 74 indicates the relative stabilities of the various species we have been discussing and shows why sodium chloride exists as a solid at room temperature.

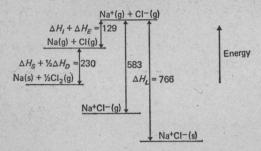

$$\Delta H_I + \Delta H_E = 129$$
$$Na^+(g) + Cl^-(g)$$
$$Na(g) + Cl(g)$$
$$\Delta H_S + \tfrac{1}{2}\Delta H_D = 230$$
$$Na(s) + \tfrac{1}{2}Cl_2(g)$$
583
$$\Delta H_L = 766$$
Energy
$$Na^+Cl^-(g)$$
$$Na^+Cl^-(s)$$

FIG. 74.—*Stabilisation energy for the Na^+Cl^- system in kJ mol^{-1}*

10. The Born–Haber cycle for Na^+Cl^-(s). The reaction

$$Na(s) + \tfrac{1}{2}Cl_2(g) = NaCl(s) \quad \Delta H_f \quad . \quad . \quad . \quad (XVI, 4)$$

may be considered to take place directly or through a series of intermediate reactions which comprise a *Born–Haber cycle*.

NOTE: This cycle is a special case of Hess's law of heat summation (*see* VII, 3).

$$Na(s) = Na(g) \qquad \Delta H_S = 109 \text{ kJ}$$
$$. \quad . \quad . \quad (XVI, 5)$$
$$\tfrac{1}{2}Cl_2(g) = Cl(g) \qquad \tfrac{1}{2}\Delta H_D = 121 \text{ kJ}$$
$$. \quad . \quad . \quad (XVI, 6)$$
$$Na(g) = Na^+(g) + e^- \qquad \Delta H_I = 494 \text{ kJ}$$
$$. \quad . \quad . \quad (XVI, 7)$$
$$e^- + Cl(g) = Cl^-(g) \qquad \Delta H_E = -365 \text{ kJ}$$
$$. \quad . \quad . \quad (XVI, 8)$$
$$Na^+(g) + Cl^-(g) = NaCl(s) \qquad \Delta H_L = -766 \text{ kJ}$$
$$. \quad . \quad . \quad (XVI, 9)$$

Addition of (XVI, 5), (XVI, 6), (XVI, 7), (XVI, 8) and (XVI, 9) yields (XVI, 4) and

$$\Delta H_f = \Delta H_S + \tfrac{1}{2}\Delta H_D + \Delta H_I + \Delta H_E + \Delta H_L$$
$$= 109 + 121 + 494 - 365 - 770 = -411 \text{ kJ mol}^{-1}$$
$$. \quad . \quad . \quad (XVI, 10)$$

NOTE: ΔH_S and ΔH_D are the heats of sublimation and dissociation respectively.

In this equation ΔH_E and ΔH_L are sufficiently negative to overcome the contribution from the other three positive terms and to make ΔH_f large and negative, indicating that NaCl(s) is more stable than its component atoms. ΔH_f may also be determined experimentally (see VII, 12) and is in good agreement with the above value. Of the quantities given in (XVI, 10), ΔH_f, ΔH_S, ΔH_D and ΔH_I may be determined experimentally and ΔH_L calculated using (XVI, 3). Hence the equation is often used to determine ΔH_E, a quantity which is not easy to measure directly.

11. The Born–Haber cycle for multiply charged ions. The formation of cations such as $Ca^{2+}(g)$ and $Al^{3+}(g)$ requires the expenditure of a much larger amount of energy (1738 kJ mol^{-1} for $Ca^{2+}(g)$) than is required in the formation of $Na^+(g)$. The major factor which makes ΔH_f sufficiently negative for $CaCl_2$ and AlF_3 to exist as stable ionic solids at room temperature is the lattice energy, which is very much greater for multiply charged than for singly charged ions.

12. The Born–Haber cycle for hypothetical ionic compounds. The Born–Haber cycle also enables us to understand why di- and trivalent metals fail to form stable ionic compounds with low valency states, e.g. sub-halides CaCl, sub-oxides AlO, etc. Question 1 of Progress Test 16 is an example of a Born–Haber calculation for just such a hypothetical ionic compound.

13. The limitations of ionic bonding. Ionic bonding is limited to those atoms that achieve a stable electronic configuration by losing one to three electrons or by gaining one or two. This is because the energy required for cation formation increases while the energy released in anion formation decreases with the increasing charge of the ion, and the lattice energy cannot outweigh the positive contributions to the heat of formation. Simple cations and anions rarely have charges of more than $+3$ and -2 respectively.

Ionic bonding is not confined to those elements which in gaining or losing electrons attain noble gas electronic configurations. The important considerations are the energy changes involved. Thus the elements of the first transition series readily form M^{2+} and M^{3+} ions, although the noble gas electronic configuration is not attained.

COVALENT AND IONIC RADII

X-ray diffraction techniques (XIX, 24–33) enable inter-nuclear distances to be measured in crystals and such distances may be thought of as representing the sum of the radii of the individual atoms or ions. Because of the diffuse nature of the electron cloud it is not possible to state precisely the size of an atom or ion. Nevertheless, it is very convenient to have self-consistent atomic and ionic radii, even though the values for a particular atom or ion will vary slightly from compound to compound. The size of an atom or ion is determined by two factors which are closely related: (a) the electronic configuration, (b) the effective nuclear charge.

14. Covalent or atomic radii. In general, the radii of neutral atoms decrease from left to right along a period in the periodic table, until a noble gas is reached, when there is a sudden increase, e.g.

Li 0·134, Be 0·090, B 0·082, C 0·077, N 0·075, O 0·073, F 0·072, Ne 0·131.

All radii are given in nm. While an outer shell is being filled, the principal effect is the increase in the effective nuclear charge which causes the decrease in the radius. However, the completion of the outer shell produces a marked increase in the radius.

Covalent radii increase on descending any group in the periodic table as might be expected from the increasing number of electron shells.

15. Ionic radii. When a neutral atom gains an electron to yield the corresponding anion the size increases because of the decrease in the effective nuclear charge. Conversely, a cation is smaller than the neutral atom from which it was formed because of the increase in the effective nuclear charge. This effect can be readily seen from the following comparison:

O   0·073, F   0·072, Ne 0·112, Na   0·154, Mg   0·130
O^{2-} 0·140, F^- 0·136, Na^+ 0·095, Mg^{2+} 0·065

The electron(s) added to form the anions complete a partially filled outer shell, and it is the completion of this shell rather than the decrease in the effective nuclear charge which determines the size of the anions. The formation of Na^+ and Mg^{2+} from the corresponding neutral atoms results in the disappearance of a whole shell. The remaining electrons are held more tightly to the nucleus and hence cations are much smaller than their neutral atoms. Although a sodium atom is more than twice as large as a fluorine atom, in sodium fluoride the fluoride ion is much larger than the sodium ion.

The effect of nuclear charge on the size of an ion can be most readily seen by comparing the members of an *isoelectronic* series. An example of such a series is O^{2-}, F^-, Ne, Na^+, Mg^{2+}, where each member has ten electrons but the nuclear charge increases regularly from $+8$ for O^{2-} to $+12$ for Mg^{2+}. The decrease in radius with increase in nuclear charge is shown above.

PROGRESS TEST 16

1. Using the following data, indicate whether or not CaF will exist as an ionic solid at room temperature:

$\Delta H_S(Ca) = 153 \cdot 5$, $\Delta H_D(F_2) = 151 \cdot 4$, $\Delta H_I(Ca) = 590$, $\Delta H_E(F) = -349$, $\Delta H_L(CaF) = -803$. All energy changes are in kJ mol^{-1}. (10)

2. Explain the meaning of valency. (2)
3. Outline the elementary electronic theory of valency and give examples of the two types of valency. (3–6)
4. Using sodium chloride as an example, explain what energetic factors are important in ionic bond formation. (7–10)
5. Discuss covalent and ionic radii. (14, 15)

CHAPTER XVII

THE COVALENT BOND

The simple electrostatic approach used in the previous chapter cannot explain the strong bonding in diatomic molecules formed from like atoms. Lewis considered that such *homonuclear* molecules were held together by covalent bonds formed by the sharing of electrons, but there is no inherent reason why the sharing of electrons should give rise to strong chemical bonds. This problem had to await the advent of wave mechanics for its resolution.

THE GENERAL PRINCIPLES

To illustrate the general principles involved, we shall consider the application of wave mechanics to the simplest of all molecules, the hydrogen molecule. Even in this case the mathematics involved is extremely complex and resort must be made to approximate methods.

1. The potential energy curve. The potential energy curve in Fig. 75, which represents the approach of two hydrogen atoms to form the hydrogen molecule, is similar to that in Fig. 73, even though we are not now concerned with charged species. In the hydrogen molecule the lowering of potential energy is caused by the electrostatic interaction of the electron on one nucleus with the other approaching nucleus.

NOTE: In the case of ions electron cloud interactions caused an increase in the potential energy of the system.

The electrons are considered as belonging to the system as a whole and the repulsive forces between the positively charged nuclei limit the distance of closest approach. Wave mechanics indicates that molecules, like atoms, can exist only in certain states, and the smallest amount of energy a diatomic molecule can possess is $h\nu_0/2$, the zero point energy (*see also* IV, **24**). This zero point energy is shown in Fig. 75, along with other

permitted vibrational energy levels. Because of this inherent
vibration the separation of atoms is not constant but is usually
represented by d_e, the separation at the minimum in the potential energy curve. The energy required to break the bond, the

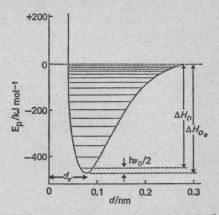

FIG. 75.—*The potential energy diagram for the hydrogen
molecule*

bond dissociation energy, ΔH_D, measures the bond strength
and corresponds to the vertical distance between the zero
point energy and the energy at infinite separation ($E_p = 0$).
The depth of the potential energy well is given the symbol
ΔH_{De} and $\Delta H_{De} = \Delta H_D + h\nu_0/2$.

Any theory of covalent bonding must be able to predict
ΔH_{De} and d_e. For the hydrogen molecule these are 455 kJ
mol^{-1} ($\Delta H_D = 435$ kJ mol^{-1}) and 0·074 nm respectively. Two
main approaches have been made to this problem, but before
going on to discuss these we must mention two important
properties of wave functions which greatly simplify the mathematics involved.

2. The variation principle. The exact wave function describing the behaviour of electrons in molecules is very complex.
However, if we can obtain simpler wave functions which
approximate to the exact wave function, then the *variation
principle* states that the best simple wave function will be that
which gives rise to the lowest energy. These simple wave

functions may often be obtained by intuitive guesswork, and then adjusted until the associated energy is a minimum.

3. The linear combination principle. If Ψ_1, Ψ_2, Ψ_3, etc., are approximate solutions of the molecular wave function, then the *linear combination principle* states that any linear combination is also a solution, *i.e.*

$$\Psi = N(c_1\Psi_1 + c_2\Psi_2 + c_3\Psi_3 + \ldots)$$

where c_1, c_2, etc., are coefficients specifying the amounts of each wave function to be incorporated in the linear combination and are adjusted by the variation method to give the state of lowest energy. N is a normalising constant and allows for the fact that the total probability of finding the electron in all space must be unity, *i.e.*

$$\int \Psi^2 dv = 1$$

Hence $$N = 1/(c_1{}^2 + c_2{}^2 + c_3{}^2 + \ldots)^{\frac{1}{2}}$$

The two methods which have been used to explain the formation of a covalent bond are the *valence bond method* and the *molecular orbital method*.

THE VALENCE BOND METHOD

In this approach, the formation of the hydrogen molecule is achieved by bringing together the constituent atoms, which are initially separated by an infinite distance. As the atoms approach one another their electron clouds interact and this may result in the formation of a stable molecule.

4. The volume bond wave function. Let A and B represent the two hydrogen nuclei, and e_A and e_B their associated electrons. If the two hydrogen atoms are separated by an infinite distance, then for A the wave function of the $1s$ electron will be $\Psi_A(e_A)$ and for B, $\Psi_B(e_B)$. The wave functions for the complete system may be represented by

$$\Psi_1 = \Psi_A(e_A)\Psi_B(e_B)$$

where e_A is specifically associated with nucleus A, and e_B with nucleus B. The valence bond approach assumes that

Ψ_1, the valence bond wave function, is a reasonably good approximation even when the atoms are close together.

Although we have labelled the electrons, we cannot in fact distinguish between them, and an equally probable wave function to Ψ_1 is Ψ_2

$$\Psi_2 = \Psi_A(e_B)\Psi_B(e_A)$$

where the electrons have changed partners. The linear combination principle (3 above) indicates that a better approximation is given by combining Ψ_1 and Ψ_2

$$\Psi = N(c_1\Psi_1 + c_2\Psi_2) \quad . \quad . \quad . \text{(XVII, 1)}$$

Ψ_1 and Ψ_2 represent states of equal energy and therefore the coefficients c_1 and c_2 are equal in magnitude. However, since it is Ψ^2 rather than Ψ which is proportional to the electron density, $c_1{}^2 = c_2{}^2$ and $c_1 = \pm c_2$. There are two possible wave functions

$$\Psi_s = (2)^{-\frac{1}{2}}(\Psi_1 + \Psi_2) \quad \text{and} \quad \Psi_A = (2)^{-\frac{1}{2}}(\Psi_1 - \Psi_2)$$

where $(2)^{-\frac{1}{2}}$ is the normalising constant (3 above).

5. The energy of the hydrogen molecule.

From these wave functions it is possible to calculate the energy of the system as a function of the internuclear distance, and the resulting curves are shown in Fig. 76. The *antisymmetric* function, Ψ_A, represents the situation where no stable molecule is formed, since it always gives a higher energy for the system than the energies of the separate atoms. The *symmetric* function, Ψ_s, passes through a minimum, indicating the formation of a stable molecule. For Ψ_s the wave functions of the electrons reinforce one another in the region between the two nuclei, which is equivalent to a build-up in electron density. The two positively charged nuclei are attracted towards the electron cloud localised between them. This attractive force more than compensates for the repulsions between (a) the two nuclei and (b) the two electrons, and a stable covalent bond results. On the other hand, for Ψ_A the electron density is depleted in the region between the two nuclei and no bond formation is possible.

NOTE: The Ψ_s wave function is symmetric since its sign is unchanged if the electrons e_A and e_B exchange partners. The Ψ_A wave function changes sign under the same operation and so is anti-symmetric.

6. Spin considerations. Because of the Pauli exclusion principle the electrons in the separate hydrogen atoms may have either parallel or antiparallel spins. The Ψ_A wave function represents the state in which the two electrons have parallel spins, while the Ψ_s wave function represents the state in which the electron spins are anti-parallel and hence can be paired. Thus the sharing of electrons between nuclei does not automatically lead to bond formation.

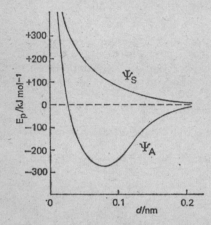

Fig. 76.—*The calculated potential energy diagrams for the hydrogen molecule*

7. Calculated values. From Fig. 76 the minimum energy in the potential energy curve is 289 kJ mol^{-1}, *i.e.* about 60% of the experimental value, and the internuclear distance 0·087 nm. These values, obtained by Heitler and London in 1927, have since been refined by making allowance for:

(*a*) Mutual shielding of the electrons by each other.

(*b*) An ionic contribution to the covalency due to the fact that both electrons may be associated with the same nucleus, *i.e.*

$$H_A^- : H_B^+ \qquad \Psi_3 = \Psi_A(e_A)\Psi_A(e_B)$$
$$H_A^+ : H_B^- \qquad \Psi_4 = \Psi_B(e_A)\Psi_B(e_B)$$

$\Psi_A(e_A)\Psi_A(e_B)$ represents the state where both electrons are associated with nucleus A. As before, the linear combination principle indicates that a better solution than Ψ_3 or Ψ_4 separately will be

$$\Psi_{\text{ionic}} = N(c_3\Psi_3 + c_4\Psi_4) \quad \ldots \text{(XVII, 2)}$$

Combination of (XVII, 1) and (XVII, 2) leads to a much better wave function for the whole molecule

$$\Psi = N(c_1\Psi_1 + c_2\Psi_2 + c_3\Psi_3 + c_4\Psi_4) \ldots \text{(XVII, 3)}$$

By applying these refinements, ΔH_{D_e} is raised to 376 kJ mol^{-1} and d_e is lowered to 0·075 nm. Subsequent refinements have made the calculated value of ΔH_{D_e} even closer to the experimentally observed value.

RESONANCE

8. Canonical forms. In the case of the hydrogen molecule, greater stability is attained if the wave function is considered to represent three separate structures or *canonical forms*.

$$\text{H} \colon \text{H} \longleftrightarrow \text{H}^- \colon \text{H}^+ \longleftrightarrow \text{H}^+ \colon \text{H}^-$$
$$c_1\Psi_1 + c_2\Psi_2 \qquad\quad c_3\Psi_3 \qquad\qquad c_4\Psi_4$$

The double-pointed arrows indicate that the forms are in *resonance* with each other and that the actual structure of the hydrogen molecule may be considered to be a *resonance hybrid* or weighted average of the three forms. This does not mean that at any one time the hydrogen molecule has one of the three structures and that the molecule oscillates between the structures. The hydrogen molecule itself has a definite structure, but this cannot be represented by a conventional bond diagram. The combined wave function in (XVII, 3) gives rise to a lower energy and therefore a more stable molecule than is obtained from any of the composite wave functions. The difference in energy between the actual structure and the most stable of the canonical forms is called the *resonance energy*.

9. Resonance in more complicated molecules. Carbon dioxide may be thought of as a resonance hybrid of the following canonical structures

$$:\!\overset{..}{\text{O}}\!::\!\text{C}\!::\!\overset{..}{\text{O}}\!: \qquad :\!\text{O}\!:::\!\text{C}\!:\!\overset{..}{\underset{..}{\text{O}}}\!: \qquad :\!\overset{..}{\underset{..}{\text{O}}}\!:\ \text{C}\!:::\!\text{O}\!:$$
$$\text{I} \qquad\qquad\qquad \text{II} \qquad\qquad\qquad \text{III}$$

The resonance energy is 155 kJ mol^{-1}, this being the extent by which the actual structure is more stable than I, the most stable of the canonical structures.

Perhaps the best known example of resonance is the benzene molecule, where the five canonical structures (two Kekulé forms and three Dewar forms) result in a resonance energy of 151 kJ mol^{-1}.

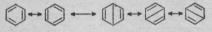

Kekulé forms   Dewar forms

The fact that the benzene molecule consists of a resonance hybrid formed from the five canonical structures provides an explanation of the fact that the length of the carbon–carbon bond in benzene (0·140 nm) is intermediate between that found for single bonds (0·153 nm in ethane) and double bonds (0·133 nm in ethylene).

10. Rules governing resonance. The following conditions must be satisfied if a particular structure is to be included in the resonance hybrid:

(a) The atoms must be in the same relative positions in all the canonical forms.

(b) The number of unpaired electrons must be the same in each canonical form.

In addition, the lower the energy of a particular canonical form the greater is its contribution to the overall structure. In benzene the Dewar forms are of higher energy and are therefore less important than the Kekulé forms.

THE MOLECULAR ORBITAL METHOD

The valence bond method considers only the electron pair which constitutes the chemical bond, and pays no attention to the remainder of the electrons in the atoms of the molecule. This pair of bonding electrons is assumed to reside in *atomic* orbitals. On the other hand, the molecular orbital method considers all the electrons in the combining atoms as belonging to the molecule as a whole. The atomic orbitals are replaced by *molecular* orbitals, which are associated with all the nuclei in the molecule.

11. The general principles. The molecular orbital approach first considers the nuclei to be in their equilibrium positions as in the stable molecule. Each electron in the molecule is then assigned to a molecular orbital. The filling up of the molecular orbitals proceeds in an analogous way to that used for atomic orbitals, namely:

(a) The molecular orbitals are filled one at a time in order of increasing energy.

(b) Because of the Pauli exclusion principle, each molecular orbital may contain no more than two electrons, the spins of which must be paired.

The method will again be illustrated using the hydrogen molecule.

12. The molecular orbital wave function. An electron moving in the molecular orbital in the neighbourhood of one of the hydrogen nuclei (A), will be influenced almost entirely by A and very little by the other nucleus, B. Under these conditions the molecular orbital in the vicinity of each atom will resemble an atomic orbital. Consequently, a reasonable approximation to a molecular orbital will be a *linear combination of atomic orbitals* (LCAO). If Ψ_A and Ψ_B are the atomic orbitals of the two separate hydrogen atoms, then the molecular orbital wave function Ψ is given by

$$\Psi = N(c_A\Psi_B + c_B\Psi_B)$$

where N and c have the same significance as in **3** above. From **4** above, $c_A = \pm c_B$ and two molecular orbitals are possible

$$\Psi_+ = (2)^{-\frac{1}{2}}(\Psi_A + \Psi_B) \quad \text{and} \quad \Psi_- = (2)^{-\frac{1}{2}}(\Psi_A - \Psi_B)$$

13. The molecular orbitals. The formation of the two molecular orbitals is shown diagrammatically in Fig. 77.

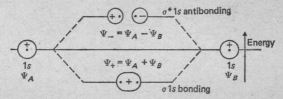

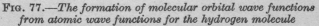

Fig. 77.—*The formation of molecular orbital wave functions from atomic wave functions for the hydrogen molecule*

(a) Ψ_+, *the bonding molecular orbital.* The positive sign in the $(\Psi_A + \Psi_B)$ wave function indicates that there is some concentration of electron density between the positively charged nuclei which diminishes the repulsive forces and leads in fact to a net force of attraction.

(b) Ψ_-, *the anti-bonding molecular orbital.* The negative sign in the $(\Psi_A - \Psi_B)$ wave function indicates depletion of electron density between the two nuclei. Under such conditions the two nuclei repel one another and no stable molecule is formed.

Early calculations based on the molecular orbital treatment gave $\Delta H_{De} = 259$ kJ mol^{-1} and $d_e = 0.085$ nm, and these values have since been refined.

NOTE: For the hydrogen molecule the molecular orbital method is similar to the valence bond method in that two atomic wave functions have been combined to give two molecular wave functions. Also, both methods predict an increase in electron density between the nuclei when a bond is formed.

APPLICATION OF THE MOLECULAR ORBITAL METHOD TO OTHER DIATOMIC MOLECULES

14. Classification of molecular orbitals. Molecular orbitals may be formed by linear combinations of atomic orbitals other than $1s$ orbitals. Some of these are shown in Fig. 78. As a consequence of the Pauli exclusion principle, *n atomic orbitals will form n molecular orbitals.* It is convenient to classify

Atomic orbitals	Molecular orbitals
$\cdot\odot \cdot \cdot\odot$ $2s$ $2s$	$(\ominus\cdot)$ $(\cdot+)$ σ^*2s $(\cdot + \cdot)$ $\sigma 2s$
$\ominus\!\!+$ $+\!\!\ominus$ $2p_x$ $2p_x$	$(\ominus\cdot)$ $(\cdot+)$ σ^*2p_X $(\cdot + \cdot)$ $\sigma 2p_X$
$\overset{+}{\underset{-}{\bigcirc}}$ $\overset{+}{\underset{-}{\bigcirc}}$ $2p_z$ $2p_z$	$\overset{+}{\underset{-}{\bigcirc}}$ $\overset{-}{\underset{+}{\bigcirc}}$ π^*2p_z $\overset{+}{\underset{\sim}{\bigcirc}}$ $\pi 2p_z$

FIG. 78.—*The linear combination of atomic orbitals to form molecular orbitals*

molecular orbitals in terms of their symmetry about the inter-nuclear axis.

(a) σ orbitals. Orbitals which are completely symmetrical about the internuclear axis are termed σ (sigma) orbitals. Thus the combination of two $2s$ orbitals gives two σ orbitals, $\sigma 2s$ and $\sigma^* 2s$, where the asterisk indicates the anti-bonding orbital. The $2p_x$ orbitals, which extend along the internuclear axis, may also combine to give two σ orbitals, $\sigma 2p_x$ and $\sigma^* 2p_x$.

(b) π orbitals. The lateral overlap of two $2p_y$ or two $2p_z$ orbitals gives rise to charge clouds above and below a plane containing the internuclear axis, with zero charge density in this plane. Such charge clouds are known as π orbitals, and the bonding $\pi 2p_y$ and $\pi 2p_z$ orbitals consist of two "sausage-shaped" charge lobes. If the $2p_x$ orbitals combine to form σ orbitals, the remaining $2p_y$ and $2p_z$ orbitals must form π orbitals. π orbitals are extremely im-portant in dealing with the bonding in unsaturated organic compounds.

15. The order of stability of molecular orbitals.

Just as it was possible to arrange atomic orbitals in order of decreasing stability (increasing energy), so it is possible to arrange molecu-lar orbitals in a similar manner:

$$\sigma 1s, \; \sigma^* 1s, \; \sigma 2s, \; \sigma^* 2s, \; \sigma 2p_x, \; \pi 2p_y = \pi 2p_z,$$
$$\pi^* 2p_y = \pi^* 2p_z, \; \sigma^* 2p_x.$$

The $\sigma 2p_x$ and $\pi 2p_y$, $\pi 2p_z$ orbitals are very similar in energy, and the actual order of stability is in some cases the reverse of that shown (Progress Test 17, Question 2).

16. Homonuclear diatomic molecules.

With the above order of stability of molecular orbitals in mind, it is possible to build up the electronic configurations of some simple diatomic molecules in the same way as we did for atoms. The hydrogen molecule may be written $H_2(\sigma 1s)^2$. The addition of two further electrons will weaken the bond, since these electrons must go into the $\sigma^* 1s$ orbital. Thus the molecule He_2, with the con-figuration $He_2(\sigma 1s)^2(\sigma^* 1s)^2$, does not exist because the bonding effect of $\sigma 1s$ is more than offset by the anti-bonding effect of $\sigma^* 1s$. He_2 in the ground state is unknown, but excited states have been observed spectroscopically, where one of the $\sigma^* 1s$ electrons is promoted to a higher energy bonding molecu-lar orbital such as $\sigma 2s$.

Other stable homonuclear diatomic molecules include

$N_2(\sigma1s)^2(\sigma*1s)^2(\sigma2s)^2(\sigma*2s)^2(\sigma2p_x)^2(\pi2p_y)^2(\pi2p_z)^2$ and
$O_2(\sigma1s)^2(\sigma*1s)^2(\sigma2s)^2(\sigma*2s)^2(\sigma2p_x)^2(\pi2p_y)^2(\pi2p_z)^2(\pi*2p_y)^1$
$$(\pi*2p_z)^1.$$

The inner $(\sigma1s)^2(\sigma*1s)^2(\sigma2s)^2(\sigma*2s)^2$ orbitals effectively cancel each other out and play no part in the bonding. Thus in nitrogen there are six bonding electrons which constitute a triple bond made up of one σ and two π bonds. The molecular orbital description of the oxygen molecule indicates the presence of two unpaired electrons, which is in agreement with experimental evidence. The simple valence bond approach leads to the prediction that all the electrons are paired.

17. Heteronuclear diatomic molecules. For heteronuclear diatomic molecules such as HF, the situation is more complicated because the separate atomic wave functions no longer contribute equally to the molecular wave function. The electronic configuration of the fluorine atom is $1s^22s^22p^5$, indicating that there is one singly occupied $2p$ orbital which can combine with the singly occupied $1s$ orbital of the hydrogen atom to form a bonding σ orbital containing two paired electrons. These electrons are not equally shared by the two nuclei as in the case of homonuclear diatomic molecules. Ψ_F contributes more to the molecular wave function than does Ψ_H, and as a result the electron cloud is distorted, with a larger region of electron density around the fluorine than around the hydrogen atom. The HF bond is said to be *polarised* and HF is a *polar* molecule (*see* XVIII, 2–4).

MOLECULAR GEOMETRY

18. Orbital overlap. Both the valence bond and molecular orbital methods for the hydrogen molecule indicate that bond formation increases the electron density between the nuclei, which can be visualised in terms of overlap of $1s$ atomic orbitals. The concept of atomic orbital overlap is extremely useful in explaining the shapes of more complicated molecules. In this approach the orbitals of the inner shell electrons are assumed to remain localised around their respective nuclei, and only those of the outer electrons are considered in bond formation. For bond formation between two atoms, each must have a singly occupied orbital (normal covalency) or alternatively one may have an unoccupied orbital and the other a doubly occupied orbital (dative covalency).

Care must be taken in selecting atomic orbitals which overlap as effectively as possible, since the strength of a covalent bond is approximately proportional to the amount of overlap. In this respect the atomic orbitals should have similar energies and possess the same symmetry properties with respect to the internuclear axis. The symmetry condition may be understood by considering the overlap of $1s$ and $2p_x$ orbitals and $1s$ and $2p_z$ orbitals. As shown in Fig. 79 (a), the $1s$ and $2p_x$ orbitals

1s   2p_x
(a)

1s 2p_z
(b)

FIG. 79.—*Overlap of 1s and 2p orbitals.*
(a) *Maximum overlap.*   (b) *Zero overlap*

overlap effectively along the x axis, and the regions of overlap have the same sign for the wave function. Since the two regions of overlap of the $1s$ and $2p_z$ orbitals are equal in size but opposite in sign, they cancel each other out and no bond results (Fig. 79 (b)).

Diatomic molecules are of necessity linear, but polyatomic molecules can be linear, as in N_2O and CO_2, or non-linear, as in H_2O, H_2S and NH_3. The chemical bonds in non-linear molecules involve p and d orbitals which possess directional characteristics.

19. Hybridisation. The bonding orbitals in molecules are not usually formed from pure s, p or d atomic orbitals but rather from a mixture, or *hybrid*, of two or more different types of atomic orbital. The hybrid atomic orbitals formed by mixing s and p atomic orbitals can be illustrated by reference to the simple hydrocarbons.

20. sp³ hybridisation in carbon. In its ground state the carbon atom has the electronic configuration $1s^2 2s^2 2p^2$ and since there are only two unpaired electrons a valency of two would be expected. In order to account for the observed valency of four, one of the $2s$ electrons must be promoted to a

$2p$ orbital giving $1s^2 2s^1 2p_x^1 2p_y^1 2p_z^1$, *i.e.* four unpaired electrons. This promotion requires the expenditure of about 416 kJ mol^{-1} of energy, but this is more than compensated by the ability to form four bonds rather than two. It might be expected that the three bonds formed by utilising the $2p$ orbitals would differ from that formed utilising the $2s$ orbital. However, it is known that in CH_4 all four C–H bonds are equivalent and moreover are directed towards the apexes of a regular tetrahedron with bond angles of $109 \cdot 5°$. The apparent anomaly is explained using the concept of hybrid atomic orbitals.

The four wave functions for the carbon atom Ψ_{2s}, Ψ_{2px}, Ψ_{2py} and Ψ_{2pz} are obtained by solution of the Schrödinger wave equation. Equally satisfactory solutions are given by linear combinations of these four wave functions to yield four equivalent combined or hybrid wave functions. Thus we say that one $2s$ and three $2p$ atomic orbitals combine together to form four equivalent sp^3 hybrid atomic orbitals. These new hybrid orbitals have strong directional characteristics, and the lobes of the orbitals point towards the four apexes of a regular tetrahedron (Fig. 80).

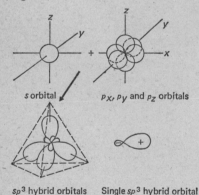

s orbital p_x, p_y and p_z orbitals

sp^3 hybrid orbitals   Single sp^3 hybrid orbital

FIG 80.—*Formation of sp^3 hybrid orbitals*

The bonding in CH_4 is understood in terms of the overlap of the four sp^3 hybrid orbitals of the carbon atom with the $1s$ orbitals of the four hydrogen atoms, giving rise to four equivalent σ bonds. Because of the very effective overlap, these bonds are particularly strong.

21. sp³ hybridisation in other molecules. Other examples of sp^3 hybridisation are to be found in NH_3 and H_2O. The sp^3 hybrid orbitals are arranged tetrahedrally around the nitrogen and oxygen atoms, but in the case of nitrogen one of the hybrids, and in the case of oxygen two, already contain pairs of electrons which are not involved in bonding with other atoms (Fig. 81). These *lone pair* hybrid orbitals have an important influence on the shape of the molecule (*see* 22 below) and on its dipole moment (*see* XVIII, 5).

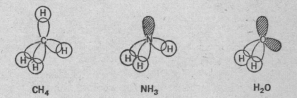

CH₄   NH₃   H₂O

FIG. 81.—*The bonding in CH_4, NH_3 and H_2O. The shaded orbitals are the lone-pair orbitals*

22. The effect of the lone pair hybrids. CH_4, which forms a regular tetrahedron, has bond angles of 109·5°, but these are reduced in NH_3 and H_2O to 107·3° and 104·5° respectively. These slight deviations from the regular tetrahedral structure are attributed to repulsive forces between the pairs of electrons around the central atom. The magnitude of these repulsive forces depends on whether the electrons are lone pairs (LP) or are involved in bonding (bonding pairs, BP). The lone pairs are closer to the nucleus than the bond pairs and hence lone pair orbitals may be visualised as being short and fat, while bond pair orbitals are long and thin. Hence repulsion between the electron pairs decreases in the order LP — LP > LP — BP > BP — BP. The effect will be greatest in water, with its two lone pairs.

23. sp² hybridisation. An explanation for the bonding in ethylene is provided by sp^2 hybridisation brought about by the mixing of one s and two p orbitals. The most energetically favourable arrangement is for the three hybrid atomic orbitals to be in the same plane at 120° to one another. The orbital overlap of the sp^2 hybrid atomic orbitals in ethylene is shown in Fig. 82 (*a*) and indicates the formation of one σ C–C bond and four σ C–H bonds. The remaining *p* orbitals on the carbon atom which do not take part in the hybridisation are perpendicular to the plane of the sp^2 hybrid orbitals and overlap

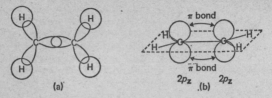

FIG. 82.—*The bonding in* C_2H_4.
(*a*) *Formation of C–C* σ *C–H bonds.*   (*b*) *Formation of C–C* π *bond*

laterally to form a second C–C bond — a π bond (Fig. 82 (*b*)). This lateral overlap is less than that of the two sp^2 hybrid orbitals and hence the π bond is weaker than the σ bond, which explains the greater reactivity of ethylene compared with ethane.

sp^2 hybridisation accounts for the plane triangular shape of molecules of the type MX_3, where M is B, Al, Ga, In or Tl and X is a halogen.

24. sp hybridisation. The mixing of one *s* and one *p* orbital to form two *sp* hybrid orbitals explains the bonding in acetylene. On energetic grounds, the two *sp* hybrid orbitals on each carbon atom are at 180° to one another and are utilised in forming one σ C–C bond and two σ C–H bonds, giving the linear H–C–C–H skeleton. The two remaining $2p$ orbitals overlap laterally to provide two further C–C π bonds, and the net result is a C–C triple bond.

The linear molecules $BeCl_2$, BeF_2 and $Hg(CN)_2$ also utilise *sp* hybrid orbitals.

25. Other types of hybrid orbital. It should not be supposed that hybrid orbitals are unique to the carbon atom. Hybridisation is important in many inorganic molecules (water and ammonia have already been mentioned). Where electrons are present in *d* orbitals, hybrid orbitals may be formed by the mixing of *s*, *p* and *d* orbitals:

(*a*) dsp^2 hybridisation results in four orbitals directed towards the corners of a square, *e.g.* PCl_4^-.

(b) dsp^3 hybridisation results in five orbitals directed towards the apices of a trigonal bipyramid, e.g. PF_5.

(c) d^2sp^3 hybridisation results in six orbitals directed towards the apices of a regular octahedron, e.g. SF_6.

DELOCALISED ORBITALS

In the valence bond approach to molecular structure the bonding orbitals formed between pairs of nuclei are *localised*. Some molecular structures are better described using the molecular orbital approach in which the bonding orbitals are delocalised, *i.e.* spread over several nuclei.

26. The benzene molecule. The benzene molecule, which we previously considered in terms of a resonance hybrid, may be elegantly described using the molecular orbital approach. The atomic orbitals of each carbon atom are hybridised to form three sp^2 hybrid orbitals, leaving one unhybridised $2p$ orbital (say $2p_z$). Two sp^2 hybrid orbitals on each carbon atom are used to form the σ-bonded carbon skeleton, while the re-

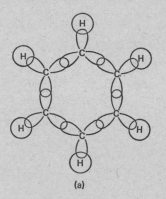

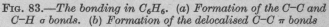

(a)   (b)

FIG. 83.—*The bonding in* C_6H_6. (a) *Formation of the C–C and C–H* σ *bonds*. (b) *Formation of the delocalised C–C* π *bonds*

maining sp^2 hybrid orbital forms a σ bond by overlap with a hydrogen $1s$ atomic orbital (Fig. 83(a)). The six unhybridised $2p_z$ orbitals (one per carbon atom) each contain a single electron and are perpendicular to the benzene ring. Although these orbitals may overlap laterally in pairs, there is no reason why the overlap should be limited in this way. Thus the bonding may be considered in terms of delocalised molecular orbitals which are spread right round the benzene ring. The combination of six atomic orbitals gives rise to six molecular orbitals—three bonding and three anti-bonding. The six p electrons are accommodated in three delocalised bonding orbitals which have π symmetry and one of these is illustrated in Fig. 83 (*b*).

27. Other molecules. Delocalised molecular orbitals may be used to describe the bonding in inorganic molecules such as NO_2, N_2H_4 and CO_2, and also in metals (*see* XVIII, **11**).

PROGRESS TEST 17

1. Draw all possible canonical structures for the following molecules: C_2N_2, C_3O_2, CS_2 and SO_3. The first three molecules are linear, while SO_3 is planar, with the S–O bonds pointing towards the corners of an equilateral triangle. (**7–9**)

2. Using the molecular orbital theory, predict the stability of the following diatomic molecules and ions: Li_2, Li_2^+, Li_2^{2+}, Be_2, B_2, C_2. (**14–16**)

3. Explain why the bond angle in H_2Se is 90° while that in H_2O is 104·5°. (**18–22**)

4. Draw and label the potential energy diagram for the hydrogen molecule. (**1**)

5. Explain the meaning of (*a*) the variation principle (**2**), (*b*) the linear combination principle (**3**).

6. Outline the application of (*a*) the valence bond approach (**8–10**), (*b*) the molecular orbital approach (**11–13**) to the bonding in the hydrogen molecule.

7. What are (*a*) σ orbitals, (*b*) π orbitals ? (**14**)

8. How can the concept of hybridisation explain the shapes of molecules ? Illustrate your answer with examples. (**19–25**)

9. What are delocalised orbitals ? (**26**)

CHAPTER XVIII

OTHER TYPES OF BOND

The description of a particular bond as either wholly ionic or wholly covalent is an over-simplification: covalent bonds possess a degree of ionic character arising from unequal sharing of the bonding electrons and ionic bonds involve some covalent contribution.

POLARISATION EFFECTS

1. Polarisation of ions. When two oppositely charged ions approach one another, in addition to coulombic interaction the smaller cation attracts the outermost electrons of the larger anion. This distortion of the initially symmetrical electron cloud of the anion is termed *polarisation* (*see* Fig. 84). The

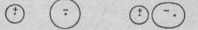

FIG. 84.—*The polarisation of a large anion by a small cation*

cation is polarised by the anion to a much smaller extent, and may be neglected in most cases. Polarisation will be favoured when:

(*a*) *The cation is small and highly charged.* The electric field strength at the surface of the cation is given by $E = ze/4\pi\epsilon_0 r^2$ and will obviously be greater the higher z (the ionic charge) and the smaller r (the ionic radius).

(*b*) *The anion is large.* The outer shell electrons of a large anion are well shielded from the nuclear charge by the inner shell electrons and are readily polarisable. Among isoelectronic anions, that with the highest negative charge will be most polarisable, since it has the largest ionic radius (*see* XVI, **15**).

(*c*) *The cation does not have the noble gas electronic configuration.* Such a configuration results in a low electric field strength at the surface of the cation because of effective shielding by the electrons. Transition metal ions have incomplete d sub-shells which do not provide effective shielding and hence possess high polarising powers.

The above conditions constitute *Fajan's rules*. Polarisation gives rise to a certain amount of electron sharing or covalent character in ionic bonds and as a result no bond can be completely ionic.

2. Covalent bond polarisation.

The pair of bonding electrons in homonuclear diatomic molecules are shared equally by the two atoms. This is not the case in heteronuclear diatomic molecules, where internal polarisation causes the electrons to be displaced towards one of the two atoms. The power of an atom in a molecule to attract electrons to itself is called its *electronegativity*. There are several methods available for determining relative electronegativities. The Pauling electronegativity scale is based on the difference between the measured bond energy in a heteronuclear diatomic molecule and that calculated assuming the bond to be completely covalent. On the Mullikan electronegativity scale the electronegativity of an atom is the mean of its ionisation energy and its electron affinity. The two scales are not equal but lead to the same relative values for electronegativities.

Electronegativity values generally decrease on descending a particular group in the periodic table and increase from left to right along a period. The most electronegative atoms are F, O, N and Cl respectively.

3. Dipole moments.

In the molecule XY the bonding electrons are displaced towards the more electronegative atom, say Y, and some charge separation takes place which may be represented by $\overset{\delta+}{X} - \overset{\delta-}{Y}$. Such molecules are said to be *polar*, since they possess an *electric dipole*, and the degree of polarity is represented by the *dipole moment*, p, which is defined by the relationship

$$p = \delta e \times l$$

where l is the distance between the centres of equal and opposite effective charges of magnitude δe.

NOTE: p is a vector quantity and the XY molecule may be represented by $X \overset{\longrightarrow}{-} Y$, where the vector arrow points to the negative charge and the length of the arrow represents the magnitude of the dipole moment.

4. The dipole moment of the hydrogen chloride molecule. The dipole moment of $\overrightarrow{H-Cl}$ is $3\cdot46 \times 10^{-30}$ C m and may be used to estimate the amount of ionic character possessed by the H–Cl bond. If HCl were completely ionic, then a charge of $\pm e$, *i.e.* $\pm1\cdot602 \times 10^{-19}$ C, would reside on each nucleus separated by an internuclear distance of $0\cdot127$ nm. The resulting dipole moment would be $1\cdot602 \times 10^{-19} \times 1\cdot27 \times 10^{-10} = 2\cdot03 \times 10^{-29}$ C m. The measured dipole moment is only 17% of this value, and we say that HCl has 17% ionic character.

5. Dipole moments in polyatomic molecules. Dipole moments may also be exhibited by polyatomic molecules, and in some cases conclusions concerning the structure of molecules can be made from the magnitudes of their dipole moments. Thus carbon dioxide has a zero dipole moment, indicating a linear molecule (Fig. 85 (*a*)). Water, on the other hand, has a dipole moment of $6\cdot2 \times 10^{-30}$ C m, indicating that it is not linear and in fact possesses a bent structure (Fig. 85 (*b*)). This large

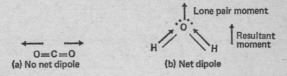

$$O=C=O$$
(a) No net dipole   (b) Net dipole

FIG. 85.—*Addition of dipole moment vectors for carbon dioxide and water*

dipole moment is partly due to the polar nature of the O–H bond and partly to the lone pair of electrons on the oxygen atom. Lone pair moments may considerably influence the resultant dipole moment and must be taken into consideration before any conclusions are drawn with regard to molecular structures.

HYDROGEN BONDING

6. General principles. When a hydrogen atom is bonded to a highly electronegative atom (X, say), a strong dipole will be produced $\overset{\delta-}{-X}\overset{\delta+}{-H}$. Suppose a second highly electronegative atom B in another molecule forms the negative end of a second dipole. Dipole–dipole interaction will result (*see* **8** below), with the hydrogen atom forming a second weaker bond with B $-\overset{\delta-}{X}-\overset{\delta+}{H}\cdots\overset{\delta-}{B}-$. This bond, represented by the dotted line, is the hydrogen bond and is a direct consequence of bond polarity. The electronegative atoms are usually F, O or N.

7. Evidence for hydrogen bonding. Although hydrogen bond strengths are not very great (less than 40 kJ mol^{-1}), the bonds are extremely important. Molecular association through hydrogen bonding is responsible for the following:

(a) The abnormally high boiling points, dielectric constants and low vapour pressures of water, sulphuric acid, ammonia and various organic compounds such as alcohols, carboxylic acids and amines containing $-OH$ and $-NH_2$ groups. The structures of liquid water and ice are considered in XX.

(b) The polymeric molecules which exist in gaseous hydrogen fluoride at room temperature (up to about $(HF)_5$). In the solid state endless polymer chains are formed.

(c) The structures of many hydrates, solid acids and acid salts are largely determined by hydrogen bonding, e.g. in solid orthoboric acid, H_3BO_3, the molecules are joined in infinite parallel sheets by hydrogen bonds, while weak van der Waals bonds (see below) hold the layers together.

(d) The stability and configuration of almost every large biologically important molecule depends on hydrogen bonding, e.g. the double helices in deoxyribonucleic acid (DNA) are linked by hydrogen bonds. Because of the weakness of the hydrogen bond, biochemical processes involving the making and breaking of these bonds occur readily at ordinary temperatures.

THE VAN DER WAALS BOND

This type of bond exists between all atoms and molecules, irrespective of whether or not they are joined by other bonds. The bond strength is low (less than 40 kJ mol^{-1}) but it is, for example, the sole binding force between noble gas atoms in the liquid and solid states. Unlike covalent bonds, van der Waals bonds are largely non-directional, and the packing of noble gas atoms in the solid state is determined solely by geometrical considerations (see XIX, 18, 19).

8. The nature of the van der Waals bond. There are three distinct types of force which comprise the van der Waals bond.

Two are significant only for polar molecules, while the third is important for all atoms and molecules.

(a) *Dipole–dipole force.* This type of force exists between polar molecules such as ammonia and water, where the positive end of one dipole attracts the negative end of another.

(b) *Dipole-induced dipole force.* A polar molecule can induce a dipole in a non-polar molecule by polarising the latter's electron cloud. An attractive force can then exist between the dipole and the induced dipole.

(c) *Dispersion force.* The origin of this force can be fully understood only from wave mechanical considerations, but a simple picture will serve to illustrate its nature.

Although the average behaviour of the electrons in an atom or non-polar molecule results in a zero dipole moment, at any given moment in time the electronic arrangement will be such as to produce a transient dipole. This dipole will induce further dipoles in adjacent atoms or molecules, which will fluctuate in phase with the first dipole, giving rise to an attractive force between the molecules. The dispersion force is the only force which exists between atoms and between non-polar molecules.

The three contributions to the van der Waals bond all vary with the reciprocal of the sixth power of the distance and are appreciable only over short distances. Repulsion between electron clouds limits the distances of closest approach.

9. Van der Waals radius. The distance between atoms in solids where only van der Waals forces are present may be represented as the sum of the *van der Waals radii* of the individual atoms. These radii may be estimated if the distance of closest approach of non-bonded atoms (*i.e.* no ionic or covalent bonding) can be found from X-ray diffraction studies (*see* XIX, 24–33). For example, in solid iodine the shortest distance between atoms in adjacent molecules is 0·454 nm and hence the van der Waals radius is 0·227 nm. The covalent and ionic radii of iodine are 0·133 and 0·219 nm respectively. The van der Waals radius for a given element will vary slightly from compound to compound and is adjusted to give the best overall agreement.

METALLIC BONDING

10. Properties of metals. Metallic elements, which are crystalline solids built up of identical atoms, comprise three-quarters of the elements in the periodic table. Two important properties give a guide to the nature of the bonding:

(a) Metals are good conductors, indicating that some of the electrons are weakly bound and are able to move under the influence of an applied electric field.

(b) Each metal atom in the crystal is surrounded by eight to twelve nearest neighbours, i.e. the co-ordination number is eight to twelve, but the number of valency electrons, usually one to three, is insufficient to allow normal covalent bond formation with all the nearest neighbours.

11. The band model. Metallic bonding may be conveniently understood in terms of the molecular orbital theory, and as an example we will consider the formation of molecular orbitals in sodium. Because of their low energy the inner electrons in the sodium atom, electronic structure $1s^2 2s^2 2p^6 3s^1$, are assumed to be localised around the sodium nucleus, and only the single $3s$ valency electron participates in the bonding. When two sodium atoms come together the two $3s$ atomic orbitals combine to give two molecular orbitals (one bonding and one antibonding), while a three-atom system will give three molecular orbitals, and so on. In general, for N atoms there will be N molecular orbitals, each with a distinct energy. When N is very large (as it will be in a metallic crystal) the energy levels merge together to form an *energy band* with an essential continuous distribution of energies (*see* Fig. 86). A similar operation can be performed with the empty $3p$ orbitals, giving rise to an unoccupied $3p$ energy band.

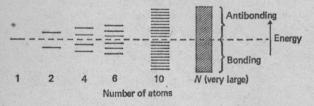

FIG. 86.—*The formation of an energy band*

Since each atomic orbital contributes one level to the band and each level may accommodate two electrons, an energy band formed from N atoms will accommodate a total of $2N$ electrons. In sodium the $3s$ energy band will be only half filled, *i.e.* half the available energy levels will be doubly occupied while the remaining anti-bonding energy levels will be empty. The completely filled 1s, 2s and 2p bands are essentially localised, while the half-filled 3s band is delocalised over the whole crystal. The band model for sodium is represented in Fig. 87 (*b*) together with the discrete energy levels of an isolated sodium atom for comparison.

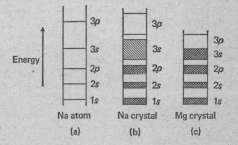

Fig. 87.—*The energy levels in isolated atoms, (a), and the energy bands in metal crystals, (b) and (c). The cross-hatching represents filled bands and the single oblique lines half-filled bands*

When an electric field is applied the 3s electrons acquire sufficient energy to be raised to unoccupied anti-bonding 3s levels and move through the crystal in the direction of the field. Electrical conduction is observed only when unoccupied energy levels are available to the conduction electrons. From Fig. 87 (*c*) it can be seen that the 3s and 3p energy bands in magnesium overlap considerably, and even though the 3s level is completely filled the 3s electrons still have vacant energy levels in the 3p band available to allow them to flow under an applied electric field.

12. Metallic radii. As shown above, a metal can be pictured as a collection of positive ions permeated by a mobile "electron gas," the valency electrons being a property of the metal as a whole rather than of the individual atoms. However, fundamentally the bonding is covalent, *not* ionic, and metallic radii are identical with covalent radii.

13. Insulators. Insulators, such as solid sodium chloride, have all the lower energy bands completely filled and a wide energy gap exists (the *forbidden band*) between the highest filled band (the *valency band*) and the lowest empty band (the *conduction band*) which cannot be traversed under normal conditions (*see* Fig. 88 (*b*)). Since the electrons in the valency band cannot acquire sufficient energy to reach the conduction band, insulators do not conduct electricity.

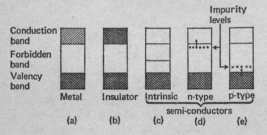

FIG. 88.—*Energy bands in various crystalline solids. The cross-hatching represents filled bands and the single oblique lines half-filled bands*

14. Semiconductors. These are substances with properties intermediate between those of conductors and insulators.

(*a*) *Intrinsic semiconductors* are characterised by having a relatively narrow forbidden band (*see* Fig. 88 (*c*)) such that thermal energy alone is sufficient to excite some electrons across the gap. In some instances the gap may be crossed by exciting the electrons with light, and this gives rise to the phenomenon of *photoconductivity*.

(*b*) *Extrinsic semiconductors* rely on the introduction of small amounts of impurity (doping) into the insulator which give rise to energy levels in the forbidden band. There are two types of extrinsic semiconductors, depending on whether the impurity energy levels are filled or empty.

(*i*) *n-type semiconductors* are formed when impurity atoms containing more valency electrons than the parent insulator are introduced, resulting in filled energy levels in the forbidden band close to the conduction band (*see* Fig. 88 (*d*)). Thermal energy is sufficient to promote some of these extra electrons into the conduction band, enabling a current to flow under an applied field. The term "n-type" originated because the current carriers are negatively charged. An example is tetravalent silicon doped with pentavalent arsenic.

(ii) *p-type semiconductors* contain impurity atoms with fewer valency electrons than the parent insulator, resulting in vacant energy levels in the forbidden band close to the valency band (*see* Fig. 88 (*e*)). Electrons may be thermally excited from the valency band to the impurity energy levels, leaving behind electron vacancies, or *positive holes*, which may then migrate under an applied field. Since the current carriers are positively charged, such semiconductors are termed "p-type" semiconductors. Tetravalent germanium doped with trivalent aluminium is an example.

PROGRESS TEST 18

1. The interhalogen molecules, FCl, BrCl and ICl, have dipole moments of $2\cdot93 \times 10^{-30}$, $1\cdot90 \times 10^{-30}$ and $1\cdot80 \times 10^{-30}$ C m, with internuclear separations of $0\cdot163$, $0\cdot214$ and $0\cdot234$ nm respectively. Calculate the percentage ionic characters. (3–4)

2. The measured dipole moment of H_2S is $3\cdot10 \times 10^{-30}$ C m. Calculate the dipole moment of each S–H bond, assuming that the bond angle is 90° and neglecting the important contribution from the lone pair of electrons on the sulphur atom. (3–5)

3. Explain the meaning of the term "polarisation" when applied to ions, and list Fajan's rules. (1)

4. Show how dipole moments arise in heteronuclear molecules. (2–5)

5. Give the experimental evidence for hydrogen bonding and explain the circumstances under which such bonds are formed. (6–7)

6. What is the nature of the van der Waals bond? (18)

7. Describe the band model of metallic bonding and use it to explain the properties of (*a*) metals, (*b*) insulators, (*c*) semiconductors. (10–14)

PART SEVEN

THE STRUCTURES OF SOLIDS AND LIQUIDS

The next two chapters are concerned with the molecular structures of solids and liquids.

Solids are considered first in Chapter XIX, since the very high degree of order present in most solids allows an adequate physical/mathematical description of their structures. Much of this chapter is devoted to a study of the technique of X-ray diffraction, which has been of such immense importance in the determination of crystal structures.

Liquids are considered in Chapter XX because their much greater complexity makes it desirable to attempt to describe their structures in terms of the structure of an imperfect solid. In the final part of this chapter some substances are considered which show certain resemblances to associated liquids but which do not fall neatly into a solid–liquid–gas classification.

CHAPTER XIX

SOLIDS

THE SOLID STATE

1. Properties of solids. The most obvious feature distinguishing a solid from a liquid or gas is its characteristic shape and rigidity. This distinction is due to the fact that in a solid the structural units are fixed in a characteristic ordered arrangement by large intermolecular forces. We describe this situation by saying that solids have both short-range and long-range order, *i.e.* a minute sample of a crystalline solid containing only a few atoms, ions or molecules is representative of the entire solid.

Crystalline solids (other than cubic crystals) have anisotropic properties because the ordering of the structural units is different in different directions, *e.g.* silver iodide, in which the coefficient of thermal expansion is positive along one of the crystal directions and negative along another.

NOTE: *Isotropic* means "the same in all directions"; *anisotropic* means the reverse.

2. Amorphous solids. Amorphous solids (*e.g.* glass) are best considered as very viscous liquids, having indistinct melting points and isotropic properties owing to the lack of any long-range order.

Some apparently amorphous solids are actually polycrystalline, being composed of minute randomly orientated crystals. Although the random orientation confers isotropic properties on the sample as a whole, single crystals may often be picked out or grown which then exhibit anisotropic properties. Metals often occur in the polycrystalline condition, although the individual crystals can usually be shown up on etching.

CRYSTAL SYSTEMS

3. The external shape of crystals. Individual crystals can often be recognised by their characteristic external shape. Thus crystals of sodium chloride grown from aqueous solution are invariably cubic with well defined faces and with characteristic angles between these faces (each 90°). The constancy of interfacial angles in different specimens of a particular crystal was first noted by Stern in 1669. This is not always apparent from the external form of the crystal, *e.g.* sodium chloride crystallises from water as cubes but from 15% aqueous urea as octahedra. This might suggest that the detailed arrangement of the sodium and chloride ions is different in the two cases. However, X-ray diffraction studies (*see* **24–32** below) have shown that this is not so. The cube and the octahedron are related geometrically (Fig. 89) and both belong to the same cubic crystal system (*see* **6** below).

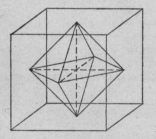

Fig. 89.—*The geometrical relationship between the cube and the octahedron*

4. Unit cells and crystal lattices. In 1784 Hauy postulated that the external regularity of the crystal was a reflection of the regularity of its constituent structural units, and this has since been confirmed using X-ray diffraction. These structural units are now called *unit cells* and contain small numbers of atoms, ions or molecules at definite positions within the cell. The complete crystal consists of unit cells placed side by side in a three-dimensional pattern.

It is helpful to idealise crystal structures by representing each atom, ion or group within the crystal by a point on a diagram. The regular array of points so obtained is called a *point lattice*. A three-dimensional point lattice is illustrated in Fig. 90 and the unit cell is also shown. The unit cells for the

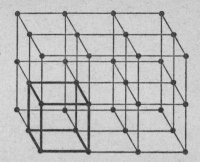

FIG. 90.—*A point lattice, indicating the unit cell*

three types of cubic lattice are shown in Fig. 91. These are derived from the unit cube by placing one atom at each vertex (simple cubic), one atom at each vertex and one at the centre of the cube (body-centred cubic) and one atom at each vertex and one at the centre of each face (face-centred cubic).

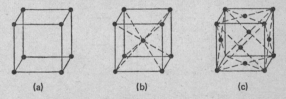

(a)   (b)   (c)

FIG. 91.—*Unit cells of the cubic lattice. (a) Simple cube. (b) Body-centred cube. (c) Face-centred cube*

NOTE: The atoms in the unit cell are actually in "contact" and are not separated by appreciable distances as is implied in Fig. 91. However, such diagrams will be extensively used for clarity.

5. Miller indices. Crystal faces, and planes through the crystal parallel to these faces, may be described in terms of their points of intersection with three conveniently chosen co-ordinate axes. Three such axes and a plane intersecting them are shown in Fig. 92. Unit distances, a, b and c, are marked along these axes (these distances need not necessarily be the same) and if the plane makes intercepts, m, n and p, with these axes, then the ratio

$$a/m : b/n : c/p = h : k : l$$

is such that h, k and l are either small integers or zero.

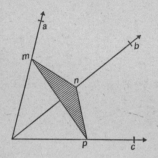

FIG. 92.—*The Miller notation: crystal axes*

The various crystal faces or planes are characterised by their values of h, k and l, as shown in Figs. 93 (a) and (b). Thus in Fig. 93 (a) the plane does not intersect the vertical

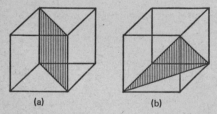

(a)   (b)

FIG. 93.—*The Miller notation: Miller indices.*
(a) (0, 1, 1) plane. (b) (1, 1, 1) plane

axis (or intersects it at infinity), so that $m = \infty$ and $h = 0$. The other two intercepts are of unit length so that $k = l = 1$. The *Miller index* of this plane is therefore (0, 1, 1). Similarly, in Fig. 93 (b) $h = k = l = 1$, giving a Miller index (1, 1, 1).

6. Crystal systems. The very large number of crystal systems may be classified according to the lengths of the crystal axes and the angles between them. If a, b and c represent the lengths of the three crystal axes and the angle between b and c is α, that between c and a is β, and that between a and b is γ (*see*

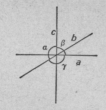

FIG. 94.—*Crystal axes and angles*

Fig. 94), then seven situations are generally recognised as shown in Table 22 overleaf and illustrated in Fig. 95.

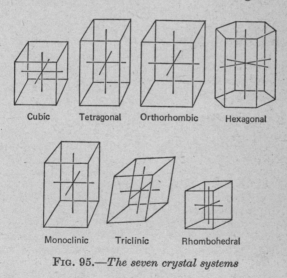

Cubic   Tetragonal   Orthorhombic   Hexagonal

Monoclinic   Triclinic   Rhombohedral

FIG. 95.—*The seven crystal systems*

7. Bonding in crystals. In order to simplify the following detailed account of crystal structure we classify crystals according to the type of bonding present into *metallic*, *ionic*, *molecular* (or van der Waals) and *covalent networks*.

TABLE 22 THE SEVEN CRYSTAL SYSTEMS

System	Lengths of axes	Angles between axes	Examples
Cubic	$a = b = c$	$\alpha = \beta = \gamma = 90°$	NaCl
Tetragonal	$a = b \neq c$	$\alpha = \beta = \gamma = 90°$	TiO_2-rutile
Orthorhombic	$a \neq b \neq c$	$\alpha = \beta = \gamma = 90°$	rhombic sulphur
Hexagonal	$a = b \neq c$	$\alpha = \beta = 90°\, \gamma = 120°$	PbI_2
Monoclinic	$a \neq b \neq c$	$\alpha \neq \beta = \gamma = 90°$	monoclinic sulphur
Triclinic	$a \neq b \neq c$	$\alpha \neq \beta \neq \gamma \neq 90°$	$CuSO_4.5H_2O$
Rhombohedral	$a = b = c$	$\alpha = \beta = \gamma$	$NaNO_3$

METALLIC STRUCTURES

8. The hard sphere model. The structures of substances such as metallic crystals in which the bonding is non-directional are determined by purely geometrical considerations. To a good approximation, the individual atoms in a metallic crystal can be thought of as hard spheres. The structure of the crystal is then based on the most efficient packing of identical spheres so as to achieve the minimum of unoccupied volume. The two ways of doing this, both of which leave 26% of unoccupied volume, are known as *cubic* and *hexagonal close-packing* (ccp and hcp).

9. Cubic and hexagonal close-packing. Fig 96 (*a*) shows a single layer of close-packed spheres. Each sphere is surrounded by six others and the spaces left, though entirely identical, have been divided into two sets (marked *x* and *y*) for convenience.

The crystal is now built up by placing a second identical layer on top of the first, each of the spheres in the second layer lying directly over a space in the first. Since the *x* and *y* spaces are identical, the second layer can lie directly over either. However, if spheres are placed over all the *x* spaces, it is not then possible also to place spheres over the *y* spaces, and vice versa.

If the *x* spaces are covered, as in Fig. 96 (*b*), then for the third layer there are again two types of space, but these are not now equivalent. Spaces marked *z* lie directly over the centres of *spheres* in the first layer and spaces marked *y* lie directly above the *y spaces* in the first layer.

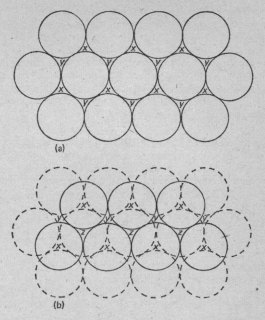

Fig. 96.—*Close packing of spheres.* (*a*) *Single layer.*
(*b*) *Double layer*

Covering the z spaces gives layers repeating as *AB AB AB*, etc., which corresponds to hexagonal close-packing (Fig. 97 (*a*)). Covering the y spaces gives layers repeating as *ABC ABC ABC*, etc., corresponding to cubic close-packing.

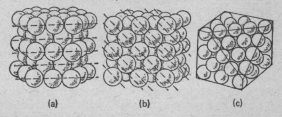

Fig. 97.—*Close packing of spheres.* (*a*) *Hexagonal close packing.*
(*b*) *Cubic close packing.* (*c*) *Cubic close packing with corner spheres removed. The dotted lines indicate the close-packed layers*

Fig. 97 (*b*) illustrates this type of packing and it can be seen that the unit cell is a face-centred cube. The close-packed layers are not parallel to the base of the cube, but rather are parallel to the diagonal across one face (*i.e.* to the (1, 1, 1) plane). In Fig. 97 (*c*) a corner has been cut away to show the close-packing in the (1, 1, 1) plane. In both arrangements each sphere is surrounded by twelve others, six in the same close-packed layer, three above and three below, *i.e.* the *co-ordination number* is 12.

10. Body-centred cubic structure. In this arrangement, based on the body-centred cubic unit cell (Fig. 91 (*b*)), the spheres are not close-packed and each sphere has a co-ordination number of eight, four in the layer above and four in the layer below (Fig. 98). There are six other spheres only slightly farther away than the eight nearest neighbours. The unoccupied volume is now 32%, and because of the less efficient packing the body-centred cubic (bcc) structure gives rise to a lower density than the ccp or hcp structures.

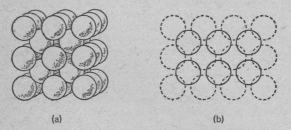

(a)   (b)

FIG. 98.—(*a*) *Body-centred cubic packing of spheres.*
(*b*) *Alternate layers of spheres*

11. Examples. The most malleable and ductile metals, *e.g.* copper, silver and gold, crystallise with the ccp structure, where the layers of atoms can very readily move with respect to one another without disrupting the bonding forces. Beryllium, magnesium and calcium are examples of metals which crystallise with the hcp structure, while many other metals, including the alkali metals, crystallise in a bcc arrangement. Below 906°C iron has the bcc structure, while above this temperature the lattice changes to the ccp structure. Metallic crystals have high lattice energies, *e.g.* iron -415 kJ mol^{-1}.

12. Interstices in close-packed structures. The holes or *interstices* formed between the layers of spheres in close-packed structures are of two different types, (a) *tetrahedral* interstices and (b) *octahedral* interstices.

   (a) *Tetrahedral interstices.* These are formed between three spheres in one layer and either one in the layer above or one in the layer below. The interstice is thus surrounded by four spheres whose centres form the apices of a regular tetrahedron (Fig. 99). The spaces marked x in Fig. 96 (*b*) are tetrahedral

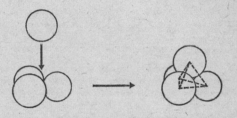

FIG. 99.—*Formation of a tetrahedral interstice between close-packed layers of spheres*

interstices and there are two per sphere. These interstices may be occupied by smaller spheres, but in order to do so the radius of the smaller sphere must be less than 0·225 times that of the larger sphere (*see* Progress Test 19, 3).

   (b) *Octahedral interstices.* The spaces marked y in Fig. 96 (*b*) are surrounded by six spheres whose centres lie at the apices of a regular octahedron, and there is one octahedral interstice per sphere. The geometry of an octahedral interstice is more clearly shown by considering layers of spheres in a face-centred cubic lattice. In Fig. 97 (*b*) the horizontal layers of spheres, *i.e.* those parallel to the faces of the unit face-centred cube, are not close-packed and two are shown in plan form in Fig. 100 (*a*). The octahedral interstice is formed between four spheres in one layer with one sphere in the layer above and one in the layer below. The radius of the sphere which just fills this interstice can be found by taking a cross section through the interstice (Fig. 100 (*b*)). If r_1 and r_2 are the radii of the smaller and larger spheres respectively, then

$$r_2{}^2 + r_2{}^2 = 2r_2{}^2 = (r_1 + r_2)^2$$
$$(2)^{\frac{1}{2}}r_2 = r_1 + r_2$$
$$r_1/r_2 = (2)^{\frac{1}{2}} - 1 = 0 \cdot 414$$

i.e. in order to occupy an octahedral interstice the radius of the smaller sphere must be less than 0·414 times that of the larger sphere.

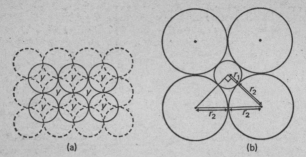

FIG. 100.—(*a*) *Positions of octahedral interstices* (*y*) *formed between layers of spheres parallel to the face of a face-centred lattice.* (*b*) *Cross-sectional, representation of octahedral interstice*

13. Interstitial compounds. The interstices present in close-packed structure may be occupied by small foreign atoms such as H, B, C and N to form *interstitial* metallic hydrides, borides, carbides and nitrides respectively. The smaller tetrahedral interstices may be occupied by H atoms, while B, C and N can occupy only octahedral interstices. In interstitial compounds the interstices are occupied in a random manner and not all the interstices need be filled. Steel is an interstitial compound of iron and carbon, and the type of steel depends on the number of octahedral interstices occupied by carbon atoms. In all metallic interstitial compounds the metal atoms determine the structure and the small non-metallic atoms are inserted between them.

IONIC STRUCTURES

14. Properties of ionic crystals (*see also* XVI). Ionic crystal lattices are made up of repeating units of positively and negatively charged ions arranged so as to achieve the maximum possible stabilisation. Each of the positive ions occupies an equivalent position to every other positive ion, and likewise the negative ions. Every ion in the crystal interacts with every other ion so that the whole crystal can be considered as a single giant molecule.

The crystals are usually hard and brittle, since distortion may bring ions of the same charge close together and the repulsive forces then cause the crystal to fracture. Ionic crystals are insulators because in the solid state there is no mechanism by which electrons or ions can migrate under the influence of an applied electric field. In the fused state, however, the structure is more disordered and the ions conduct the current.

Ionic crystals have high lattice energies, *e.g.* sodium chloride, -766 kJ mol^{-1}.

15. Co-ordination numbers in ionic crystals. Since ionic bonding is non-directional, the structures of ionic crystals are determined by geometrical considerations and are therefore related to the metallic structures described in **9** and **10** above. A given ion tends to surround itself with as many ions of opposite sign as possible, but since anions are usually larger than cations the maximum co-ordination number of twelve for the close packing of equal spheres is never attained. The usual co-ordination numbers in simple ionic crystals are eight, six and four, and these give rise to cubic, octahedral and tetrahedral arrangements of one ion around another. With such symmetrical arrangements, the repulsions between neighbouring ions of the same sign are minimised.

16. Radius ratios. The factor which determines the co-ordination number and hence the structure of an ionic crystal is the *radius ratio*, r_+/r_-, where r_+ is the radius of the cation and r_- that of the anion. The problem is entirely analogous to the filling of interstices in the close-packed structures of metallic crystals (*see* **12** above). Here the larger anions may be regarded as providing the close-packed structure, with the smaller cations occupying the interstices. Thus in sixfold co-ordination the cation is surrounded octahedrally by six anions, *i.e.* it is in an octahedral interstice. In order to occupy this interstice the ratio r_+/r_- must be at least 0·414 (*see* **12** above).

Similarly, in fourfold co-ordination the anions are arranged tetrahedrally around the central cation, and r_+/r_- must be at least 0·225 (*see* Progress Test 19, 3).

The maximum co-ordination, eightfold co-ordination, is obtained when the anions form a cube around the central cation and calculations similar to those used earlier indicate that r_+/r_- must be greater than 0·732.

NOTE: $r_+/r_- = 1$ corresponds to close-packing of equal spheres and gives rise to a co-ordination number of 12.

For a radius ratio from 1 to 0·732, eightfold co-ordination is favoured, but below 0·732 oppositely charged ions no longer touch and the structure changes to sixfold co-ordination. The three situations which arise are summarised below:

Eightfold co-ordination, cubic $1 > r_+/r_- > 0·732$
Sixfold co-ordination, octahedral $0·732 > r_+/r_- > 0·414$
Fourfold co-ordination, tetrahedral $0·414 > r_+/r_- > 0·225$

These rules are not completely rigorous, owing to the fact that ions are not hard spheres and that ions of opposite charge are not in contact. Nevertheless they do provide a reasonable guide, as the following examples show.

17. Some typical ionic structures

(a) *Eightfold co-ordination, the caesium chloride structure*, e.g. CsCl, CsBr, CsI. These ionic solids have the bcc structure, in which each ion is surrounded by eight others of opposite sign (Fig. 101 (a)).

(b) *Sixfold co-ordination, the rock salt structure*, e.g. alkali metal halides other than those mentioned in (a) above. The alkali metal ion is too small to allow eightfold co-ordination, and only six-fold co-ordination is possible. The halide ions form a ccp structure, with alkali metal ions in all the octahedral interstices. As shown in Fig. 101 (b), the structure can be thought of as two interpenetrating face-centred cubic lattices.

(c) *Fourfold co-ordination, the zinc blende structure*, e.g. ZnS and many 1:1 ionic (and partly ionic) compounds. In ZnS the large S^{2-} ions form a ccp structure, with Zn^{2+} ions occupying alternate tetrahedral interstices. Fig. 101 (c) shows an expanded diagram of the unit cell, which is composed of a face-centred cubic lattice of S^{2-} ions interpenetrated by a simple cubic lattice of Zn^{2+} ions. Each Zn^{2+} ion is surrounded by four S^{2-} ions, and vice versa.

(d) *The fluorite structure*, e.g. CaF_2 and certain other XY_2 compounds. In CaF_2 the Ca^{2+} ions form a ccp structure, with F^- ions in all the tetrahedral interstices. As shown in Fig. 101 (d), each F^- ion is tetrahedrally surrounded by four Ca^{2+} ions, and each Ca^{2+} ion is surrounded by eight F^- ions at the corners of a cube. Thus the co-ordination number of the Ca^{2+} ion is twice that of the F^- ion, as must be the case if electrical neutrality is to be maintained. The unit cell comprises a face-centred cubic lattice of Ca^{2+} ions interpenetrated by a simple cubic lattice of F^- ions.

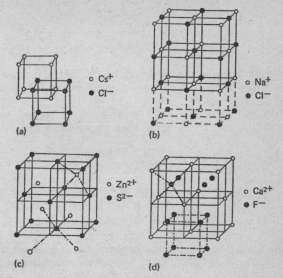

○ Cs⁺
● Cl⁻
(a)

○ Na⁺
● Cl⁻
(b)

○ Zn²⁺
● S²⁻
(c)

○ Ca²⁺
● F⁻
(d)

Fig. 101.—(a) *The caesium chloride structure, showing the eightfold co-ordination of each ion.* (b) *The rock salt structure, showing the sixfold co-ordination of each ion.* (c) *The zinc blende structure, showing the fourfold co-ordination of each ion.* (d) *The fluorite structure, showing the eightfold co-ordination of the cation, and the fourfold co-ordination of the anion*

MOLECULAR (VAN DER WAALS) CRYSTALS

18. Properties of molecular crystals. In molecular crystals the fundamental particle of the unit cell is an atom or molecule carrying no net charge. The binding in the crystal is due solely to the weak van der Waals forces (*see* XVIII, 8), so that the crystals show little resistance to distortion, are readily volatile and have low lattice energies.

19. Close-packing in molecular crystals. Since the van der Waals forces are non-directional, the structures are usually close-packed, *e.g.* the high temperature forms of solid hydrogen, nitrogen and oxygen crystallise with the hcp structure. The noble gases, with the exception of helium, crystallise with the ccp structure.

COVALENT NETWORKS

20. Properties of covalent networks. Covalent networks are crystals in which the geometry is determined by the formation of well defined covalent bonds.

Distortion of the crystal involves breaking strong covalent bonds, and the hardest substances known are built up of three-dimensional covalent networks.

Lattice energies are high (*e.g.* diamond, -711 kJ mol^{-1}), and the crystals have very high melting points.

Though often similar in form to ionic crystals, three-dimensional covalent networks can be distinguished by the fact that their very low electrical conductivity does not increase in the fused state (if they can be fused).

21. Three-dimensional covalent networks. The most familiar example of a covalent network is the diamond crystal, in which the carbon atoms are joined tetrahedrally in a three-dimensional network by sp^3 carbon–carbon bonds. Diamond has a zinc blende structure (Fig. 101 (*c*)) in which both zinc and sulphur ions are replaced by carbon atoms. In effect the entire crystal is a single giant molecule. Carborundum, SiC, also has a diamond structure, in which a silicon atom is tetrahedrally surrounded by four carbon atoms, and vice versa.

22. Two-dimensional networks. An example of a two-dimensional network is provided by the layer structure of graphite (Fig. 102).

Each carbon atom in a particular layer is covalently bonded to three other carbon atoms by sp^2 carbon–carbon bonds (0·142 nm long), the π electrons being delocalised over the entire layer very much as in benzene. Each layer is effectively a giant covalently bonded flat molecule. The bonding between successive layers is by the weak van der Waals attraction, the bond distance being 0·340 nm—much larger than any normal carbon–carbon bond.

The electrical conductance of graphite is explained by the mobility of the π electrons, and its lubricating properties by the weak van der Waals forces allowing layers to slide easily over one another.

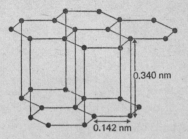

FIG. 102.—*The graphite structure*

23. One-dimensional networks. Atoms with only two valency electrons form chains (*e.g.* monoclinic plastic sulphur—spiral chains) or rings (*e.g.* rhombic sulphur—puckered eight-membered rings). Individual chains or rings are held together by van der Waals forces.

X-RAY CRYSTALLOGRAPHY

24. The discovery of X-ray diffraction. Von Laue (1912) suggested that if the wavelengths of X-rays were of the same order as the spacings between atoms, then it should be possible to observe diffraction patterns when X-rays passed through, or were reflected by, crystals. This was very soon confirmed using copper sulphate crystals, and the technique was taken up for systematic study by W. H. and W. L. Bragg in England as well as by von Laue's group in Munich.

The description of the theory and technique which follows is mainly due to the Braggs.

25. The reflection of X-rays. The scattering of X-rays by crystals is due to reflection by successive planes of particles, the angle of reflection θ being equal to the angle of incidence, as shown in Fig. 103. Here we consider reflection from two planes, the interplanar spacing being d, so that X-rays reflected from the second plane travel a distance $2d \sin \theta$ farther than those

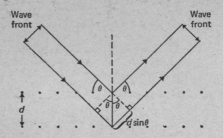

FIG. 103.—*Reflection of X-rays by parallel planes of atoms*

reflected from the first plane. If the two reflections are to reinforce as in Fig. 104 (*a*) (constructive interference), instead of destroying each other as in Fig. 104 (*b*) (destructive interference), then the extra distance travelled by the X-rays re-

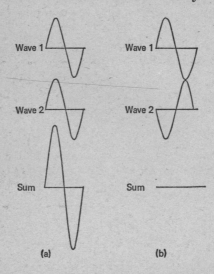

FIG. 104.—(*a*) *Constructive and* (*b*) *destructive interference between waves*

flected from the second plane must be an integral number of wavelengths, *i.e.*

$$2d \sin \theta = n\lambda \qquad \text{. . . (XIX, 1)}$$

where λ is the wavelength of the X-rays and n is an integer called the order of the reflection.

26. The Bragg apparatus. A schematic representation of the Bragg apparatus is shown in Fig. 105. The detector was

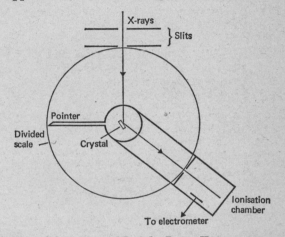

FIG. 105.—*Schematic model of the Bragg X-ray spectrometer.*

originally an ionisation chamber filled with methyl bromide, though nowadays photographic recording is normally used. The wavelength of the X-rays depends on the target material used in the X-ray source, *e.g.* Cu—0·1537 nm, Cr—0·2285 nm, Mo — 0·0708 nm, and is usually selected from prior knowledge of the likely spacings in the crystal.

27. The Bragg experiment. The crystal is set at an arbitrary angle to the incident beam and rotated until one set of planes satisfies the Bragg relation. At this point the reflected waves reinforce and a strong signal is registered by the detector. As the crystal is rotated further the signal disappears, then on further rotation another strong signal is registered as another set of planes satisfies the Bragg relation, and so on.

By matching up the observed reflections with those calculated for postulated structures, the structure of the crystal under investigation can be fixed, as shown in **28** and **29** below.

DETERMINATION OF THE ROCK SALT STRUCTURE BY X-RAY CRYSTALLOGRAPHY

28. Determination of the lattice type. From its external form, sodium chloride is obviously based on a cubic lattice.

Fig. 106 shows the distances between various successive planes of atoms in the three types of cubic lattices. Taking a simple cubic lattice of side a as an example, the distance between successive $(1, 0, 0)$ planes is a (*i.e.* $d_{100} = a$), that

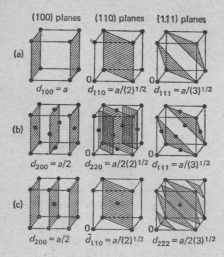

FIG. 106.—*Atomic planes in (a) simple, (b) face-centred, (c) body-centred cubic lattices*

between successive $(1, 1, 0)$ planes $a/(2)^{\frac{1}{2}}$ (*i.e.* $d_{110} = a/(2)^{\frac{1}{2}}$) and between successive $(1, 1, 1)$ planes $a/(3)^{\frac{1}{2}}$ (*i.e.* $d_{111} = a/(3)^{\frac{1}{2}}$). Similarly, in the body-centred and face-centred lattices the distances are: $d_{100} = a/2$, $d_{110} = a/(2)^{\frac{1}{2}}$, $d_{111} = a/2(3)^{\frac{1}{2}}$ and $d_{100} = a/2$, $d_{110} = a/2(2)^{\frac{1}{2}}$, $d_{111} = a/(3)^{\frac{1}{2}}$ respectively.

Thus for the three types of cubic lattices the ratios of the distances between successive planes are:

Simple cubic: $d_{100} : d_{110} : d_{111} = 1 : 0{\cdot}707 : 0{\cdot}577$
Face-centred: $d_{100} : d_{110} : d_{111} = 1 : 0{\cdot}707 : 1{\cdot}544$
Body-centred: $d_{100} : d_{110} : d_{111} = 1 : 1{\cdot}414 : 0{\cdot}577$

From the Bragg relation (XIX, 1), the condition for maximum reflection is:

$$d = n\lambda/2 \sin \theta$$

Thus if we confine our attention to any particular order of reflection:

$$d_{100} : d_{110} : d_{111} = 1/\sin \theta_1 : 1/\sin \theta_2 : 1/\sin \theta_3 \ldots \text{(XIX, 2)}$$

By comparing the observed ratios of $\sin \theta$ with those calculated for the three types of cubic lattices, sodium chloride can be shown to have the face-centred cubic structure.

29. Determination of the positions of sodium and chloride ions within the lattice.

For the $(1, 0, 0)$ and $(1, 1, 0)$ planes, the intensity of the reflection falls off regularly with increasing order, indicating that these planes contain equal numbers of sodium and chloride ions.

For the $(1, 1, 1)$ planes there is an alternation of intensities, reflections of odd order being weak and those of even order being strong. This indicates that the $(1, 1, 1)$ planes are composed alternately of sodium and chloride ions, the sodium ion planes lying midway between successive chloride ion planes. Thus when the crystal is correctly oriented for odd order reflections from the chloride ion $(1, 1, 1)$ planes, reflection is also achieved from the sodium ion $(1, 1, 1)$ planes, which are exactly half a wavelength behind, causing destructive interference. For even order reflection the X-rays reflected from the chloride ion $(1, 1, 1)$ planes are one wavelength behind those reflected from the sodium ion $(1, 1, 1)$ planes, so that constructive interference takes place.

30. X-ray powder photography.

X-ray crystallography can also be performed on microcrystalline powders.

Out of the infinite number of random orientations of the microcrystallites there will always be some which are at the correct angle for reflection. The reflected beam makes an angle of 2θ with the incident beam, as in Fig. 103, but since the microcrystallites are oriented in all directions the reflected X-rays will describe a cone of half-angle 2θ about the incident beam. The reflected beam is detected by a cylindrical strip of film surrounding the specimen to give a diffraction photograph such as that shown in Fig. 107. The calculation of θ (and therefore of d) from the radius of the film and the spacings between the lines is then quite straightforward.

Point where incident
beam enters ($2\theta = 180°$)

FIG. 107—X-ray powder photograph

NOTE: As the particle size of the powder decreases, the lines in the diffraction photograph become broader and more blurred.

OTHER APPLICATIONS OF X-RAY DIFFRACTION

31. Electron density maps. In many crystals the lattice sites are occupied not by simple spherical atoms or ions but by complete molecules or groups of atoms. In such cases X-ray diffraction may also be used to obtain electron density maps of the occupants of different lattice sites, since the scattering power of a particular atom in a molecule depends on the number of electrons it contains.

This technique has been applied with great success to the analysis of the structures of highly complex organic molecules.

32. Chemical analysis. A further use of X-ray diffraction is in chemical analysis, *e.g.* an oxide of cadmium can be prepared whose stoichiometry suggests a formula Cd_2O. However, the X-ray reflections from this compound are characteristic of CdO and metallic cadmium, suggesting the presence of a mixture rather than a true compound.

An allied analytical use is the "fingerprinting" technique, in which a compound or mixture is analysed by comparing its diffraction photograph with those of known compounds.

33. Limitations of X-ray diffraction. The scattering of X-rays is due solely to interactions with electrons, and the higher the atomic number of an element the greater will be its scattering power. Hydrogen atoms, having only one electron, scatter X-rays very weakly and cannot be located using this technique. However, *neutron scattering*, which is due to the interaction of neutrons with *nuclei*, is independent of the number of electrons present, and can be used to locate the position of hydrogen atoms.

THE HEAT CAPACITIES OF SOLIDS

34. The law of Dulong and Petit. This law, which was formulated in 1819, states that the molar heat capacity at constant pressure of a solid element is about 27 J K^{-1} mol^{-1}. For solids C_p and C_V are very similar.

The constancy of the molar heat capacity was explained using the principles involved in the kinetic theory of gases (*see* IV). The atoms in a solid can vibrate about their mean lattice positions, and these vibrations can be considered as taking place along three mutually perpendicular axes. Each mode of vibration will contribute R to C_V (*see* IV, **18**) and for three independent modes of vibration $C_V = 3R = 24 \cdot 9$ J K^{-1} mol^{-1}, which is in agreement with the above law.

However, some of the lighter solid elements have molar heat capacities much less than 25 J K^{-1} mol^{-1}; *e.g.* 5·8 J K^{-1} mol^{-1} for carbon (diamond), 11·7 J K^{-1} mol^{-1} for boron and 16·3 J K^{-1} mol^{-1} for beryllium. In addition, all molar heat capacities decrease with decreasing temperature and tend to zero at the absolute zero of temperature (*see* Fig. 108). According to the classical approach, the molar heat capacity should be independent of temperature.

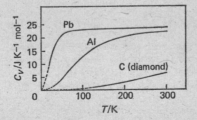

FIG. 108.—*Dependence on temperature of the molar heat capacities of three elements*

35. Quantum theories of molar heat capacities. The above anomaly was resolved by Einstein in 1907 using the quantum theory (*see* XV). Einstein assumed that all the atoms in a solid vibrate with a single frequency which is characteristic of the solid and depends on the mass of the atoms and the lattice forces. Also, the absorption of vibrational energy can take place only in discrete units in accordance with Planck's quantum hypothesis. At low temperatures most atoms will have small or zero vibrational energy and hence will contribute little to the heat capacity. As the temperature is raised, the vibrational energy will increase and C_V will tend towards $3R$.

Einstein's theory was subsequently refined by Debye, who assumed that the atoms in a solid could vibrate with any frequency from zero up to some limiting frequency ν_D. ν_D is low for solids composed of light, firmly held atoms such as carbon, and high for those composed of heavy, loosely held atoms such as metals. Debye was able to formulate an expression for C_V which gave the correct form of the temperature dependence in terms of ν_D. This expression is used to estimate the heat capacities of solids in the region 0–20 K, where experimental determinations are difficult.

PROGRESS TEST 19

1. A crystal plane intercepts the crystallographic axes at $\frac{3}{4}, \frac{3}{2}$, 1 multiples of the unit distance. Find the Miller index of the plane. (5)

2. Using the diagrams of Fig. 96, find the unoccupied volume in the ccp and body-centred cubic arrangements. (8–10)
Hint. Remember that atoms at the corners of a cube are shared by eight unit cubes and those at the centres of faces by two unit cubes.

3. Show that the radius of a tetrahedral site in a close-packed lattice formed by spheres of radius r is $0.225r$. (12)
Hint. A tetrahedral site may be generated by placing four spheres at alternate corners of a cube. The spheres are in contact along the face diagonals.

4. First-order reflections from the (1, 0, 0), (1, 1, 0) and (1, 1, 1) planes of a certain crystal are recorded at angles 7° 10′, 5° 4′ and 10° 12′ respectively. From which cubic lattice is the crystal derived? (25–29)

5. Using X-rays from a copper target ($\lambda = 0 \cdot 1537$ nm) the angle for first-order reflection from the (1, 0, 0) planes of a sodium chloride crystal is $15°$. The density of sodium chloride is $2 \cdot 17 \times 10^3 \ kg \ m^{-3}$. Calculate Avogadro's constant. (25–29)

6. Explain, with examples, the meaning of Miller indices. (5)

7. Write notes on the close-packing of equal spheres. (9)

8. Explain how radius ratios influence the structure of ionic crystals. (16)

9. Draw the following structures: (*a*) caesium chloride, (*b*) rock salt, (*c*) zinc blende, (*d*) fluorite. (17)

10. Derive the Bragg equation. (25)

11. Outline the quantum theories of molar heat capacities. (34)

CHAPTER XX

LIQUIDS

COMPARISON OF LIQUIDS WITH SOLIDS AND GASES

1. Mathematical models for gases and solids (recapitulation). In the gas, intermolecular forces are far outweighed by thermal energy, and the resulting disorder can be satisfactorily treated by statistical means.

In the crystal, the reverse situation holds. The atoms, ions or molecules are held rigidly in their lattice positions by powerful intermolecular forces, and the only motion is vibration about the mean lattice position. This situation may also be described satisfactorily by mathematical means.

2. Mathematical model for liquids. The liquid state is intermediate between the solid and gaseous states.

NOTE: This does not imply that any particular property of a liquid is the average of the corresponding properties of the solid and gaseous states. Rather, that some properties resemble those of a gas, while others resemble those of a solid.

Intermolecular forces in liquids, though strong, are not sufficient to prevent the molecules having considerable translational energy, *i.e.* in a liquid the regularity of structure induced by the cohesive forces is partially destroyed by thermal motions. The mathematical study of the liquid state is therefore extremely complex, and most treatments consider the liquid either as an imperfect gas or an imperfect solid.

3. Properties of liquids. Liquids resemble gases in their exhibition of the Brownian motion, viscous flow, and in their isotropic properties. However, in their densities and incompressibilities liquids resemble solids.

Further indications of the resemblance to solids are found in the fact that much more energy is required to convert a liquid into a gas than to convert a solid into a liquid (*e.g.* for water, latent heat of fusion $= 6 \cdot 01$ kJ mol^{-1}, latent heat of vaporisation $= 40 \cdot 6$ kJ mol^{-1}).

Liquids also exhibit surface tension, a property essentially similar to that of hardness in solids. Molecules at the liquid surface experience only attractive forces due to molecules in the same plane or below. The whole liquid surface is thus under tension and tends to contract to reduce the surface area (causing the formation of spherical drops) and to resist penetration.

The most striking and important resemblance between liquids and solids, however, is to be found in their X-ray diffraction patterns, which are considered in detail below.

THE STRUCTURES OF LIQUIDS

4. X-ray diffraction in liquids. The observation of definite maxima in the X-ray diffraction patterns of liquids shows that some sort of structural order is present. In fact the diffraction patterns very closely resemble those of microcrystalline powders (*see* Fig. 107), except that the sharp lines in the powder photograph became blurred and diffuse in the liquid. This indicates that the diffraction angle θ in the Bragg relation (XIX, 1) $n\lambda = 2d \sin \theta$ is not well defined and therefore that the distance between repeating structural units has no single value.

5. Interpretation of the diffraction pattern using the radial distribution function. The intensity at any point in the diffraction pattern can be interpreted statistically by relating it to the probability that an atom will be found at a distance r from some arbitrarily chosen reference atom. Fig. 109 shows

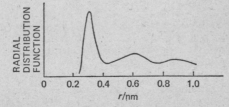

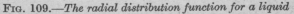

FIG. 109.—*The radial distribution function for a liquid*

a plot of the probability that an atom will be found in a spherical shell of thickness dr at a distance r from the central atom (the radial distribution function, *see* XV, 29).

NOTE: The radial distribution shown in Fig. 109 and its interpretation in 6–8 below are characteristic of unassociated liquids. Associated liquids (in which the cohesive forces are strongly directional), such as water, are considered later in this chapter.

6. Interpretation of the radial distribution function in terms of the liquid structure.
As shown in Fig. 109, the probability of finding an atom at values of r less than the van der Waals radius (*see* XVIII, 9) is zero.

The probability rises to a maximum for values of r around the van der Waals distance (showing that the first co-ordination shell around the reference atom is approximately close-packed) and then falls to a minimum. This is because the presence of a nearest neighbour at the van der der Waals distance tends to prevent another molecule being present at, say, one and a half times the van der Waals distance.

The next maximum occurs at about twice the van der Waals distance, but has greatly reduced intensity, and so on.

NOTE: Notice that the probability does not drop to zero between maxima as it would in the corresponding crystal, where the atoms are rigidly fixed.

7. The hole theory.
The very presence of well defined maxima in Fig. 109 shows that regular co-ordination is present, though their small number and relative sizes indicate that departure from an ordered arrangement becomes greater as the distance from the reference atom increases. This lack of long-range order in a liquid can be pictured as caused by the introduction of randomly moving holes of variable size and shape into the regular structure characteristic of a solid.

NOTE: Just as a gas may be thought of as mainly empty space through which a few molecules move at random, so a liquid may be thought of as the inverse, *i.e.* mainly matter through which a few holes move randomly.

8. Evidence supporting the hole theory.
The presence of holes nicely explains the flow and diffusional properties of liquids, since the molecules can move past each other by

utilising the "extra" volume of the holes. Also, liquids are fairly compressible over the first 3% or so of their volumes, but at higher pressures the compressibility drops markedly. This initial high compressibility can be thought of as being due to molecules occupying the free volume of the holes.

9. Order–disorder phenomena. The sharp melting point characteristic of the transformation of a solid to a liquid occurs because it is not possible to introduce disorder into an ordered structure gradually. Long-range order is a property associated with the positions of many molecules, and it is not possible to introduce small regions of disorder into a crystal without at the same time destroying the long-range order.

THE WATER STRUCTURE

The structures of associated liquids are illustrated here by a fairly detailed study of the structure of water.

10. Structure of ice. As shown in XVI, **20**, a water molecule has two lone pairs in sp^3 hybrid orbitals on the oxygen, and two σ O–H bonds formed by overlap of a hydrogen $1s$ orbital with an sp^3 orbital on oxygen.

In ice each water molecule is tetrahedrally linked to four others by hydrogen bonds (*see* XVIII, **6, 7**). The four hydrogen bonds associated with a particular water molecule are formed by each of the sp^3 lone pairs hydrogen bonding to the hydrogen of another water molecule, while each of the hydrogens forms a hydrogen bond with the lone pair on another water molecule. The structure of ice thus consists of a three-dimensional network held together by hydrogen bonds, each water molecule being surrounded tetrahedrally by four others.

11. Structure of liquid water. The structure of water is best thought of as retaining the tetrahedral co-ordination of ice over short ranges and for short periods of time.

When ice melts, some of the hydrogen bonds are broken, permitting closer packing, and the open structure characteristic of ice collapses to some extent. This effect increases with increasing temperature, each water molecule being surrounded on average by 4·4 molecules at 1·5°C and by 4·9 at 60°C. The combination of this effect with the normal decrease in density with increasing temperature accounts for the fact that water has its maximum density at 4°C.

Hydrogen bonding is still present in water vapour, steam being approximately 10% hydrogen bonded at 100°C.

12. Solutions of non-polar gases in water. Simple non-polar gases dissolve in water with a loss of entropy. Since the entropy of a system measures the disorder prevailing (*see* VIII, **12, 13**), this indicates that the water structure becomes more ordered under the influence of the dissolved molecules. The water molecules can be thought of as building a minute iceberg around the non-polar molecule.

13. Solutions of ions in water. The gain in entropy on dissolution of an ionic substance in water indicates that the structure has become less ordered. This arises because of the orientation of the water molecules around the dissolved ions. Thus water molecules in the first layer around a cation cannot all participate in the normal tetrahedral arrangement, since they are very strongly held ("frozen"), with all their hydrogens pointing outwards, as in Fig. 110 (*a*); the normal arrangement

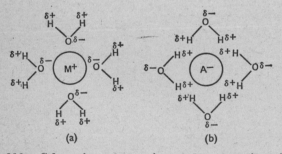

Fig. 110.—*Schematic representation of the orientation of the first layer of water molecules around* (a) *a cation,* (b) *an anion*

requires two of the hydrogens to be pointing inwards. In a similar fashion, water molecules in the first layer around an anion are oriented with all their hydrogens pointing inwards, as in Fig. 110 (*b*). Polyvalent monatomic ions (*e.g.* Al^{3+}) have a still greater effect, since the "frozen" region extends to layers beyond the first.

OTHER STATES OF MATTER

Not all materials fall neatly into a solid–liquid–gas classification. The chief offenders in this category are considered in 14–16 below.

14. Liquid crystals. In some substances the tendency towards an ordered arrangement is so great that, before melting to a true liquid, the compound passes through a *mesomorphic* or *liquid crystal* state in which it exhibits some of the properties of both solids and liquids. Thus between 116° and 135°C p-azoxyanisole flows like a liquid but shows anisotropic reflection and refraction of light.

Molecules of such compounds are parallel to one another in the crystal and remain parallel in the liquid crystal over limited regions, though the regular disposition within rows or layers is lost and the population of any given region is not stationary.

15. Glasses. The formation of a glass such as silica can be pictured as follows. Most liquids can be cooled at least 5°–10°C below their freezing points before crystals form. The temperature then rises to the freezing point, since this is the only temperature at which solid and liquid can co-exist. If for some reason crystallisation is difficult, then as the temperature is lowered translational motions become smaller and smaller until, within a small temperature range, all translational motion ceases. The glass so formed is not crystalline, but is effectively a "frozen" liquid configuration. Crystallisation does not normally take place from the glassy state because the necessary translational motions cannot occur.

The same bonds are present in a glass as in the corresponding liquid, but these are not all of the same length. Thus a glass softens gradually over a wide temperature range rather than melting sharply—because there is no single temperature at which all the bonds are simultaneously loosened.

16. Rubbers. Rubbers consist of polymerised hydrocarbons forming long, linear chains. Figs. 111 (a) and (b) show idealised models of the structure of a rubber in the stretched and unstretched states. The characteristic elasticity of rubber is a measure of its tendency to revert to the more disordered structure of Fig. 111 (*b*) on release of the tension, and X-ray diffraction shows quite a high degree of crystallinity in stretched rubbers.

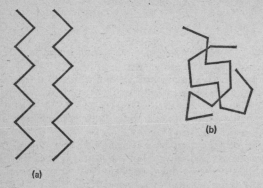

(a)

(b)

Fig. 111.—*Idealised models of the rubber structure.*
(a) *stretched,* (b) *unstretched*

PROGRESS TEST 20

1. List some of the properties of liquids in which they resemble (*a*) gases, (*b*) solids. (**3**)
2. What is meant by the hole theory of liquids? (**7**)
3. Describe the structures of ice and liquid water. (**10, 11**)
4. Describe three states of matter which do not fall neatly into the solid, liquid, gas classification. (**14–16**)

PART EIGHT

CHEMICAL KINETICS

Chapter XXI begins with the definition of some important terms and follows with an outline of the methods used for measuring reaction rates. The characteristic rate equations for zero-, first-, second- and third-order reactions are derived, and many examples are given of these types of reaction. More complicated systems, in the form of reversible reactions, consecutive reactions and chain reactions, are also considered, and the chapter closes with a list of the methods available for the determination of the order and the rate constant of a particular reaction.

The dependence of the rate constant on temperature is discussed in Chapter XXII. The two theories which have been postulated to explain the temperature dependence of the rate constant, the collision theory and the transition state theory, are compared and contrasted. Finally the general properties of catalytic reactions are outlined and an account is given of homogeneous catalysis.

CHAPTER XXI

ELEMENTARY CHEMICAL KINETICS

We have seen that thermodynamics enables us to deduce whether or not a particular reaction is feasible but gives no information on how rapidly the reaction will proceed. Thus for the reaction

$$H_2(g) + \tfrac{1}{2}O_2(g) \longrightarrow H_2O(l)$$

$\Delta G_{289}^{\ominus} = -285\cdot8$ kJ mol^{-1}, but the reaction is not observed when hydrogen and oxygen are mixed in the correct proportions at room temperature. A reaction path must be provided by introducing free radicals (*see* **26** below) or by performing the reaction on a catalyst surface (*see* XXIII). Reaction kinetics involves a quantitative treatment of the rates of chemical reactions and allows deductions to be made about the reaction paths.

NOTE: The arrow in the above equation indicates that only the reaction from left to right is important.

DEFINITIONS

1. The rate of reaction. The rate of a chemical reaction is the rate at which the concentration of any reactant or product varies with time. The rate of the general reaction

$$aA + bB \longrightarrow cC + dD \ . \ . \ . \ (XXI, 1)$$

may be expressed either as the rate of disappearance of a reactant or as the rate of appearance of a product. Thus $-d[A]/dt$ is the rate of disappearance of A and $+d[C]/dt$ is the rate of appearance of C. The relationship between the various rates is given by

$$-d[A]/adt = -d[B]/bdt = d[C]/cdt = d[D]/ddt$$

The rate of reaction (XXI, 1) is given by any of these expressions, the quantity which is actually measured being a matter of experimental convenience. Thus if A is a gas, while B, C and D are all solids, it may be most practical to follow the reaction by measuring the decrease in pressure (which is proportional to the decrease in concentration) of A.

2. The law of mass action. The rate of a chemical reaction is dependent on the concentrations of the reactants, and this dependence is given by the *law of mass action*. One statement of this law is: *the rate of any chemical reaction is proportional to the concentration of the reactants participating in the rate-determining step*. The concept of a rate-determining step is discussed in **4** below.

3. The rate law. The rate law is an explicit equation for the mass action effect in a particular reaction, *i.e.* it describes the dependence of the rate of the reaction on concentration, and for the reaction

$$2A + B \longrightarrow 3C$$

is given by

$$-d[A]/2dt = -d[B]/dt = d[C]/3dt = k[A]^m[B]^n$$
$$\ldots \text{ (XXI, 2)}$$

where m and n are constants having integral or fractional values. The *order of reaction* with respect to A is m, with respect to B is n, and the *overall order of reaction* is $m + n$. The overall order may range from zero to three, with first-order ($m + n = 1$) and second-order ($m + n = 2$) being the most common. The proportionality constant k in (XXI, 2) is called the *rate constant* and is numerically equal to the rate of reaction when all reactants have unit concentration.

4. Order and molecularity. m and n in (XXI, 2) are not necessarily equal to the stoichiometric coefficients of A and B (2 and 1 respectively) in the overall reaction. Thus for the reactions:

$$2N_2O_5(g) \longrightarrow 4NO_2(g) + O_2(g) \quad -d[N_2O_5]/dt = k[N_2O_5]$$
$$\ldots \text{ (XXI, 3)}$$
$$2NO(g) + 2H_2(g) \longrightarrow N_2(g) + 2H_2O(g)$$
$$-d[NO]/dt = k[NO]^2[H_2]$$
$$CO(g) + Cl_2(g) \longrightarrow COCl_2(g) \quad -d[CO]/dt = k[CO][Cl_2]^{\frac{3}{2}}$$

It is apparent that the order for each reactant cannot be predicted from the equation for the reaction and must be found experimentally.

The term *molecularity* is often confused with order. The *order* of a reaction is the number of chemical species (molecules, atoms, etc.) whose concentrations affect the rate of the reaction. The molecularity of a reaction is the number of chemical species taking part in the *rate-determining step*. The difference may readily be seen by reference to an example. The decomposition of N_2O_5, (XXI, 3), does not take place in a single step but rather by a sequence of *elementary* reactions which form a complex reaction mechanism.

NOTE: An elementary reaction is one which occurs in a single stage.

$$N_2O_5 \rightleftharpoons NO_2 + NO_3 \qquad \ldots \text{ (XXI, 4)}$$
$$NO_2 + NO_3 \longrightarrow NO_2 + O_2 + NO \ldots \text{ (XXI, 5)}$$
$$NO + NO_3 \longrightarrow 2NO_2 \qquad \ldots \text{ (XXI, 6)}$$

Equilibrium (XXI, 4) is rapidly attained and produces very small amounts of NO_2 and NO_3. (XXI, 5), the slowest of the three steps, determines the rate of the reaction and is known as the rate-determining step. (XXI, 6) is also fast. (XXI, 5) may be thought of as acting as a bottleneck to the other two faster reactions. The rate-determining step (XXI, 5) is *bimolecular*, while the overall reaction is first-order. *Unimolecular* and *termolecular* reactions are also known. Molecularity is meaningless when applied to complex reactions and is reserved for elementary rate-determining steps, since it is only for such reactions that the order can be inferred directly from the stoichiometric coefficients. Molecularity is a theoretical concept which is applied once the mechanism of a reaction is known, whereas order is empirical and must be determined experimentally.

MEASUREMENT OF REACTION RATES

In order to formulate the rate equation it is necessary to collect data on the rate of the reaction at some specified temperature as a function of the concentrations of the reacting species. This is achieved by determining the concentration of one of the constituents of the reacting system at various times during the course of the reaction. There are two techniques in general use.

5. Sampling techniques. In this method a sample is removed from the reaction mixture at various times and analysed. For reactions which are fast relative to the time involved in sampling and analysis there is an uncertainty in fixing this time. This effect may be minimised by *quenching* or *freezing* the sample, thus abruptly slowing down or stopping the reaction. Quenching techniques include:

(a) Rapid cooling of the sample to a suitable low temperature.
(b) Diluting the sample.
(c) Adding a reagent which chemically stops the reaction, *e.g.* the addition of a base to an acid-catalysed reaction.

NOTE: A catalyst is a substance which alters the rate of a chemical reaction but remains unchanged in quantity at the end of the reaction (*see* XXII, 12–21).

6. Continuous techniques. A more elegant way of following the course of a reaction is by continuous observation of some physical property which changes as the reaction proceeds. The following physical methods have been used:

(a) *Optical rotation.* The course of a reaction involving optically active compounds may be followed by polarimetric measurement of the degree of rotation of plane-polarised light passing through the solution. The amount of rotation depends, among other things, on the concentration of the optically active components.

NOTE: The wave motion of a beam of light can be described in terms of the components of its vibration in two mutually perpendicular planes containing the beam. It is possible to isolate the component in one plane, giving rise to a beam of plane-polarised light.

(b) *Spectrophotometry.* If one or more of the constituents in a reaction absorbs light at a particular wavelength, the reaction may be followed by monitoring the change in intensity of the light transmitted by the reaction mixture.
(c) *Dilatometry.* The small volume changes which often accompany a reaction in solution can be followed using a dilatometer. This is an extremely sensitive instrument which allows precise measurement of small volume changes.

(d) *Electrical methods.* These include measurements of the dielectric constant, the electrical conductance and the pH.

(e) *Pressure measurements.* Gas phase reactions or reactions in which a gas is consumed or produced can often be followed by observing the pressure change in the reaction vessel.

(f) *Radioactive isotopes.* Certain reactions can only be followed using radioactive isotopes. The fate of a radioactive isotope can be followed during the course of the reaction by virtue of the particles or electromagnetic radiation it emits.

ZERO-ORDER REACTIONS

7. The rate equation. Zero-order reactions are those which proceed at the same rate regardless of concentration. The general reaction may be represented by

$$A \longrightarrow \text{Products}$$

If the reaction is zero-order with respect to A, the rate law is

$$-da/dt = k_0$$

where a is the concentration of A at time t and k_0 is the zero-order rate constant. This expression may be integrated between the limits $a = a_0$ at $t = 0$ and $a = a$ at $t = t$:

$$-\int_{a_0}^{a} da = k_0 \int_{0}^{t} dt$$
$$-[a]_{a_0}^{a} = k_0[t]_{0}^{t}$$
$$\therefore a_0 - a = k_0 t \quad \ldots \quad (XXI, 7)$$

NOTE: In the remainder of this chapter a_0, b_0, etc., represent the concentration of species A, B, etc., at zero time, whilst a, b, etc., represent those at time t.

The dimensions of k_0 are (concentration) (time)$^{-1}$. Equation (XXI, 7) indicates that a plot of a against t will give a straight line of slope $-k_0$ and intercept a_0.

8. The half-life. The half-life $(t_{\frac{1}{2}})$ of a chemical reaction is the time taken for the concentration of any reactant to fall to half its initial value. Thus substituting the condition $a = a_0/2$ when $t = t_{\frac{1}{2}}$ into (XXI, 7) gives

$$t_{\frac{1}{2}} = a_0/2k_0$$

For a zero-order reaction, the half-life is proportional to the initial concentration of the reactant.

9. Examples. Zero-order reactions are not common and usually involve heterogeneous processes occurring on surfaces. A typical example is the decomposition of ammonia gas on a hot tungsten wire catalyst. The explanation for the zero-order behaviour appears to be that the decomposition can occur only on the surface of the catalyst. Once the surface of the tungsten wire is covered at a given concentration of ammonia, further increase in concentration in the gas phase cannot further increase the surface concentration. Thus, beyond a certain concentration, the rate of reaction will be constant and therefore of zero-order.

FIRST-ORDER REACTIONS

10. The rate equation. A first-order reaction is one in which the rate is proportional to the concentration of a single reacting substance. The rate law for the reaction

$$A \longrightarrow Products$$

may be written $\qquad -da/dt = k_1 a \qquad . \quad . \quad . \quad (XXI, 8)$

where k_1 is the first-order rate constant. This expression may be integrated between the limits $a = a_0$ at $t = 0$ and $a = a$ at $t = t$ (*see* II, 24) to give

$$\ln(a/a_0) = -k_1 t$$
or $\qquad a = a_0 \exp(-k_1 t)$

In terms of logarithms to the base 10

$$\lg(a/a_0) = -k_1 t/2\cdot303 \quad \text{or} \quad \lg a = k_1 t/2\cdot303 + \lg a_0$$
$$. \quad . \quad . \quad (XXI, 9)$$

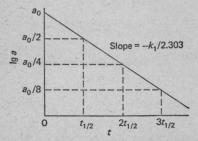

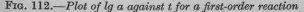

FIG. 112.—*Plot of lg a against t for a first-order reaction*

k_1 is independent of concentration (since it depends on the *ratio* of two concentrations) and has dimensions $(\text{time})^{-1}$. Equation (XXI, 9) indicates that a plot of $\lg a$ against t will give a straight line of slope $-k_1/2 \cdot 303$ and intercept $\lg a_0$ as shown in Fig. 112.

Fig. 113 shows a plot of a against t, and tangents are drawn to this curve at various times. These tangents are equal to the rate of the reaction $(-\mathrm{d}a/\mathrm{d}t)$ and illustrate the general phenomenon that the rate decreases as the reaction proceeds.

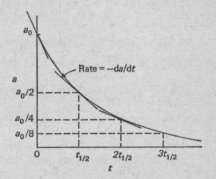

FIG. 113.—*Plot of a against t for a first-order reaction*

The first-order rate law may be written in another form. If x represents the decrease in the concentration of A after time t, then a, the concentration of A remaining after time t, is replaced by $(a_0 - x)$ and (XXI, 8) becomes

$$-\mathrm{d}(a_0 - x)/\mathrm{d}t = \mathrm{d}x/\mathrm{d}t = k_1(a_0 - x)$$

since a_0 is a constant.

Separating the variables and integrating between the limits $x = 0$ at $t = 0$ and $x = x$ at $t = t$

$$\int_0^x \frac{\mathrm{d}x}{a_0 - x} = k_1 \int_0^t \mathrm{d}t$$

$$[-\ln(a_0 - x)]_0^x = k_1[t]_0^t$$

$$\therefore \ln\{a_0/(a_0 - x)\} = k_1 t \quad \text{or} \quad \lg\{a_0/(a_0 - x)\} = k_1 t/2 \cdot 303$$
$$\qquad \qquad \ldots \text{(XXI, 10)}$$

A plot of $\lg\{a_0/(a_0 - x)\}$ against t will give a straight line of slope $k_1/2 \cdot 303$ passing through the origin.

11. The half-life. If a or x is put equal to $a_0/2$ in (XXI, 9) and (XXI, 10) respectively while $t = t_{\frac{1}{2}}$, we have

$$\lg\tfrac{1}{2} = -k_1 t_{\frac{1}{2}}/2\cdot 303$$
$$\therefore t_{\frac{1}{2}} = 2\cdot 303 \lg 2/k_1 = 0\cdot 693/k_1$$

For any first-order reaction, $t_{\frac{1}{2}}$ is a constant, independent of the initial concentration. $t_{\frac{1}{2}}$ is indicated in Figs. 112 and 113, from which it can be seen that a_0 falls to $a_0(\tfrac{1}{2})$ in $t_{\frac{1}{2}}$, to $a_0(\tfrac{1}{2} \times \tfrac{1}{2})$ in $2t_{\frac{1}{2}}$, $a_0(\tfrac{1}{2} \times \tfrac{1}{2} \times \tfrac{1}{2})$ in $3t_{\frac{1}{2}}$, etc.

EXAMPLE 1: The half-life of a first-order reaction is 100 s. Calculate the rate constant, and determine what fraction will have reacted after 250 s.

$$k_1 = 0\cdot 693/t_{\frac{1}{2}} = 0\cdot 693/100 = 6\cdot 93 \times 10^{-3}\,\mathrm{s}^{-1}$$

The fraction remaining after 250 s may be calculated using (XXI, 9)

$$\lg(a_0/a) = k_1 t/2\cdot 303 = (6\cdot 93 \times 10^{-3} \times 250)/2\cdot 303 = 0\cdot 752$$
$$\therefore a_0/a = 5\cdot 65$$

Fraction which has reacted $= (a_0 - a)/a_0 = 1 - a/a_0$
$$= 1 - 1/5\cdot 65$$
$$= 1 - 0\cdot 177 = 0\cdot 823$$

EXAMPLE 2: Show from the following data that the reaction $SO_2Cl_2 \longrightarrow SO_2 + Cl_2$ follows first-order kinetics. Calculate (a) the rate constant, (b) the half-life.

$t/10^4$s	0	1	2	3	4	5
p/mmHg	300	362	406	442	475	501

(a) Since two moles of product are formed per mole of reactant decomposed, the total pressure at any time t is given by

$$p_T = p + 2(p_0 - p)$$

where p_0 and p refer to the pressures of SO_2Cl_2 at time $t = 0$ and $t = t$ respectively. Thus $p = 2p_0 - p_T$.

Now the pressure of the gas will be proportional to its concentration. Thus a/a_0 may be replaced by p/p_0 (the constant of proportionality will cancel out), and (XXI, 9) becomes

$$\lg(p/p_0) = \lg\{(2p_0 - p_T)/p_0\} = -k_1 t/2\cdot 303$$

Since the reaction is first-order, a plot of $\lg\{(2p_0 - p_T)/p_0\}$ against t will give a straight line of slope $-k_1/2\cdot 303$.

$t/10^4$ s	0	1	2	3	4	5
$(2p_0 - p_T)/p_0$	1	0·794	0·646	0·525	0·417	0·331
$\lg\{(2p_0 - p_T)/p_0\}$	0	−0·10	−0·19	−0·28	−0·28	−0·48

The plot is shown in Fig. 114 and has a slope given by

$$-k_1/2{\cdot}303 = -0{\cdot}38/4 \times 10^4$$

Hence
$$k_1 = 2{\cdot}19 \times 10^{-5}\,s^{-1}$$

(b) The half-life is given by

$$t_{\frac{1}{2}} = 0{\cdot}693/2{\cdot}19 \times 10^{-5} = 3{\cdot}16 \times 10^4\,s$$

The half-life may also be found from the graph, being the time for p to change from p_0 to $p_0/2$. Putting $p = p_0/2$ in $\lg\{(2p_0 - p_T)/p_0\}$ gives $-\lg 2 = -0{\cdot}301$. The corresponding time from the graph is $3{\cdot}15 \times 10^4$ s.

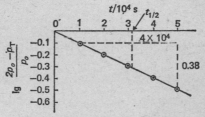

FIG. 114.—*Decomposition of SO_2Cl_2*

EXAMPLES OF FIRST-ORDER REACTIONS

12. Decomposition of N_2O_5. N_2O_5, a solid with a high vapour pressure, readily decomposes in the vapour phase:

$$2N_2O_5(g) \longrightarrow 4NO_2(g) + O_2(g)$$

Since there are different numbers of moles on the two sides of the equation, the reaction may be followed by monitoring the increase in pressure as decomposition proceeds. A correction can be made for the equilibrium

$$4NO_2 \rightleftharpoons 2N_2O_4$$

from a knowledge of the equilibrium constant. The partial pressure and hence the concentration of N_2O_5 ($n/V = p/RT$ for an ideal gas) may be found in terms of the total pressure of the gaseous constituents at a series of times. The reaction is first-order in the gas phase and also in solvents such as carbon tetrachloride and chloroform. N_2O_5, N_2O_4 and NO_2 are soluble in these solvents, while O_2 is insoluble and the reaction may be followed by making measurements on the O_2 evolved.

13. Inversion of sucrose. This reaction was the earliest first-order reaction to be quantitatively studied by polarimetry.

$$C_{12}H_{22}O_{11} + H_2O \xrightarrow[\text{catalyst}]{H_3O^+} C_6H_{12}O_6 + C_6H_{12}O_6$$

sucrose　　　　　　　　　　　fructose　　glucose

The reactant and the products rotate the beam of polarised light in opposite senses. If R_0, R and R_∞ represent the rotations at time $t = 0$, $t = t$ and $t = \infty$ (*i.e.* at a time when the reaction is complete for all practical purposes), then $R_0 - R_\infty$ is proportional to the initial concentration of sucrose (a_0), and $R_0 - R_t$ the change in rotation after time t is proportional to the decrease in concentration of sucrose (x). Thus the concentration of sucrose remaining at time $t = (a_0 - x) = (R_0 - R_\infty) - (R_0 - R_t) = (R_t - R_\infty)$. Thus from (XXI, 10)

$$\lg\{(R_0 - R_\infty)/(R_t - R_\infty)\} = k_1 t/2 \cdot 303$$

The inversion of sucrose is an example of a *pseudo* first-order reaction. If the concentrations of the three reactants (sucrose, water and catalyst) all change appreciably, third-order kinetics should be observed. However, the concentration of catalyst is unchanged during the reaction, and the concentration of solvent is so high that it may be regarded as constant during the reaction. Thus the third-order kinetics reduce to pseudo first-order kinetics.

$$-d[\text{sucrose}]/dt = k[\text{sucrose}][H_2O][H_3O^+] = k'[\text{sucrose}]$$

14. Other examples

(*a*) The acid-catalysed hydrolysis of an ester, *e.g.*

$$CH_3COOCH_3 + H_2O \xrightarrow{H_3O^+} CH_3COOH + CH_3OH$$

In the initial stages this is a pseudo first-order reaction where the concentration of catalyst is unchanged and water is present in large excess. As the concentration of products builds up, so a reverse reaction sets in (*see* 24 below).

(*b*) Isomerisation reactions, *e.g.* vinyl allyl ether to allyl-acetaldehyde

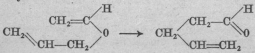

(c) The hydrogenation of ethylene on a nickel catalyst

$$C_2H_4 + H_2 \xrightarrow{\text{Ni}} C_2H_6$$

(d) Radioactive decay of unstable nuclei.

SECOND-ORDER REACTIONS

Second-order reactions may be divided into two categories, those in which the rate is proportional to:

(a) The square of the concentration of one reacting species.

(b) The product of the concentrations of two different reacting species each raised to the first power.

15. One reacting species. The general reaction may be represented by

$$2A \longrightarrow \text{Products}$$

and the rate law is

$$-da/dt = k_2 a^2 \quad \text{. . . (XXI, 11)}$$

where k_2 is the second-order rate constant. This expression may be integrated as follows:

$$-\int_{a_0}^{a} \frac{da}{a^2} = k_2 \int_{0}^{t} dt$$

$$[1/a]_{a_0}^{a} = k_2[t]_{0}^{t}$$

$$\therefore 1/a - 1/a_0 = k_2 t \quad \text{. . . (XXI, 12)}$$

k_2 has the dimensions $(\text{concentration})^{-1}(\text{time})^{-1}$. From (XXI, 12) it can be seen that a plot of $1/a$ against t will give a straight line of slope k_2 and intercept $1/a_0$.

As with first-order reactions, it is possible to express the rate law in terms of the amount of A which has reacted. Thus (XXI, 11) becomes

$$-d(a_0 - x)/dt = dx/dt = k_2(a_0 - x)^2 \quad \text{. . . (XXI, 13)}$$

(XXI, 13) may be integrated:

$$\int_{0}^{x} \frac{dx}{(a_0 - x)^2} = k_2 \int_{0}^{t} dt$$

Hence $\qquad 1/(a_0 - x) - 1/a_0 = k_2 t$ . . . (XXI, 14)

or $\qquad k_2 t = x/a_0(a_0 - x)$ . . . (XXI, 15)

(XXI, 14) is identical with (XXI, 12) if $a = a_0 - x$.

The half-life of the reaction may be found from (XXI, 12) or (XXI, 14), using the condition that when a or $x = a_0/2$, $t = t_{\frac{1}{2}}$:

$$t_{\frac{1}{2}} = 1/k_2 a_0 \qquad \text{. . . (XXI, 16)}$$

For a second-order reaction, $t_{\frac{1}{2}}$ is inversely proportional to the initial reactant concentration.

16. Two reacting species. The rate law for the general reaction

$$A + B \longrightarrow \text{Products}$$

is

$$-da/dt = -db/dt = k_2 ab \text{ . . . (XXI, 17)}$$

a and b in (XXI, 17) may be replaced by $a_0 - x$ and $b_0 - x$ (*see* **10** above) and (XXI, 17) becomes

$$-d(a_0 - x)/dt = dx/dt = k_2(a_0 - x)(b_0 - x)$$

This expression has been integrated in II, 24 to give

$$\frac{1}{a_0 - b_0} \ln\left\{\frac{b_0(a_0 - x)}{a_0(b_0 - x)}\right\} = k_2 t \text{ . . . (XXI, 18)}$$

k_2 may be found from the slope of a plot of

$\ln\left\{\dfrac{b_0(a_0 - x)}{a_0(b_0 - x)}\right\}$ against t.

EXAMPLES OF SECOND-ORDER REACTIONS

17. The alkaline hydrolysis or saponification of an ester. Consider the reaction

$$C_2H_5COOC_2H_5 + OH^- \longrightarrow C_2H_5COO^- + C_2H_5OH$$

This differs from acid hydrolysis (*see* **14** above) in that water plays no part in the reaction, and the two reactants are present in similar amounts. The reaction may be followed in one of two ways:

(*a*) *Sampling method.* In this method the concentration of alkali in a sample is determined by titration with standard acid. In practice the reaction is quenched by mixing the sample with a known excess of standard acid, which is then back titrated with standard alkali. If V_0 and V represent the volumes of acid required at times $t = 0$ and $t = t$ respectively then, for equal initial concentrations of ester and alkali

$$a_0 = b_0 = fV_0$$
$$x = f(V_0 - V) \quad \text{and} \quad (a_0 - x) = (b_0 - x) = fV$$

where f is the factor required to convert the volumes of standard acid to concentrations of alkali. Equation (XXI, 15) may now be used, giving

$$k_2t = (V_0 - V)/fV_0V$$

(b) *Conductivity method.* In this method the decrease in conductivity which results from the highly conducting OH^- ions being converted into ester ions of lower conductivity is monitored. If Λ_0, Λ_t and Λ_∞ are the molar conductivities at times $t = 0$, $t = t$ and $t = \infty$, then as in 13 above for equal initial concentrations,

$$a_0 = f(\Lambda_0 - \Lambda_\infty); \; x = f(\Lambda_0 - \Lambda_t); \; (a_0 - x) = f(\Lambda_t - \Lambda_\infty)$$

where f is the factor for converting conductivities to concentrations and must be determined experimentally. Equation (XXI, 15) becomes

$$k_2t = (\Lambda_0 - \Lambda_t)/f(\Lambda_0 - \Lambda_\infty)(\Lambda_t - \Lambda_\infty)$$

18. Dimerisation of alkenes and dienes. For example, the dimerisation of butadiene

$$2C_4H_6(g) \longrightarrow C_8H_{12}(g)$$

The course of this reaction may be followed by monitoring the decrease in pressure as the reaction proceeds. The method of calculating the rate constant from the pressure changes is given in Progress Test 21, 2.

19. Other examples. Reactions of alkyl halides with tertiary amines and with sodium phenoxide in alcoholic solution are second-order and have been monitored by titration of the halide ions produced with silver nitrate.

THIRD-ORDER REACTIONS

20. The rate equation. The rate equation for the general reaction

$$A + B + C \longrightarrow \text{Products}$$

is
$$-da/dt = -db/dt = -dc/dt = k_3abc$$

Since $a = a_0 - x$, $b = b_0 - x$ and $c = c_0 - x$, the above equation may be written

$$dx/dt = k_3(a_0 - x)(b_0 - x)(c_0 - x) \quad . . . \text{(XXI, 19)}$$

In the special case when the three initial concentrations are equal, (XXI, 19) becomes

$$dx/dt = k_3(a_0 - x)^3$$

which on integration yields

$$2kt = 1/(a_0 - x)^2 - 1/a_0^2 \quad . . . \text{(XXI, 20)}$$

21. Examples. Among the few examples of third-order reactions are those between nitric oxide and other gases. For example:

$$2NO + Cl_2 \longrightarrow 2NOCl \qquad -d[NO]/dt = k'[NO]^2[Cl_2]$$
$$2NO + O_2 \longrightarrow 2NO_2 \qquad -d[NO]/dt = k''[NO]^2[O_2]$$
$$2NO + 2H_2 \longrightarrow N_2 + 2H_2O \quad -d[NO]/dt = k'''[NO]^2[H_2]$$

These reactions undoubtedly take place in stages, and the last reaction, for example, is known to be a two-stage process:

$$2NO + H_2 \longrightarrow N_2 + H_2O_2 \quad \text{slow}$$
$$H_2O_2 + H_2 \longrightarrow 2H_2O \qquad \text{fast}$$

OPPOSING OR REVERSIBLE REACTIONS

In principle all reactions are reversible (in a kinetic sense). However, if the position of equilibrium lies well on the side of the products, only the forward reaction need be considered. When this is not the case, complications arise because the products initiate a reaction which reforms the reactants. The rate of this reverse reaction increases as the products accumulate and at equilibrium becomes equal to the rate of the forward reaction (*see* XI, 4). The simplest case of a reversible reaction is one which is first-order in both directions.

22. Reversible first-order reactions. Such a reaction may be represented by

$$A \underset{k_{-1}}{\overset{k_1}{\rightleftharpoons}} B$$

where k_1 and k_{-1} are the rate constants for the forward and reverse reactions respectively. If only A is present initially at concentration a_0, and x is the concentration which has reacted after time t, then the concentration of A at time t is $a_0 - x$. The net rate of the forward reaction is

$$-\mathrm{d}(a_0 - x)/\mathrm{d}t = \mathrm{d}x/\mathrm{d}t = k_1(a_0 - x) - k_{-1}x$$

$$\dots \text{(XXI, 21)}$$

At equilibrium the net rate is zero

$$\therefore k_1(a_0 - x_e) = k_{-1}x_e \dots \text{(XXI, 22)}$$

or $$k_{-1} = k_1(a_0 - x_e)/x_e \dots \text{(XXI, 23)}$$

where x_e is the amount of A decomposed or B formed at equilibrium.

NOTE: $K_c = k_1/k_{-1} = x_e/(a_0 - x_e) = ([\text{B}]/[\text{A}])_{\text{equil}}$ (see XI, 4).

Substituting the value for k_{-1} into (XXI, 21) gives

$$\mathrm{d}x/\mathrm{d}t = k_1(a_0 - x) - k_1x(a_0 - x_e)/x_e = k_1a_0(x_e - x)/x_e$$

$$\dots \text{(XXI, 24)}$$

Equation (XXI, 24) may be integrated between the limits $x = 0$ at $t = 0$ and $x = x$ at $t = t$:

$$\int_0^x \frac{\mathrm{d}x}{x_e - x} = \frac{k_1a_0}{x_e}\int_0^t \mathrm{d}t$$

$$[-\ln(x_e - x)]_0^x = \frac{k_1a_0}{x_e}[t]_0^t$$

$$\ln\{x_e/(x_e - x)\} = k_1a_0t/x_e$$

$$\therefore k_1 = \frac{x_e}{a_0t}\ln\left(\frac{x_e}{x_e - x}\right)$$

$$\dots \text{(XXI, 25)}$$

The reaction may be followed in a suitable manner and x_e found by allowing sufficient time for the system to reach equilibrium. Once k_1 has been found (from a plot of $\ln\{x_e/(x_e - x)\}$ against t) k_{-1} may be calculated from (XXI, 23).

23. Examples

(a) *Intramolecular rearrangements and isomerisation reactions.* The isomerisation of α-d-glucose to β-d-glucose may be followed polarimetrically, since the two substances differ in the degree to which they rotate plane-polarised light. The reaction is first-order in both directions.

(b) *The hydrolysis of methylacetate.* This reaction, which was assumed to be pseudo first-order (*see* 14 above), is in fact reversible and should be written

$$CH_3COOCH_3 + H_2O \underset{}{\overset{H_3O^+}{\rightleftharpoons}} CH_3COOH + CH_3OH$$

The rate law is

$$-d[CH_3COOCH_3]/dt = k'_1[CH_3COOCH_3] \\ - k'_{-2}[CH_3COOH][CH_3OH]$$

where k'_1 is the pseudo first-order rate constant of the forward reaction and k'_{-2} is the pseudo second-order rate constant of the reverse reaction.

NOTE: The reverse reaction is also catalysed by H_3O^+ ions.

In the early stages of the hydrolysis the concentrations of CH_3COOH and CH_3OH are so small that the term involving them is negligible and the forward reaction predominates.

(c) *The dissociation of hydrogen iodide.* Both the dissociation and the formation are second-order

$$2HI \underset{k_{-2}}{\overset{k_2}{\rightleftharpoons}} H_2 + I_2$$

NOTE: The formation reaction has recently been shown to be termolecular, *i.e.* $H_2 + 2I \longrightarrow 2HI$, where the I atoms are formed by the reaction $I_2 \rightleftharpoons 2I$. The system as a whole exhibits second-order kinetics.

If only HI is present initially (concentration a_0) and x is the amount decomposed after time t, then the concentrations of H_2 and I_2 present after time t will be $x/2$. The rate expression is

$$dx/dt = k_2(a_0 - x)^2 - k_{-2}(x/2)^2 \quad . \quad . \quad . \quad (XXI, 26)$$

At equilibrium $\quad k_2(a_0 - x_e)^2 = k_{-2}(x_e/2)^2$

$$K_c = k_2/k_{-2} = x_e^2/4(a_0 - x_e)^2$$

K_c may be determined from a knowledge of a_0 and x_e. $k_{-2} = k_2/K_c$ can be substituted into (XXI, 26)

$$dx/dt = k_2(a_0 - x)^2 - k_2(x/2)^2/K_c$$

$$= k_2 \left\{ a_0^2 - 2a_0x + \left(1 - \frac{1}{4K_c}\right)x^2 \right\}$$

This expression may be integrated to yield k_2.

CONSECUTIVE REACTIONS

24. Consecutive reactions. A product of one reaction may be the reactant in a second reaction, and successive stages may follow. Consider the simple case of two consecutive first-order reactions

$$A \xrightarrow{k_1} B \xrightarrow{k'_1} C$$

If only A is present initially, the rate of disappearance of A is given by

$$-da/dt = k_1 a$$

which on integration gives

$$a = a_0 \exp(-k_1 t) \qquad (see \ \mathbf{10} \ above)$$

Rate of increase of B = Rate of formation — Rate of removal, *i.e.*
$$db/dt = k_1 a - k'_1 b$$
$$= k_1 a_0 \exp(-k_1 t) - k'_1 b$$

This expression may be integrated to give

$$b = \frac{a k_1}{k'_1 - k_1} \left\{ \exp(-k_1 t) - \exp(-k'_1 t) \right\}$$

A simple example of a consecutive reaction is the acid hydrolysis of the esters of dibasic acids, where the two ester groups are hydrolysed in turn, providing two successive pseudo first-order reactions.

25. Chain reactions. These constitute a special type of consecutive reaction and may be illustrated by reference to the gas phase reaction between hydrogen and chlorine. In the dark, in clean containers, no reaction occurs when the two gases are mixed. Reaction does take place in the presence of a trace of sodium vapour (in which the sodium is present as atoms) or in light. In each case the initial step in the reaction is the dissociation of the chlorine molecules

$$Cl_2 + Na \longrightarrow NaCl + Cl \quad \cdot \ \cdot \ \cdot \ (XXI, 27)$$

$$Cl_2 \xrightarrow{light} 2Cl \qquad \cdot \ \cdot \ \cdot \ (XXI, 28)$$

(XXI, 27) and (XXI, 28) are called *initiation steps*. The chlorine atoms then react with hydrogen molecules

$$Cl + H_2 \longrightarrow HCl + H \quad \cdot \ \cdot \ \cdot \ (XXI, 29)$$

and the hydrogen atoms in turn react with other chlorine molecules to produce more chlorine atoms, which then react as in (XXI, 29)

$$H + Cl_2 \longrightarrow HCl + Cl \quad . \quad . \quad . \quad (XXI, 30)$$

(XXI, 29) and (XXI, 30) are known as the *chain-propagating* steps and H and Cl are called *chain carriers*. The chain reaction does not continue indefinitely and is terminated when two atoms come together in the presence of a third body, usually the container wall.

$$\left.\begin{array}{l} H + Cl + M \longrightarrow HCl + M \\ H + H + M \longrightarrow H_2 + M \\ Cl + Cl + M \longrightarrow Cl_2 + M \end{array}\right\} \begin{array}{l} \textit{chain termination} \\ \textit{steps} \end{array}$$

The third body, M, is required to absorb the excess energy of the colliding atoms and thus to stabilise the resulting molecule. In its absence the two atoms would fly apart without forming a bond.

26. Free radicals. Chain reactions usually involve *free radicals* (*i.e.* molecular, atomic or ionic species which contain unpaired electrons) as the chain carriers.

NOTE: Strictly speaking, molecules such as O_2 and NO which contain unpaired electrons are free radicals, but the term is usually restricted to unpaired electron species possessing high chemical reactivity.

In the example in 25 above the chlorine atom is a free radical and the existence of the unpaired electron may be emphasised with a dot, *e.g.* Cl·, CH_3·, CHO·.

The thermal decomposition of acetaldehyde occurs through the methyl radical chain carrier:

$$\begin{array}{ll} CH_3CHO \longrightarrow CH_3· + CHO· & \text{initiation} \\ CH_3CHO + CH_3· \longrightarrow CH_4 + CO + CH_3· & \text{propagation} \\ 2CH_3· \longrightarrow C_2H_6 & \text{termination} \end{array}$$

The overall reaction is

$$CH_3CHO \longrightarrow CH_4 + CO$$

Small traces of C_2H_6 are formed.

No third body is required in this termination step, since the excess energy given out in bond formation can be taken up in internal forms of motion such as vibration and rotation which are not possible for atoms.

THE DETERMINATION OF ORDER AND RATE CONSTANT

27. Differential method. If the reaction obeys the general rate equation

$$\text{Rate} = -dc/dt = k_n c^n$$

Taking logarithms, $\lg \text{Rate} = \lg k_n + n \lg c$

$$\text{. . . (XXI, 31)}$$

Hence a plot of $\lg$ Rate against $\lg c$ should give a straight line of slope n and intercept $\lg k_n$.

28. Initial rate method. This is a variation on the differential method. In a sufficiently slow reaction the rate may be determined before any appreciable change has occurred in the reactant concentrations. Under such conditions it is permissible to assume that all reactant concentrations are effectively constant at their initial values. The rate law for the reaction

$$A + B + C \longrightarrow \text{Products}$$

may therefore be written

$$\text{Rate} = dx/dt = k(a_0 - x)^{n_1}(b_0 - x)^{n_2}(c_0 - x)^{n_3}$$

If x is small $\text{Rate} = k\, a_0^{n_1} b_0^{n_2} c_0^{n_3}$ . . . (XXI, 32)

If b_0 and c_0 are held constant while a_0 is varied

$$\text{Rate} = k' a_0^{n_1} \quad \text{where } k' = k b_0^{n_2} c_0^{n_3}$$

Taking logarithms, $\lg \text{Rate} = \lg k' + n_1 \lg a_1$

n_1 may be found from the slope of the plot of $\lg$ Rate against $\lg a_1$. The procedure may be repeated to find n_2 and n_3. Once the individual orders are known, k may be found from (XXI, 32).

29. Integral method. The experimental data may be substituted into the various integrated rate equations (e.g. (XXI, 7), (XX, 9), (XXI, 12), etc.) until a constant value of k is obtained. Alternatively, a graphical method may be used and the functions of concentrations relevant to the various orders plotted against time until a linear relationship is obtained. This is essentially a trial-and-error method and is not very sensitive.

30. Half-life method. Reaction orders and rate constants may be determined by finding the variation of the half-life with the initial concentration. Thus if $t_{\frac{1}{2}}$ is independent of concentration the reaction is first-order.

31. Isolation method. If all reactants save one are present in large excess and therefore have effectively constant concentrations, the apparent order will be the order with respect to the reactant whose concentration does change. The general rate law

$$dx/dt = k(a_0 - x)^{n_1}(b_0 - x)^{n_2}(c_0 - x)^{n_3}$$

reduces to

$$dx/dt = k'(a_0 - x)^{n_1}$$

if b_0 and c_0 are large with respect to a_0. n_1 may then be found by application of one of the above methods; n_2 and n_3 are found by repeating the above procedure with the relevant reactants present in large excess. Once the individual orders have been determined, the rate constant may be found.

EXAMPLE 3: The following data were obtained for the gas phase decomposition of HI:

% decomposition	Rate (arbitrary units)
0	4·58
5	4·07
10	3·71
15	3·37
20	2·95

Determine the order of the reaction.

From (XXI, 3), lg Rate = lg k_n + n lg[HI]
Instead of [HI] we may use (% HI remaining). A plot of lg Rate against lg (% HI remaining) will give a straight line of slope n.

% HI remaining	lg (% HI remaining)	lg Rate
100	2·000	0·661
95	1·978	0·610
90	1·954	0·569
85	1·929	0·526
80	1·903	0·470

These data are plotted in Fig. 115. The line has a slope of 1·96, indicating the reaction to be second-order.

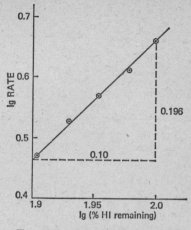

FIG 115.—*Decomposition of HI*

PROGRESS TEST 21

1. Using the following data on the inversion of sucrose, show that the reaction is first-order and determine the rate constant.

t/hr	0	4	12	24	48	∞
angle of rotation/deg	87·3	74·25	53·75	31·6	7·19	−12·0

(10–11)

2. The gaseous dimerisation reaction, $2A \longrightarrow A_2$ follows second-order kinetics at 400°C. From the following data determine the rate constant in (a) $mmHg^{-1} s^{-1}$, (b) $dm^3 mol^{-1} s^{-1}$.

t/s	0	100	200	300	400	500
p/mmHg	600	500	450	420	400	385

(15)

3. The reaction $CH_3CH_2NO_2 + NaOH \longrightarrow CH_3CH = NO_2Na + H_2O$ is second-order, with a rate constant of $6·5 \, dm^3 \, mol^{-1} \, s^{-1}$ at 0°C. If the initial concentrations of nitromethane and sodium hydroxide are $4 \times 10^{-3} \, mol \, dm^{-3}$ and $5 \times 10^{-3} \, mol \, dm^{-3}$ respectively, how long will it take for 80% of the nitromethane to react? (16)

4. A substance, A, may decompose by either first- or second-order kinetics. If both reactions have the same half-life for the same initial concentration of A, calculate the ratio of the first-order rate to the second-order rate, (a) initially, (b) at the time of one half-life. (7–8, 15)

5. The rate law for the reaction A $\longrightarrow$ Products may be represented by $dx/dt = k_n(a_0 - x)^n$, where a_0 is the initial concentration, x is the decrease in concentration after time t, k_n is the rate constant and n is the order of reaction. If $n \neq 1$ integrate the expression to determine k_n and derive the relationship between the half-life and the rate constant. Indicate how the latter expression could be used to determine n. (20)

6. A substance decomposes according to a third-order rate law. If the rate constant is $1 \cdot 2 \times 10^{-3}$ dm^6 mol^{-2} s^{-1}, calculate $t_{\frac{1}{2}}$ if the initial concentration is (a) 1 mol dm^{-3}, (b) 10^{-1} mol dm^{-3}. (20 and question 5 above)

7. A reversible first-order reaction may be represented by A $\rightleftharpoons$ B. The percentage composition of the mixture of A and B is shown below as a function of time. Calculate the equilibrium constant and the rate constants for the forward and reverse reactions.

t/s	0	100	200	∞
%A	100	75·6	62·1	40

(22)

8. The decomposition of acetaldehyde gave the following data at 500°C:

Initial pressure/mmHg	200	400	600
$t_{\frac{1}{2}}$/s	800	400	267

Determine the order of the reaction and the rate constant. (8, 15)

9. The rate of a reaction may be represented by: Rate $= k[A]^m[B]^n$. From the following data deduce the values of k, m and n.

Experiment No.	[A]/10^3 mol dm^{-3}	[B]/10^3 mol dm^{-3}	Rate/10^{-3} mol dm^{-3} s^{-1}
1	1·2	4·4	12·1
2	2·4	4·4	24·3
3	1·2	1·1	3·0

(10, 15–16)

10. Distinguish between the molecularity and the order of a reaction. (3–4)

11. Outline the methods by which the course of a reaction may be followed. (5–6)

12. Derive the characteristic equations for (a) zero-order, (b) first-order, (c) second-order reactions and give the units of the rate constant in each case. (7, 10, 15–16)

13. Distinguish between first-order and pseudo first-order reactions and give examples to illustrate your answer. (12–14)

14. Derive the rate equation for a reversible reaction which is first-order in both directions. (22)

15. Explain by reference to the gas phase reaction between hydrogen and chlorine the meaning of the following terms: (a) initiation step, (b) chain-propagating step, (c) chain termination step, (d) chain carrier, (e) free radical. (25–26)

16. Indicate methods by which the order of a reaction may be determined. (27–31)

CHAPTER XXII

THEORETICAL ASPECTS

THE EFFECT OF TEMPERATURE

An increase in temperature almost invariably increases the rate of a chemical reaction, a 10° rise in temperature around room temperature often resulting in a twofold or threefold increase.

1. The theory of activation. It is reasonable to suppose that reaction between two molecules will occur only when they collide. If collision alone determines the reaction rate, then the collision number and the reaction rate should be equal. However, for gaseous reactions the calculated collision rate exceeds the reaction rate by many factors of ten. Also, from (V, 4) and (V, 5) it can be seen that the collision number is proportional to $T^{\frac{1}{2}}$. Thus for a gaseous reaction carried out at 300 K and 310 K

$$Z_{310}/Z_{300} = (310/300)^{\frac{1}{2}} = 1.02$$

i.e. there is only a 2% increase in Z, while there is a 200–300% increase in reaction rate.

Arrhenius explained these anomalies by assuming that reaction could occur only between molecules possessing more than the normal amount of energy. Such molecules are said to be *activated*, and the excess energy they possess is called the *activation energy*.

2. The Arrhenius equation. The dependence of the concentration equilibrium constant on temperature is given by the equation d ln $K_c/\mathrm{d}T = \Delta U/RT^2$ (*see* XIII, 33), and since the equilibrium constant is equal to a ratio of rate constants (*see* XI, 4) Arrhenius suggested that the variation of the rate constant with temperature should be of the form

$$\mathrm{d}\ln k/\mathrm{d}T = E^{\ddagger}/RT^2 \quad . \quad . \quad . \quad \text{(XXII, 1)}$$

where $E^{\ddagger}$ is the activation energy. If $E^{\ddagger}$ is independent of temperature, (XXII, 1) may be integrated to give

$$\ln k = -E^{\ddagger}/RT + \ln A \quad . \quad . \quad . \text{ (XXII, 2)}$$

where $\ln A$ is the constant of integration. Taking antilogarithms

$$k = A \exp(-E^{\ddagger}/RT) \quad . \quad . \quad . \text{ (XXII, 3)}$$

A is known as the *Arrhenius, frequency* or *pre-exponential factor*, and the complete equation is known as the *Arrhenius equation*. Thus molecules must acquire a certain critical energy $E^{\ddagger}$ before they can react, and the Boltzmann factor, $\exp(-E^{\ddagger}/RT)$, gives the fraction of molecules possessing this energy. Equation (XXII, 2) indicates that a plot of $\ln k$ against $1/T$ should give a straight line of slope $-E^{\ddagger}/R$ and intercept $\ln A$. The vast majority of rate constants have been found to conform to this behaviour.

The frequency factor has the same units as k, *e.g.* s^{-1} for a first-order reaction and dm^3 mol^{-1} s^{-1} for a second-order reaction. For bimolecular reactions, A usually lies between 10^9 and 10^{11} dm^3 mol^{-1} s^{-1}, while $E^{\ddagger}$ usually lies in the range 40–300 kJ mol^{-1}.

$E^{\ddagger}$ may also be found from a knowledge of the rate constants at two temperatures. If k_1 and k_2 are the rate constants at temperatures T_1 and T_2 respectively, substitution of these values into (XXII, 2) leads to

$$\ln(k_2/k_1) = \frac{E^{\ddagger}}{R}\left(\frac{1}{T_1} - \frac{1}{T_2}\right) \quad . \quad . \quad . \text{ (XXII, 4)}$$

3. The relationship between ΔU and $E^{\ddagger}$. Chemical reaction proceeds via the formation of an active species, now called an *activated complex*, which subsequently breaks down to yield the products. Thus a reversible reaction may be represented by

$$A + B \underset{k_b,\, E_b^{\ddagger}}{\overset{k_f,\, E_f^{\ddagger}}{\rightleftharpoons}} X^{\ddagger} \rightleftharpoons C + D \quad \Delta U. \quad . \quad . \text{ (XXII, 5)}$$

Equation (XXII, 5) indicates that the forward reaction between A and B and also the reverse reaction between C and D both proceed through the same activated complex, $X^{\ddagger}$. In both cases the rate-determining step is the formation of the activated complex. If k_f, $E_f^{\ddagger}$ and k_b, $E_b^{\ddagger}$ represent the rate constants and activation energies of the forward and reverse processes, then from (XXII, 1)

$$\mathrm{d}\ln k_f/\mathrm{d}T = E_f^{\ddagger}/RT^2 \quad \text{and} \quad \mathrm{d}\ln k_b/\mathrm{d}T = E_b^{\ddagger}/RT^2$$
$$\cdots \text{(XXII, 6)}$$

Hence $\mathrm{d}\ln k_f/\mathrm{d}T - \mathrm{d}\ln k_b/\mathrm{d}T = \mathrm{d}\ln(k_f/k_b)/\mathrm{d}T = \mathrm{d}\ln K_c/\mathrm{d}T$
$$= (E_f^{\ddagger} - E_b^{\ddagger})/RT^2 \cdots \text{(XXII, 7)}$$

Now $\qquad \mathrm{d}\ln K_c/\mathrm{d}T = \Delta U/RT^2 \qquad \cdots \text{(XXII, 8)}$

Thus comparing (XXII, 7) and (XXII, 8) leads to

$$\Delta U = E_f^{\ddagger} - E_b^{\ddagger}$$

i.e. the internal energy change for the forward reaction is equal to the difference between the activation energies of the forward and reverse reactions. The concept of an activated complex and of the energy relationships discussed above can be demonstrated on a plot of the energy of the system against the reaction co-ordinate. The latter is a qualitative description of the progress of a reaction from reactants to products. Fig. 116 shows a smooth curve which is drawn through the three known energies ΔU, $E_f^{\ddagger}$ and $E_b^{\ddagger}$.

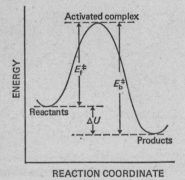

FIG. 116.—*The relation between the activation energies for a reversible reaction and the internal energy change*

The Arrhenius theory leads to an improvement in our understanding of reaction processes, but it does not give any information on A, nor does it predict the value of $E^{\ddagger}$. Two later theories, the *collision theory* and the *transition state theory*, allow a better interpretation of A and $E^{\ddagger}$. The collision theory is applicable only to gaseous reactions, whereas the transition state theory also holds for reactions in solution.

THE COLLISION THEORY

4. Modification of the collision number. The collision theory postulates that reaction can take place between two molecules only when they collide with an energy greater than or equal to $E^{\ddagger}$. For *gaseous* reactions, in which the reacting molecules can be considered as hard spheres possessing only translational energy, the collision number for unlike molecules is given by (V, 5), namely

$$Z_{12} = N_1 N_2 d_{12}{}^2 (8\pi RT/\mu)^{\frac{1}{2}}$$

At any given temperature the speeds and therefore the energies of molecules are distributed about an average value, and there are always a few molecules with energies greatly in excess of the average value (*see* IV, **13–14**). As the temperature increases, the number of molecules with energies in excess of the average greatly increases, although the number of collisions is not changed appreciably. The Maxwell–Boltzmann distribution of energies for gas molecules at two temperatures is shown in Fig. 117.

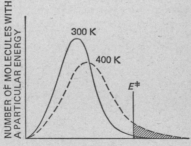

Fig. 117.—*Maxwell–Boltzmann distribution of energies among the molecules of a gas*

NOTE: The Maxwell–Boltzmann distribution of molecular speeds and therefore energies was considered in IV, **13–14**.

The fraction of molecules having energies greater than a particular value such as $E^{\ddagger}$ is given by the ratio of the area under the curve beyond $E^{\ddagger}$ (shaded in Fig. 117) to the total area under the curve. This fraction increases by a large factor as the temperature is increased, although the total area under the curve increases only slightly. The fraction of collisions of molecules with energies greater than or equal to $E^{\ddagger}$ is given by

$$N^{\ddagger}/N_0 = \exp(-E^{\ddagger}/RT)$$

where $N^{\ddagger}$ and N_0 represent the number of molecules with energies greater than or equal to $E^{\ddagger}$ and the total number of molecules respectively.

5. Calculation of the rate constant. The number of molecules reacting per second = (number of collisions per second) × (fraction of collisions with energies greater than or equal to $E^{\ddagger}$), i.e.

$$dN/dT = Z_{12} \exp(-E^{\ddagger}/RT) \quad . \quad . \quad . \quad (XXII, 9)$$

Before the above expression can be compared with the experimental rate, the units of dN/dt which are molecule $m^{-3} s^{-1}$ must be converted to the usual units of rate, namely mol $dm^{-3} s^{-1}$. The experimental rate is given by

$$dc/dt = k_2 c_1 c_2 \quad . \quad . \quad . \quad (XXII, 10)$$

where c_1 and c_2 are the concentrations of the reacting species in mol dm^{-3}. The relationship between c and N (in molecules m^{-3}) is

$$c = N/10^3 L \quad . \quad . \quad . \quad (XXII, 11)$$

Equation (XXII, 11) may be differentiated

$$dc = dN/10^3 L$$

Substituting for dc, c_1 and c_2 in (XXII, 10) yields

$$dN/dt \; 10^3 L = k_2 N_1 N_2/10^6 \; L^2$$

Hence $\quad k_2 = \dfrac{10^3 \; L}{N_1 N_2} \times \dfrac{dN}{dt} = \dfrac{10^3 \; L \; Z_{12}}{N_1 N_2} \exp(-E^{\ddagger}/RT)$

$$. \quad . \quad . \quad (XXII, 12)$$

The term in front of the exponential in (XXII, 12) has the same units as those of k_2 (dm^3 mol^{-1} s^{-1}) and is given the symbol z_{12}.

$$\therefore k_2 = z_{12} \exp(-E^{\ddagger}/RT) \quad . \quad . \quad . \quad (XXII, 13)$$

NOTE: The above approach is restricted to bimolecular reactions in the gas phase.

6. Comparison with the Arrhenius equation.

Equation (XXII, 13) may be compared with the Arrhenius equation, (XXII, 3), and it can be seen that $A = z_{12}$. Since the dependence of Z_{12} and therefore of z_{12} on temperature is very small compared with the exponential dependence, (XXII, 13) gives the correct value for the temperature dependence of the rate constant.

k_2 may be calculated from (XXII, 13) using the experimentally determined values of $E^{\ddagger}$ and the collision diameter d_{12}, and compared with the value of k_2 found by direct experimentation. For reactions between gaseous atoms and between simple gaseous molecules the two values of k_2 usually agree within a factor of ten.

7. The failure of the collision theory.

For many reactions, particularly those involving more complex molecules, the calculated and experimental rate constants may differ by many factors of 10. This has been allowed for by incorporating a steric factor P in (XXII, 13)

$$k_2 = Pz_{12} \exp(-E^{\ddagger}/RT). \quad . \quad . \quad (XXII, 14)$$

There is no way of estimating P, which is simply a measure of the discrepancy between the simple collision theory and the experimental results. A value of P less than unity allows for the fact that some collisions having the requisite energy may not produce reaction. In such cases P is usually interpreted in terms of an orientation requirement, *i.e.* the molecules may have to collide in a specific way before reaction will take place. However, P may be as low as 10^{-10} and is then much too small to be accounted for by this simple picture.

The collision theory can produce no explanation for steric factors greater than unity.

8. Unimolecular reactions in the gas phase.

It is difficult to explain the existence of unimolecular first-order reactions using the collision theory, since if two molecules must collide in order to provide the necessary activation energy a second-order rate law should result. This anomaly was resolved by Lindemann (1922), who assumed that there was a time lag between activation and reaction, during which the activated molecules could either react or be de-activated.

$$A + A \longrightarrow A + A^{\ddagger} \qquad \text{activation} \quad . \ . \ . \ \text{(XXII, 15)}$$
$$A + A^{\ddagger} \longrightarrow A + A \qquad \text{deactivation} \quad . \ . \ . \ \text{(XXII, 16)}$$
$$A^{\ddagger} \longrightarrow \text{Products} \qquad \text{reaction} \quad . \ . \ . \ \text{(XXII, 17)}$$

If the time lag is long so that the rate of deactivation is very much greater than the rate of reaction—*i.e.* (XXII, 17) is the slow step—then the reaction should follow first-order kinetics. If, however, A reacts immediately it is formed, *i.e.* (XXII, 16) is the slow step, then the reaction should be second-order. This mechanism may be readily tested by observing the effect of change of pressure (and hence of concentration) on the reaction. At high pressure the rate of deactivation will be dominant, while at sufficiently low pressures all the activated molecules will react before they can be deactivated. Hence the kinetics should change from first-order to second-order with decreasing pressure. This behaviour has been observed in several cases.

THE TRANSITION STATE OR ABSOLUTE RATE THEORY

9. The nature of the activated complex. The transition state theory is based on the assumption that the activated complex (or transition state) acts as an intermediate in all chemical reactions.

NOTE: An alternative definition of molecularity (*see* XXI, 4) is the total number of reactant molecules which form the activated complex.

To a good approximation the reactants may be considered to be *in equilibrium* with the activated complex, *even when the reactants and products are not in equilibrium*.

$$A + B \rightleftharpoons X^{\ddagger} \xrightarrow{k_2} \text{Products}$$

The reaction rate will be dependent on two factors:

(a) The concentration of $X^{\ddagger}$, which may be found by ordinary equilibrium calculations.

(b) The rate at which $X^{\ddagger}$ breaks up.

NOTE: It should be remembered that the equilibrium constant and the rate of a particular reaction are normally quite distinct, and that a large equilibrium constant does not necessarily imply a high forward rate constant.

$X^{\ddagger}$ is regarded as a normal molecule, except that it has only a transient existence, since one of its vibrational modes is so loose that it decomposes immediately into the products. Thus, for example, the bimolecular decomposition of HI may be visualised in terms of the following sequence of reactions

$$\begin{matrix} \text{H—I} \\ \text{H—I} \end{matrix} \rightleftharpoons \begin{bmatrix} \text{H} \cdots \text{I} \\ \vdots \quad \vdots \\ \text{H} \cdots \text{I} \end{bmatrix} \rightleftharpoons \begin{matrix} \text{H} \\ | \\ \text{H} \end{matrix} + \begin{matrix} \text{I} \\ | \\ \text{I} \end{matrix}$$

The formation of the activated complex results in the partial loosening of the H–I bonds and the partial formation of the H–H and I–I bonds (shown by the dotted lines). For reaction to take place, the activated complex must pass over the activation energy barrier (see Fig. 116).

10. Calculation of the rate constant. The equilibrium constant for the reactant-activated complex equilibrium is given by

$$K^{\ddagger} = [X^{\ddagger}]/[A][B]$$

Hence $\qquad [X^{\ddagger}] = K^{\ddagger}[A][B] \quad . \quad . \quad . \quad \text{(XXII, 18)}$

From (a) and (b) in **9** above

$$-d[A]/dt = [X^{\ddagger}] \times \text{(rate at which } X^{\ddagger} \text{ breaks up)} \\ . \quad . \quad . \quad \text{(XXII, 19)}$$

Now the rate of break-up of the activated complex is controlled by the frequency of the loose vibrational mode. If this frequency is ν, then the associated energy will be $h\nu$. Since the loose vibrational mode will have a low frequency, its energy will also be approximately equal to the classical value of kT (see IV, **18–24**).

$$\therefore h\nu = kT$$
$$\nu = kT/h$$

Equation (XXII, 19) may be written

$$-d[A]/dt = [X^\ddagger]kT/h$$
$$= \frac{kT}{h} K^\ddagger[A][B] \qquad \text{(from (XXII, 18))}$$

Hence the bimolecular rate constant k_2 is given by

$$k_2 = kTK^\ddagger/h \qquad \text{. . . (XXII, 20)}$$

Now $\Delta G^\ominus = -RT \ln K = \Delta H^\ominus - T\Delta S^\ominus$ . . . (XXII 21)

Rearranging (XXII, 21),

$$K = \exp(-\Delta G^\ominus/RT) = \exp(\Delta S^\ominus/R)\exp(-\Delta H^\ominus/RT)$$

Equation (XXII, 20) may be written in terms of the thermodynamic functions of the activated complex

$$k_2 = \frac{kT}{h} \exp(-\Delta G^\ddagger/RT) = \frac{kT}{h} \exp(\Delta S^\ddagger/R)\exp(-\Delta H^\ddagger/RT)$$
$$\text{. . . (XXII, 22)}$$

At a given temperature kT/h is the same for all molecules and the rate constant is therefore determined by $\Delta G^\ddagger$. Thus the higher the value of $\Delta G^\ddagger$ the lower the rate constant at a given temperature.

NOTE: $\Delta H^\ddagger$, $\Delta G^\ddagger$ and $\Delta S^\ddagger$ cannot be determined by direct measurement because of the transient nature of the activated complex.

To a good approximation, $\Delta H^\ddagger$ may be equated with $E^\ddagger$, and (XXII, 22) becomes

$$k_2 = \frac{kT}{h} \exp(\Delta S^\ddagger/R)\exp(-E^\ddagger/RT) \text{ . . . (XXII, 23)}$$

11. Comparison with the collision theory. From (XXII, 13) and (XXII, 23)

$$A = Pz_{12} = \frac{kT}{h} \exp(\Delta S^\ddagger/R)$$

Both z_{12} and kT/h are dependent on temperature and have values of about 10^{11} and 10^{13} dm^3 mol^{-1} s^{-1} respectively for bimolecular reactions at 300 K. The steric factor, introduced as an empirical constant in the collision theory, may be interpreted in terms of the change in entropy when the activated complex is formed. Where the reacting molecules contain large numbers of atoms the formation of the activated complex is accompanied by a large decrease in entropy, since the system becomes more ordered.

The rate of a chemical reaction is largely determined by $E^\ddagger$, although $\Delta S^\ddagger$ can play an important part when the reacting species must be specifically orientated in order to form the activated complex.

CATALYSIS

The rate of a chemical reaction may be dramatically altered by the presence of small quantities of certain substances other than those appearing in the stoichiometric equation for the reaction. Such substances do not undergo any permanent chemical changes and are known as *catalysts*. We will now examine some of the more important properties of catalysts.

12. Positive and negative catalysts. A catalyst may either increase or decrease the rate of a reaction. The former are known as *positive catalysts*, or just "catalysts," while the latter are called *negative catalysts* or *inhibitors*. The term "inhibitors" is preferred, since these substances are used up in the reaction and are therefore not true catalysts. The rate of decomposition of a solution of hydrogen peroxide is increased by the addition of alkalis but decreased by the addition of acids. Similarly, the oxidation of aldehydes, which proceeds by a chain mechanism, is inhibited by chain terminators, such as anthracene, which combine with and thus destroy the chain carriers.

13. The participation of the catalyst in the reaction. Although the catalyst must be unchanged chemically at the end of the reaction, its physical form is often changed. Thus the granular manganese dioxide added to catalyse the thermal decomposition of potassium chlorate is left as a fine powder at the end of the reaction. Such physical changes indicate that the catalyst actually participates in the reaction. Catalytic reactions take place in a series of steps, the catalyst being consumed in the first step of the reaction and regenerated in a subsequent step. Thus the uncatalysed bimolecular reaction

$$A + B \longrightarrow \text{Products}$$

may, in the presence of a catalyst, X, give the following sequence of reactions:

$$A + X \rightleftharpoons AX$$
$$AX + B \longrightarrow Products + X$$

The compound AX is known as the *intermediate compound* and may have only a transitory existence, although it has been isolated as a definite chemical compound in some reactions.

NOTE: AX should not be confused with the activated complex mentioned earlier.

14. The effect of concentration. A small quantity of catalyst is usually able to bring about a considerable chemical change. The reason is that the catalyst is not used up in the reaction. For reactions that do not involve chain mechanisms, the rate is proportional to the catalyst concentration. Since the concentration of catalyst does not change in the reaction, it is often combined with the true rate constant to yield a pseudo rate constant. This may give rise to an apparently lower order of reaction (*see* XXI, 13–14).

15. The effect on the equilibrium constant. For a reversible reaction, the catalyst does not effect the position of the equilibrium. This indicates that the catalyst influences the forward and reverse reactions to the same extent. Thus a catalyst that increases the rate of hydrolysis of an ester must likewise increase the rate of esterification of the corresponding acid. The catalyst speeds up the attainment of equilibrium without affecting either the value of the equilibrium constant or the free energy change.

16. The effect on the activation energy. A catalyst is unable to initiate a reaction that is not energetically favourable. For a reaction to be accelerated by a catalyst, ΔG for the reaction must be negative, *i.e.* the reaction must be thermodynamically feasible, although the rate may be too slow to observe. A catalyst functions by making available an alternative path of lower activation energy. If the Arrhenius factor is essentially unchanged, then the decrease in activation energy leads to an increased rate of reaction. For example, the uncatalysed thermal decomposition of diethyl ether has an activation energy of 261 kJ mol^{-1}, while in the presence of a few per cent of iodine vapour the activation energy is reduced to 143 kJ mol^{-1}. This decrease in activation energy causes the rate constant to increase by a factor of 10^4.

HOMOGENEOUS CATALYSIS

A catalyst may be homogeneous or heterogeneous, depending on whether the catalyst is in the same or in a different phase to the reactants. A discussion of heterogeneous catalysis is deferred until XXIII, 12–17.

17. Gas phase catalysis. One of the best known of the few homogeneously catalysed gas phase reactions is the oxidation of sulphur dioxide to sulphur trioxide in the presence of nitric oxide:

$$NO + \tfrac{1}{2}O_2 \longrightarrow NO_2$$
$$NO_2 + SO_2 \longrightarrow NO + SO_3$$

Carbon monoxide may be oxidised to carbon dioxide in an analogous manner. The thermal decomposition of certain aldehydes and ethers is markedly increased by the presence of iodine vapour.

The most important examples of homogeneous catalysis in solution are those included under the general heading of acid–base catalysis.

18. Specific acid or specific base catalysis. Certain reactions, such as the inversion of sucrose and the hydrolysis of esters, are catalysed only by H_3O^+ ions and are known as *specific acid-catalysed reactions*. Others, such as the decomposition of diacetyl alcohol, are specifically catalysed by OH^- ions.

19. General acid or general base catalysis. Using the Brønsted–Lowry concept of acids and bases (*see* XIII, 2), general acid catalysis involves the donation of a proton from the catalyst to the reacting molecule (called the substrate) followed at a later stage by the acceptance of a proton to reform the original structure of the catalyst. A general acid-catalysed reaction involving a substrate, HS, and a Brønsted–Lowry acid, HB, may be represented by

$$HS + HB \rightleftharpoons HSH^+ + B^-$$
$$HSH^+ \longrightarrow SH + H^+$$
$$H^+ + B^- \longrightarrow HB$$

A proton is added to one part of the molecule and a different proton is subsequently removed from another part. General base catalysis involves the acceptance of a proton by the catalyst from a substrate molecule, subsequently followed by the donation of the proton to another site on the substrate molecule, with the consequent reformation of the original catalyst.

20. General acid–base catalysis. Some reactions require the presence of both acids and bases before a catalytic effect is observed. Thus the mutarotation of glucose is very slow in both cresol (a weak acid) or pyridine (a weak base), but rapid in a mixture of the two. The rate of mutarotation is increased by the addition of either acid or base to an aqueous solution, but this is because water acts as a base if acid is added and as an acid if base is added (*see* XIII, 2).

21. Other catalysed reactions in solution.

(*a*) *Electron transfer reactions.* Electron transfer reactions are often catalysed by species having two valency states differing by 1. For example the reaction

$$V^{3+} + Fe^{3+} \longrightarrow V^{4+} + Fe^{2+}$$

is catalysed by Cu^{2+} ions, the probable mechanism being:

$$V^{3+} + Cu^{2+} \longrightarrow V^{4+} + Cu^+ \quad \ldots \quad (XXII, 24)$$
$$Cu^+ + Fe^{3+} \longrightarrow Cu^{2+} + Fe^{2+}$$

Reaction (XXII, 24) is rate-determining, and hence the rate is proportional to the concentration of V^{3+} and Cu^{2+} but independent of the Fe^{3+} concentration.

(*b*) *Catalysis by iodide ions.* One drawback to using ceric ions for oxidation reactions in volumetric analysis is that the reaction often takes place too slowly to be useful. However, the presence of a small amount of iodide ion markedly speeds up the oxidation. This catalyst has been used in the oxidation of arsenious and thiosulphate ions by ceric ion. The first step in the oxidation probably involves the formation of iodine atoms, which then combine to form molecular iodine. The molecular iodine reacts with the substance being oxidised to reform the iodide ions, *e.g.*

$$Ce^{4+} + I^- \longrightarrow Ce^{3+} + I$$
$$2I \longrightarrow I_2$$
$$I_2 + 2S_2O_3^{2-} \longrightarrow S_4O_6^{2-} + 2I^-$$

PROGRESS TEST 22

1. The following rate constants were found for the decomposition of N_2O_5 at various temperatures

$t/°C$	25	45	65
k/s^{-1}	3.46×10^{-5}	4.98×10^{-4}	4.87×10^{-3}

Determine graphically the activation energy and the Arrhenius factor. (2)

2. At 283°C and a concentration of 10^3 mol m^{-3} the number of molecules of HI colliding per second is 6×10^{37} m^{-3}. If the energy of activation is 184 kJ mol^{-1}, calculate the value of the rate in (a) molecules m^{-3} s^{-1}, (b) mol dm^{-3} s^{-1}. (5)

3. The rate of the thermal isomerisation of perfluorovinyl-cyclopropane to perfluorocyclopentane has been measured over the temperature range 130°C to 185°C and found to follow first-order kinetics. At 131°C $k_1 = 1.70 \times 10^{-5}$ s^{-1}, and at 183°C $k_1 = 2.28 \times 10^{-3}$ s^{-1}. Calculate (a) the activation energy, (b) the Arrhenius factor, (c) the half-time for isomerisation at 150°C, (d) the entropy of activation at 150°C. (XXII, 2, 10–11, XXI, 11)

4. State the Arrhenius equation and indicate the significance of the various factors. (2–3)

5. Outline the collision theory of reaction rates and indicate its limitations. (4–8)

6. Outline the transition state theory and indicate why it is superior to the collision theory. (9–11)

7. Describe some of the general properties of a catalyst. (12–16)

PART NINE

SURFACE CHEMISTRY

The next two chapters are concerned with systems in which surface effects predominate, giving rise to phenomena which are not apparent in bulk material.

The adsorption of substances on liquid and solid surfaces is considered in Chapter XXIII, where special reference is made to the role of solid surfaces in the catalysis of heterogeneous gas reactions.

Colloidal systems are discussed in Chapter XXIV. Of the great variety of colloidal systems, only that of a solid dispersed in a liquid medium is discussed in detail.

SURFACE PHENOMENA AND HETEROGENEOUS CATALYSIS

The surface area of finely divided matter is very high and gives rise to properties which are not apparent in bulk matter. Such properties can be explained only in terms of surface effects. Surface effects are important for particles having dimensions of less than 1 μm, since under these conditions an appreciable fraction of the system will be at or near the surface.

All surfaces are phase discontinuities, and the fact that a surface does exist indicates that the intermolecular (or interatomic) forces at the surface must be different from the forces existing in the interior of the substance. Thus a molecule in the surface of a liquid experiences a net attraction towards the interior, since there is no compensating force on the side remote from the liquid. On the other hand, a molecule in the interior is subject to a uniform attraction in all directions (*see* V, 3 for the analogous effect in gases).

Gases or dissolved substances are attracted to and retained on surfaces in an attempt to satisfy the unsaturated molecular forces. It is necessary to distinguish between the phenomena of *adsorption* and *absorption*. The former represents a process in which a substance is attached to the surface of a solid or liquid, while in the latter process the substance is distributed throughout the solid or liquid.

LIQUID SURFACES

1. Surface tension. Molecules at the surface of a liquid experience a net attractive force into the interior, and freely suspended liquid droplets are spherical, since this is the form giving the least surface area for a given volume. In order to increase the surface area, molecules in the interior must perform work and move to the surface against the forces of molecular attraction. The force which opposes this increase in area is known as the *surface tension* and is defined as the force acting perpendicular to unit length of a line in the liquid surface. Liquids with large surface tensions are those in which the cohesive forces between molecules are large (*e.g.* water). The tension which exists at the interface between immiscible liquids is commonly called the *interfacial tension*.

2. Surface tension of a solution.

The addition of a solute to a liquid may either increase or decrease the surface tension, depending on the nature of both the solute and the solvent. Thus the surface tension of water in slightly increased by the addition of strong electrolytes, sucrose and glycerol; slightly reduced by the addition of non-electrolytes and weak electrolytes, and greatly reduced by the addition of fatty acids and soaps.

J. Willard Gibbs (1878) was able to show thermodynamically that changes in surface tension arise from unequal distribution of solute between the surface and the bulk solution. Solutes which decrease the surface tension concentrate in the surface and vice versa.

3. Surface films.

The marked decrease in the surface tension of water on the addition of small concentrations of fatty acids and soaps (the so-called *surface active agents*) may be explained as follows. The water-soluble carboxyl groups at one end of the molecule lie in the surface of the water, while the water-insoluble hydrocarbon groups at the opposite end project above the surface (*see* Fig. 118).

Fig. 118.—*Surface film of fatty acid on water*

NOTE: For clarity the hydrocarbon chains have been drawn perpendicular to the surface. This will only be the case when the molecules are tightly packed together (*see* 5 below).

As the length of the hydrocarbon chain decreases, the solubility increases, owing to the increasing influence of the carboxyl group, and the effect on the surface tension decreases, *i.e.* the molecules tend to concentrate less in the surface. The terms *hydrophilic* (water-attracting) and *hydrophobic* (water-repelling) are often used to describe the functional groups.

4. The Langmuir surface balance. Confirmation of the preceding ideas comes from the study of monomolecular films on liquid surfaces using a *Langmuir surface balance*. If a small amount of a solution of a long-chain fatty acid (*e.g.* palmitic or stearic acid) in benzene is placed on a clean water surface, the solvent evaporates, leaving behind a monomolecular film of the fatty acid spread over the surface. By confining the film between two barriers in the surface of the water, one fixed and floating and the other movable (*see* Fig. 119) it is possible gradually to reduce the area of the film and to measure the lateral force per unit length, F, exerted by the film on the fixed float. Since there is a clean water surface on the side of the float remote from the film, F is given by

$$F = \gamma^\bullet - \gamma$$

where $\gamma^\bullet$ and γ represent the surface tensions of pure water and the film respectively. F is called the *surface pressure*.

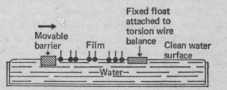

Fig. 119.—*Schematic diagram of Langmuir surface balance*

5. Molecular dimensions from the surface balance. The effect on the surface pressure of reducing the film area is shown in Fig. 120 (*a*). The surface area occupied by a single molecule (A) is readily determined from the measured surface area and the number of molecules in the weight of sample used to form the film. The initial slow increase in F with decreasing area indicates that the surface is not completely covered by the film and that considerable compression is possible. However, after reaching a certain critical area, F increases very rapidly. This is the situation when the molecules in the film are tightly packed. Any further reduction in area requires considerable force, since this would involve compressing the molecules themselves unless the structure of the single layer changes. If the area is reduced still further, the film eventually crumbles and forms a multimolecular layer.

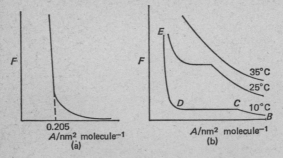

Fig. 120.—*Surface pressure–area isotherms* (a) *High surface pressure;* (b) *Myristic acid at low surface pressure*

Extrapolation of the steep portion of the curve to cut the horizontal axis gives the area occupied by one molecule in a tightly packed monolayer. This value of about 0·205 nm² is independent of the length of the hydrocarbon chain for the higher fatty acids, and is to be expected if the hydrocarbon chain is arranged at right angles to the surface such that the area measured is the cross-sectional area of the hydrocarbon chain. X-ray diffraction measurements on the solid acids confirm the above value.

6. The analogy between surface films and gases. The curves in Fig. 120 (b) represent the effect of temperature on the surface pressure–area curves for myristic acid, $CH_3(CH_2)_{12}COOH$, at low surface pressures and very high areas. The curves are strongly reminiscent of the pressure–volume isotherms for a real gas near its critical temperature (*see* Fig. 14), and surface films behave in some respects as "two-dimensional gases." Thus for the lower curve, BC represents the compression of the isolated individual surface molecules (discontinuous or gaseous phase), DE the compression of the continuous monomolecular film (continuous or liquid phase), while CD represents an intermediate situation with small islands of surface film in equilibrium with the isolated individual molecules (liquid–gas equilibrium). The upper curve may be compared with an ideal gas isotherm and follows an equation analogous to the ideal gas law

$$fA = kT$$

where k is Boltzmann's constant.

ADSORPTION OF GASES ON SOLIDS

The amount of gas adsorbed by a given amount of solid depends on:

    (a) The nature of the gas and of the solid.
    (b) The past history of the solid.
    (c) The surface area of the solid.
    (d) The temperature of the solid.
    (e) The temperature and pressure of the gas.

The adsorption may be quite specific. For example, while 1 g of charcoal will adsorb only 8 cm^3 of nitrogen at 15°C, it will adsorb almost 400 cm^3 of sulphur dioxide. The extent of adsorption may often be increased by activating the solid in various ways. Thus activated charcoal can be produced by heating wood charcoal to 500°C in a vacuum, a process which apparently distills out the hydrocarbon impurities.

7. Types of adsorption. There are two main types of adsorption, *physical adsorption* and *chemical adsorption* or *chemisorption*. In the former the attraction of the gaseous molecules to the solid is by van der Waals forces, while the latter involve chemical forces similar to those in chemical bond formation. Table 23 illustrates the differences between the two types of adsorption. Most of the characteristics of physical adsorption are readily understood on the assumption that the attractive forces are very similar to the intermolecular forces present in liquids. For example, heats of physical adsorption are of the same order as heats of vaporisation. The much greater heats of chemisorption reflect the increased strength of the bond. The bond involved in the chemisorption of oxygen on tungsten is so strong that tungsten trioxide may be distilled off from the surface by heating strongly.

NOTE: The adsorption process results in an increase in order and hence ΔS is negative. In order to make the process spontaneous (*i.e.* ΔG negative), ΔH must be negative and greater than $T\Delta S$.

TABLE 23 COMPARISON OF PHYSICAL ADSORPTION AND CHEMISORPTION

Physical adsorption	*Chemisorption*
1. Attraction due to van der Waals forces, hence heats of adsorption usually in the range 20–40 kJ mol^{-1} and activation energies small.	Attraction due to chemical bond forces, hence heats of adsorption usually in the range 40–400 kJ mol^{-1} and activation energies may be appreciable.
2. Usually occurs rapidly at low temperatures and decreases with increasing temperature.	Can occur at high temperatures.
3. Usually completely reversible.	Often irreversible.
4. The extent of adsorption is approximately related to the ease of liquefaction of the gas.	No such correlation.
5. Not very specific.	Often highly specific.
6. Forms multimolecular layers.	Forms monomolecular layers.

Adsorption may be a mixture of the two extreme types, physical adsorption being important at low temperatures and chemisorption at high temperatures. Thus at −190°C nitrogen is physically adsorbed on iron as nitrogen molecules, while at 500°C it is chemisorbed, with the formation of a surface layer of iron nitride.

8. The adsorption isotherm. The relationship between the amount of gas adsorbed at constant temperature by a given amount of solid and the pressure of the gas is represented by the *adsorption isotherm*. Two of the most common adsorption isotherms are shown in Fig. 121 (*a*) and (*b*). (*a*) is typical of chemisorption and is characterised by an initial steeply rising portion which gradually levels off as the monolayer is formed. (*b*) is typical of physical adsorption, where the first portion is similar to (*a*) and the levelling off again corresponds to monolayer formation. The subsequent rapid increase indicates multilayer formation with the eventual condensation of the gas.

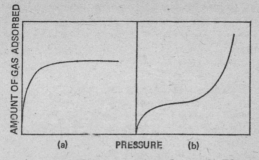

FIG. 121.—(a) *Chemisorption isotherm.* (b) *Physical adsorption isotherm*

9. The Langmuir adsorption isotherm. There is no general theory which explains the shape of all adsorption isotherms. A simple approach which has proved very successful for chemisorption is that due to Langmuir (1916), who assumed that:

(a) The surface is uniform in its adsorption properties.

(b) There is no interaction between neighbouring adsorbed molecules.

(c) The adsorbed molecules are not dissociated.

(d) The adsorbed layer is monomolecular.

The derivation of the Langmuir adsorption isotherm is as follows. The rate of desorption of gaseous molecules will be proportional to the fraction of the surface occupied by the molecules, θ, *i.e.* rate of desorption $= k_1\theta$.

The rate of adsorption will be proportional to the surface area not already covered $(1 - \theta)$ and also to the rate of collision of the gaseous molecules with the uncovered surface (directly proportional to the gas pressure, p), *i.e.* rate of adsorption $= k_2 p(1 - \theta)$.

At equilibrium the two rates will be equal, hence

$$k_1\theta = k_2 p(1 - \theta)$$
$$\theta = k_2 p/(k_1 + k_2 p) = ap/(1 + ap) \quad . \quad . \quad . \text{ (XXIII, 1)}$$

where $a = k_2/k_1$. Equation (XXIII, 1) gives the correct form for the chemisorption isotherm; in particular:

(a) At low pressures $a \gg p$ and $\theta = p/a$, so that θ increases linearly with pressure.

(*b*) At high pressures $a \ll p$ and $\theta = 1$, which is the situation when the surface is completely covered by a monolayer.

Other, more complicated, theories have been advanced which quantitatively explain the phenomena associated with multi-layer adsorption.

10. The nature of the adsorbed species. The bonding involved in the chemisorption of species on solid surfaces may be either ionic or covalent. In ionic bonding, the ease of transfer of electrons across the surface of the solid determines both the ease of formation and the strength of the bond. Ionic bonding is responsible for the bonding of alkali metal vapours to tungsten surfaces. Covalent bonds may only be formed if orbitals with unpaired electrons are available on the solid surface. Most organic and inorganic molecules are adsorbed on metal surfaces by covalent bonds.

Some gaseous molecules dissociate on adsorption, and the diatomic gases, hydrogen, nitrogen and oxygen, are known to be adsorbed as atoms on metal surfaces

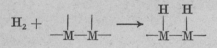

More than one adsorbed state may be possible, and experimental results suggest that methane may be adsorbed on metal surfaces in two ways:

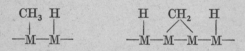

11. Adsorption from solution. The adsorption of solutes from solution is of considerable practical importance, and the technique of decolorising solutions using activated charcoal is well known. A monolayer is formed and a satisfactory empirical isotherm is

$$y = kc^{1/n} \qquad . \quad . \quad . \text{(XXIII, 2)}$$

where y is the mass of solute adsorbed per unit mass of solid, c is the solute concentration, and k and n are empirical constants. Equation (XXIII, 2), the *Freundlich isotherm*, has been applied to some gaseous systems.

HETEROGENEOUS CATALYSIS OF GASEOUS REACTIONS

Many gaseous reactions which do not proceed homogeneously at an appreciable rate can be made to proceed very rapidly in the presence of a suitable solid surface. Such effects are the result of chemisorption of the reactants on the surface. The formation of the strong chemisorption bonds brings about reorganisation of the bonds in the reactant molecules, and enables different reaction paths to be undertaken involving lower activation energies.

12. Stepwise mechanism of surface reactions. Surface reactions may be considered to involve five consecutive steps:

　(a) Diffusion of the reactants to the surface.
　(b) Adsorption of the reactants on the surface.
　(c) Reaction on the surface.
　(d) Desorption of the products from the surface.
　(e) Diffusion of the products away from the surface.

Steps (a) and (e) are usually very fast, and the rate-determining steps are either (b), (c) or (d).

13. The kinetics of surface reactions. The rate equations for surface reactions may be derived using the Langmuir adsorption isotherm, assuming that only adsorbed species are able to undergo reaction. The rate of a surface catalysed reaction is proportional to the fraction of the surface covered, i.e.

$$\text{rate} = k\theta = kap/(1 + ap) \quad . \quad . \quad . \quad \text{(XXIII, 3)}$$

The simplest example of a surface reaction is the decomposition of a single reactant. The rate of decomposition may be written $-dn/dt$, where n is the amount of gas. For a constant volume system, dn will be proportional to dp, the change in pressure of the gas ($dn = Vdp/RT$ for an ideal gas), and (XXIII, 3) may be written

$$-dp/dt = k\theta = kap/(1 + ap) \quad . \quad . \quad . \quad \text{(XXIII, 4)}$$

It is convenient to consider two extreme situations which are distinguished by the extent of adsorption of the gas.

(a) *Gas weakly adsorbed.* For a weakly adsorbed gas, a is small, and ap may be neglected in comparison with unity. Under these conditions the rate is given by

$$-dp/dt = kap = k_1p$$

This is the expression for a first-order reaction which on integration becomes

$$p = p_0\exp(-k_1t)$$

where p_0 and p are the initial pressure and the pressure after time t respectively. Examples of this type of reaction are the decomposition of nitrous oxide on gold and of hydrogen iodide on platinum. Both these reactions are second-order in the gas phase.

(b) *Gas strongly adsorbed.* When a gas is strongly adsorbed, $ap \gg 1$, thus $\theta = 1$ and (XXIII, 3) becomes

$$-dp/dt = k$$

The reaction is independent of pressure, *i.e.* is zero-order, and the integrated form is

$$p_0 - p = kt$$

Zero-order behaviour has been observed for the decomposition of ammonia on tungsten and of hydrogen iodide on gold. It is implicit in the above formulation that sufficient gas is present to completely cover the surface. The reaction will be zero-order only above a certain pressure. At low pressures the conditions described in (a) will operate and the reaction will be first-order. At intermediate pressures, where the surface is only partially covered, the reaction will have a fractional order.

14. Bimolecular reactions.

Reactions involving two gaseous reactants can take place only if the molecules are adsorbed in adjacent positions. If θ_A and θ_B represent the fractions of the surface covered by molecules A and B, then the probability of A and B occupying adjacent positions is proportional to $\theta_A\theta_B$, and the reaction rate is given by

$$-dp/dt = k\theta_A\theta_B$$

If both reactants are weakly adsorbed

$$-dp/dt = k_2p_Ap_B$$

and the reaction will be second-order.

If one reactant, say A, is strongly adsorbed, while B is only weakly adsorbed, the fraction of the surface covered by B will be small. Under such conditions the rate of the reaction will depend upon the rate at which B molecules can be adsorbed on the surface. Increase in the pressure of B will *increase* the rate, while increase in the pressure of A will *decrease* the rate. A is said to *inhibit* the reaction, since it prevents the adsorption of B. In the reaction between carbon monoxide and oxygen on platinum, carbon monoxide is strongly adsorbed and retards the reaction.

15. Inhibition by products. The products of a surface reaction may also be adsorbed, and reactants and products will compete for the available surface area. If the products are adsorbed more strongly than the reactants, the rate of reaction will decrease as the concentration of products increases, *e.g.* the decomposition of ammonia on platinum surfaces is inhibited by the hydrogen produced, since this is strongly adsorbed.

16. Active centres. The surface of a solid catalyst is not uniform in its catalytic properties, and catalysis occurs only at a few specially favoured sites or *active centres*. The rate of reaction will be proportional to the amount of gas adsorbed on the active centres. These active centres are thought to be dislocations, crystal edges and other surface irregularities. The evidence for their existence is as follows:

(a) *Poisoning of catalysts*. Small amounts of foreign substances —far less than that required to form a monolayer—can completely remove the catalytic activity by preferentially adsorbing on the active centres. Thus the hydrogenation of ethylene on a copper surface is almost completely prevented by a trace of mercury vapour.

(b) *Selective poisoning*. Two different reactions may require different active centres on a given catalyst. Thus small amounts of carbon disulphide largely inhibit the reaction between carbon dioxide and hydrogen on a platinum surface, although the surface is still able to catalyse the decomposition of nitrous oxide.

(c) *Sintering*. Many catalysts are deactivated by heating to a temperature just below their melting points. This results in the surface irregularities being "ironed out."

17. The specific nature of heterogeneous catalysis. The rate of a given reaction may alter by many factors of ten depending on the nature of the catalytic surface. This specificity is further illustrated by the fact that different surfaces may give rise to different products from the same initial reactants. For example, on a copper surface ethanol decomposes into acetaldehyde and hydrogen, while on an alumina surface the products are ethylene and water. This indicates that the two catalysts adsorb the ethanol molecule in different ways, and so are able to yield quite distinct products. In this context it should be noted that alumina surfaces have a great affinity for water and are therefore efficient at removing the elements of water from substances, while copper surfaces adsorb hydrogen very strongly (as atoms) and readily abstract hydrogen from substances.

PROGRESS TEST 23

1. Explain what is meant by surface tension. In what way is the surface tension of water affected by the addition of solutes? (1–3)

2. Describe the Langmuir surface balance. How may it be used to measure molecular dimensions? (4–5)

3. In what respects does a surface film resemble a two-dimensional gas? (6)

4. Distinguish between physical adsorption and chemisorption. (7–8)

5. Derive the Langmuir adsorption isotherm and list the assumptions made. (9)

6. What sequence of five steps is assumed to be involved in surface catalysed gaseous reactions? (12)

7. Derive the rate expression for the decomposition of a single reactant adsorbed on a surface when the reactant is (a) weakly adsorbed, (b) strongly adsorbed. (13)

8. What are active centres? List some experimental evidence for their existence. (16)

CHAPTER XXIV

COLLOIDS

Colloidal solutions are intermediate between true solutions and suspensions, and may contain either large molecules (macromolecules), such as proteins and polymers, or small particles. The diameters of colloidal particles may range from 1 to 100 nm and their unique properties are due to the high surface area to volume ratio conferred by such dimensions.

The two phases present in a colloidal system are the *disperse phase* (the phase containing the particles) and the *dispersion medium* (the medium in which the particles are distributed). Since either of these can be solid, liquid or gas, there should be nine types of colloidal system; however, the gas–gas system cannot exist. The most important type of colloidal solution, a *sol*, consists of a solid dispersed in a liquid and is the only type we shall consider in detail.

CLASSIFICATION AND PREPARATION OF SOLS

1. Classification. Sols may be classified as *lyophobic* (solvent-hating) and *lyophilic* (solvent-attracting). Where water is the solvent, the terms used are *hydrophobic* and *hydrophilic* respectively. In lyophobic sols there is little attraction between the two phases, and such sols tend to coagulate into coarse particles, particularly on the addition of small quantities of electrolyte (*see* **18** below), until finally precipitation occurs. The reverse process, *i.e.* the reformation of the sol from the precipitate on the addition of liquid, cannot occur. Examples of lyophobic sols include metals, metallic sulphides and silver halides in aqueous solution.

The strong attractive forces between the two phases in lyophilic sols give rise to stable solvated particles. The solid obtained on evaporation may be reconverted to the original sol simply by the addition of liquid. Examples of lyophilic sols include proteins in aqueous solution and high polymers in organic solvents.

Some sols possess properties intermediate between those of typical lyophilic and lyophobic sols.

2. The nature of sols. The particles in a sol may be very large single molecules or aggregates of small molecules, atoms or ions. Thus proteins, many carbohydrates and plastics consist of macromolecules, colloidal sulphur of clusters of S_8 molecules held together by van der Waals forces, colloidal gold of minute crystals, while stearate ions form clusters or *micelles* (*see* **20** below) containing about seventy individual ions.

3. Preparation. Whereas lyophilic sols may be prepared simply by warming the solid and the liquid together, lyophobic sols must be prepared by special methods, which can be subdivided into *condensation* methods and *dispersion* methods. In the former the sol particles are built up from smaller particles, while in the latter coarse particles are broken down to aggregates of colloidal size.

4. Condensation methods. In these methods careful control of the experimental conditions is essential to ensure that the building up process does not go too far and precipitation result.

(a) *Oxidation*. Aqueous solutions of hydrogen sulphide and hydrogen selenide may be oxidised by oxygen or sulphur dioxide to sulphur and selenium sols respectively.

(b) *Reduction*. Metal sols of silver, gold and platinum may be prepared by reduction of their aqueous solutions, using formaldehyde or hydrazine.

(c) *Hydrolysis*. Sols of the oxides of weakly electropositive metals such as iron, aluminium and chromium may be obtained by hydrolysis of suitable salts in aqueous solution.

(d) *Double decomposition*. Silver halide sols are obtained on mixing dilute solutions of silver salts and alkali halides.

5. Dispersion methods.

(a) *Physical grinding*. A colloid mill, consisting of a series of closely spaced metal discs rotating at high speeds in opposite directions, reduces many substances to colloidal size.

(b) *Peptisation*. The direct disintegration of coarse particles into particles of colloidal size may be possible in the presence of a peptising agent. Thus the addition of dilute hydrochloric acid to freshly precipitated silver chloride results in sol formation.

(c) *Electrical dispersion.* Sols of silver, gold and platinum may be prepared by striking an electric arc between wires of the relevant metal immersed in water. The high temperature of the arc vaporises some of the metal, which then condenses to form colloidal particles.

6. Purification. After a sol has been prepared, it often contains electrolytes, which would cause coagulation on standing and must therefore be removed. Since ordinary filtration is ineffective, resort must be made to *dialysis* or *ultrafiltration.*

(a) *Dialysis.* This technique depends on the fact that particles of colloidal size cannot pass through a membrane such as parchment or cellophane, as can small molecules and ions. The process is very slow normally, but can be speeded up by applying a potential difference across the membrane (*electrodialysis*). Under the influence of the field the ions pass through the membrane much faster.

(b) *Ultrafiltration.* The pores in a normal filter paper can be made small enough to retain colloidal particles by impregnation with collodion (a partially evaporated solution of cellulose nitrate in an ethanol–diethyl ether mixture). Increased pressure or suction is required to force the liquid through the small pores.

PROPERTIES OF SOLS

7. Colligative properties. The colligative properties of a solution depend upon the number of particles present, and are well marked for true solutions (*see* XX, 4–11). Owing to the large size of sol particles, the magnitude of these properties is much reduced and only osmotic pressure is readily measurable.

8. Optical properties. The amount of light scattered by true solutions is very small, whereas colloidal solutions are turbid, owing to intense light scattering. This phenomenon is called the *Tyndall effect.* Measurements of the intensity and angular distribution of the scattered light enable estimates to be made of the dimensions and relative molecular masses of the scattering particles. The *ultramicroscope* utilises the phenomenon of light scattering to observe colloidal particles which are too small to be seen in an ordinary microscope. Because of

Brownian motion the particles appear as dancing halos of light (*see* 9 below).

9. Kinetic properties.

(*a*) *Diffusion*. Colloidal particles diffuse much more slowly than solutes in solution. Particle dimensions can be estimated from diffusion studies.

(*b*) *Brownian motion*. Brownian motion is caused by the erratic bombardment of the colloidal particles by the molecules of the liquid.

(*c*) *Sedimentation*. Colloids tend to settle out very slowly under the influence of gravity, but the rate of sedimentation may be greatly increased using a high speed centrifuge known as an *ultracentrifuge*. Measurements of the settling rate in the ultracentrifuge provide a means of measuring the relative molecular masses of large molecules.

(*d*) *Viscosity*. Viscosity is a measure of the internal resistance offered to the relative motion of different parts of a liquid. Lyophilic sols have a high viscosity, partly due to the increase in effective particle size caused by solvation and partly due to the consequent decrease in the amount of free solvent. The viscosity of a lyophobic sol is not appreciably different from that of the solvent.

10. The charge on sols.

Sol particles tend to acquire a surface charge, either by ionisation or by adsorption of ions from the solvent.

Lyophilic sols such as proteins possess both acidic and basic functional groups. In aqueous solution at low pH, the $-NH_2$ group (basic) acquires a proton to give $-NH_3^+$, while at high pH the $-COOH$ group (acidic) transfers a proton to OH^- to give $-COO^-$. Thus the charge on the protein molecule is dependent on pH and there is an intermediate pH called the *isoelectric point* at which the molecule will be electrically neutral.

A similar mechanism operates in the case of some lyophobic sols such as metal oxides in aqueous media, the surface being positively charged at low pH and negatively charged at high pH:

$$MO^- \underset{}{\overset{H^+}{\rightleftharpoons}} MOH \underset{}{\overset{H^+}{\rightleftharpoons}} MOH_2^+$$

At some intermediate pH the metal oxide will have no net surface charge. As a general rule, however, ionic lyophobic sols have a preference for adsorbing common ions, and the surface charge will depend on the method of preparation. In the presence of excess Ag^+ ions, silver chloride sols are positively charged, while excess Cl^- ions confer a negative surface charge.

11. The electrical double layer. To counterbalance the charge on the surface of the sol, oppositely charged ions (the *counter*-ions) are attracted from the bulk solution. Originally this was thought to result in a compact *electrical double layer* or *Helmholtz double layer* of positive and negative charges (*see* Fig. 122 (a)). However, because of thermal agitation, such

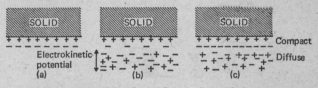

FIG. 122.—*Models of electrical double layers.*
(a) *Helmholtz double layer.* (b) and (c) *Stern double layers*

a compact arrangement is not possible. The double layer is now considered to be made up of the fixed charges on the surface of the solid, a layer of more or less tightly bound counter-ions, and a more diffuse layer extending into the solution. The diffuse layer, which is analogous to the ionic atmosphere in electrolytes (*see* XII, **18**), may have the same or opposite sign to that of the tightly bound counter-ions (*see* Fig. 122 (b) and (c)). The combination of compact and diffuse layers is called a *Stern double layer*. The diffuse layer is only loosely attached to the particle surface and moves in the opposite direction to the particle under an applied electric field. The net charge on the solid and the medium is zero. Because of the distribution of charge there is a difference in potential between the compact layer and the bulk of the solution, and this called the *electrokinetic* or *zeta potential*. The presence of the double layer accounts for the existence of the four electrokinetic phenomena described in **12—15** below. All four phenomena allow estimates to be made of the magnitude of the electrokinetic potential.

12. Electrophoresis. The migration of charged sol particles under an applied electric field is called *electrophoresis*, and is analogous to the migration of ions in an electrolyte. The sol particles are neutralised at the electrode and precipitation takes place. By noting the direction in which the particles migrate it is possible to determine the charge. For certain sols, the charge is pH-dependent (*see* **10** above), and at the isoelectric point no migration takes place. Different sol particles migrate at different rates, and this effect has been used to characterise and purify proteins and other substances of biological interest.

13. Electro-osmosis. The passage of solvent through a semi-permeable membrane under the influence of an applied electric field is known as *electro-osmosis*. The direction of the liquid flow depends on the charge on the diffuse layer, and can be used to produce an electro-osmotic pressure in a suitable apparatus.

14. The streaming potential. When a sol is forced through a porous membrane a *streaming potential* is produced across the membrane. The diffuse layer is separated from the charged particles so that opposite charges are built up on opposite sides of the membrane. The resulting potential may be measured by placing identical reversible electrodes on either side of the membrane. This phenomenon can be regarded as the reverse of electro-osmosis.

15. The sedimentation potential. The potential difference built up between the top and bottom of a container when a charged sol sediments (*i.e.* moves relative to a stationary liquid phase) is known as the *sedimentation potential*. In sedimenting, the charged particles tend to leave the diffuse layer in the upper parts of the container.

THE STABILITY OF SOLS

The stability of a sol is determined by the nature of its surface.

16. Lyophobic sols. Since a sol particle with its associated double layer is neutral, no electrostatic force exists between widely separated particles. However, as the particles approach one another the diffusive parts of their double layers interact, giving rise to a repulsive force. Thus as long as there is sufficient electrolyte present to form the double layer, aggregation will be difficult. The effect of high concentrations of electrolytes is discussed in **18** (*d*) below.

17. Lyophilic sols. The same short-range repulsive forces are also present in lyophilic sols, and in addition the particles are surrounded by a tightly bound solvent sheath. Such a molecule is thus surrounded by two protective layers which hinder coalescence. The stability of a lyophilic sol is reduced if the charge is removed, *e.g.* by the addition of an electrolyte, but this does not necessarily lead to precipitation, since the particles are still protected by a surrounding layer of solvent. For the same reason many lyophilic sols are stable at their isoelectric points.

18. Coagulation of lyophobic sols.

(*a*) *Mixing of oppositely charged sols.* If two sols of opposite charge are mixed together, both are precipitated. This is due to the short-range attractive forces which favour the building up of aggregates.

(*b*) *Electrophoresis.* In electrophoresis the charge on the sol is neutralised at the electrode and the sol precipitated.

(*c*) *Boiling.* Some lyophobic sols, such as sulphur and silver halides in aqueous media, may be coagulated by boiling. This is probably brought about by increased collisions between the solvent molecules and the sol particles removing some of the adsorbed electrolyte and hence encouraging coagulation.

(*d*) *Addition of electrolyte.* Small amounts of electrolyte are essential to the stability of lyophobic sols, since they are responsible for the electrical double layer. However, the addition of larger amounts of electrolyte results in precipitation. Oppositely charged ions are adsorbed on the surface of the particle, thus lowering the charge. Mutual repulsion between particles is reduced and coagulation takes place. The greater the charge on the oppositely charged ions, the greater the coagulating power of the electrolyte.

19. Coagulation of lyophilic sols.

(*a*) *Addition of electrolyte.* The addition of *large* amounts of electrolyte results in precipitation of the sol, and this is known as the *salting-out effect*. The salting-out effect is different from the coagulation caused by the addition of small amounts of electrolyte to lyophobic sols and is due to removal of the solvent from the sol particles. In effect, the electrolyte competes with the sol particles for the solvent. The salting-out efficiency depends on the tendency of the ions to become hydrated.

(b) *The addition of liquids in which the solvent is soluble.* The addition of alcohol and acetone to aqueous lyophilic sols often results in precipitation of the sol, since the particles become desolvated. In the presence of alcohol or acetone quite small quantities of electrolyte can bring about precipitation.

OTHER TYPES OF COLLOID

20. Associated colloids. One class of lyophilic sols behave as normal strong electrolytes at low concentrations, but at higher concentrations exhibit properties consistent with the formation of aggregated particles. Such substances are known as *associated colloids* and comprise the surface active agents such as soaps and synthetic detergents. Sodium stearate is typical of an associated colloid, having the physical properties expected for individual sodium and stearate ions in dilute aqueous solution. However, on increasing the concentration there is an abrupt change in physical properties, indicating that association of the stearate ions has taken place. The aggregates formed are called *micelles* and revert to individual ions on dilution.

The sodium stearate micelle is a spherical structure of about seventy stearate ions with the hydrocarbon chains directed towards the centre and the polar $-COO^-$ groups on the surface in contact with the aqueous medium. These polar groups are responsible for the stability of the micelles even though the net charge on the micelle is considerably reduced by the presence of the counter-ions (Na^+) bound to the surface. The hydrocarbon interior of the micelle acts as a solvent for water-insoluble organic material (oils, greases, etc.) and is responsible for the cleansing action of soaps.

21. Gels. Under certain conditions it is possible for sol particles to unite to form thread-like chains which can interlock, giving rise to a continuous three-dimensional network. The solvent becomes mechanically trapped within this network, forming a more or less rigid mass or *gel*. Gelatin is an *elastic gel* prepared by cooling a dispersion of gelatin in warm water. Silicic acid gel (or silica gel) is a *non-elastic* gel prepared by the acidification of aqueous sodium silicate solution.

Elastic gels may be completely dehydrated and then reformed by the addition of water. Dehydration of non-elastic gels gives a powdery solid which will not reform the original gel on the addition of water. This difference is attributed to the rigidity of the thread-like chains which form the three-dimensional network. In elastic gels the chains are flexible and can reaccommodate the water following dehydration. Non-elastic gels, on the other hand, possess a more rigid structure and cannot expand sufficiently to re-enclose the water.

22. Emulsions. A suspension of finely divided liquid droplets in another immiscible liquid is called an *emulsion*. These droplets are somewhat larger than the particles found in sols. Emulsions may be produced by agitating a mixture of the two liquids or, better, by passing the mixture through a colloid mill. Stable emulsions may also be prepared in the presence of small amounts of other substances called *emulsifying agents*. These are surface active agents such as soaps and synthetic detergents which coat the droplets and hinder coalescence. Emulsions may be broken down (*demulsified*) to yield the constituent liquids by physical methods such as boiling, freezing or centrifuging, and by chemical methods which destroy the emulsifying agent.

PROGRESS TEST 24

1. Explain the differences between lyophobic and lyophilic sols. (1)
2. List the methods of preparation of lyophobic sols. (3–5)
3. How may sols be separated from the excess electrolyte used in their preparation? (6)
4. Outline some of the non-electrical properties of sols. (7–9)
5. Explain the formation of the electrical double layer. (10–11)
6. What electrokinetic phenomena owe their existence to the electrical double layer? (12–15)
7. Explain how lyophobic and lyophilic sols are stabilised and how this stabilisation may be removed. (16–19)
8. What are (a) associated colloids (20), (b) gels (21), (c) emulsions (22)?

SUGGESTIONS FOR FURTHER READING

GENERAL PHYSICAL CHEMISTRY TEXTBOOKS

S. Glasstone and D. Lewis: *Elements of physical chemistry* (Macmillan, 1970).

S. H. Maron and C. F. Prutton: *Principles of physical chemistry*, 4th edition (Macmillan, 1969).

W. J. Moore: *Physical chemistry*, 3rd edition (Longman, 1964).

G. W. Castellan: *Physical Chemistry* (Addison-Wesley, 1964).

W. H. Hamill, R. R. Williams and C. MacKay: *Principles of physical chemistry*, 2nd edition (Prentice-Hall, 1966).

E. A. Moelwyn-Hughes: *A short course of physical chemistry* (Longman, 1966).

G. M. Barrow: *Physical chemistry* (McGraw-Hill, 1961).

F. Daniels and R. A. Alberty: *Physical chemistry*, 3rd edition (J. Wiley, 1966).

R. H. Cole and J. S. Coles: *Physical principles of chemistry* (W. H. Freeman, 1964).

SPECIFIC PHYSICAL CHEMISTRY TEXTBOOKS

Gas laws

W. Kauzmann: *Kinetic theory of gases* (Benjamin, 1966).

Chemical energetics and equilibrium

E. F. Caldin: *An introduction to chemical thermodynamics* (Oxford University Press, 1958).

K. Denbigh: *The principles of chemical equilibrium*, 2nd edition (Cambridge University Press, 1966).

P. G. Ashmore: *Principles of chemical equilibrium*, Monographs for teachers No. 5 (Royal Institute of Chemistry, 1961).

E. A. Guggenheim: *Elements of chemical thermodynamics*, Monographs for teachers No. 12 (Royal Institute of Chemistry, 1966).

D. C. Firth: *Elementary chemical thermodynamics* (Oxford University Press, 1969).

I. M. Kloty: *Chemical thermodynamics* (Benjamin, 1964).

J. R. W. Warn: *Concise chemical thermodynamics* (Van-Nostrand Reinhold, 1969).

F. D. Ferguson and T. K. Jones: *The phase rule* (Butterworth, 1966).

G. Hargreaves: *Elementary chemical thermodynamics*, 2nd edition (Butterworth, 1968).

Structure and bonding

A. Holden: *Bonds between atoms* (Oxford University Press, 1971).

E. Cartmell and G. W. A. Fowles: *Valency and molecular structure*, 3rd edition (Butterworths, 1966).

J. E. Spice: *Chemical bonding and structure* (Pergamon, 1966).

A. F. Wells: *Structural inorganic chemistry*, 3rd edition (Oxford University Press, 1962).

J. Barrett: *Introduction to atomic and molecular structure* (J. Wiley, 1970).

J. W. Linnett: *The electronic structure of molecules : a new approach* (Methuen, 1964).

Electrochemistry

R. A. Robinson and R. H. Stokes: *Electrolyte solutions*, 2nd edition (Butterworth, 1968).

A. R. Denaro: *Elementary electrochemistry*, 2nd edition (Butterworth, 1971).

C. W. Davies: *Principles of electrolysis*, Monographs for teachers No. 1, 2nd edition (Royal Institute of Chemistry, 1968).

A. G. Sharpe: *Principles of oxidation and reduction*, Monograph for teachers No. 2, 2nd edition (Royal Institute of Chemistry, 1968).

Chemical kinetics

P. G. Ashmore: *Principles of reaction kinetics*, Monographs for teachers No. 9, 2nd edition (Royal Institute of Chemistry, 1967).

G. L. Pratt: *Gas kinetics* (J. Wiley, 1969).

J. L. Latham: *Elementary reaction kinetics*, 2nd edition (Butterworth, 1969).

K. J. Laidler: *Reaction kinetics* (2 volumes) (Pergamon, 1963).

A. A. Frost and R. G. Pearson: *Kinetics and mechanisms* (J. Wiley, 1961).

Surface chemistry

D. L. Shaw: *Introduction to colloid and surface chemistry*, 2nd edition (Butterworth, 1970).

Miscellaneous

M. L. McGlashan: *Physical-chemical quantities and units, the grammar and spelling of physical chemistry*, Monographs for teachers No. 15 (Royal Institute of Chemistry, 1968).

F. Daniels: *Mathematical preparation for physical chemistry* (McGraw-Hill, 1956).

R. B. Heslop: *Numerical aspects of inorganic chemistry* (Elsevier, 1970).

EXAMINATION TECHNIQUE

The passing of examinations is a science in itself, and one which is neglected to a sad degree by the majority of candidates. The advice given in this appendix is intended to help you under examination conditions to make the most of the knowledge you have gained from this and other books, but remember that no amount of examination technique can replace a sound knowledge of the work on which you are being examined.

1. Planning. A common sight in an examination hall is the candidate who begins writing almost before he has had time to make contact with the chair. This is generally a sign that the candidate is nervous and flustered and will turn in an ill-thought-out paper. Remember that a few minutes spent thinking about the questions, deciding which you can best answer, and how you are going to answer them can save a considerable amount of wasted effort. A suggested sequence follows; a time for each of the various stages is not given as this will, to a large extent, depend on the individual:

(a) Read the "instructions to candidates" carefully.

(b) Read *each* question carefully and make sure you grasp the point that the examiner is making; a great deal of time can be wasted answering questions that were never asked.

(c) Choose the questions you intend to answer and mark beside each the times at which you intend to start and finish your answer.

(d) Begin on the question you feel you can answer best. This strategy is largely a morale booster; the feeling that you have competently answered one question puts you in a better frame of mind for those which you may feel less well equipped to deal with.

(e) Leave sufficient time to check through your answers and to make any necessary corrections.

2. The question. Again assure yourself that you are answering the question the examiner asked, rather than a related topic about which you happen to know. Realise that the examiner will be marking according to a mark scheme and try to weigh up the importance of each of the sections in a question that consists of several parts; if the question is not divided (*e.g.* an essay) realise that the examiner will have effectively sectionalised it when arriving at his mark scheme and will have allotted a fixed number of marks to each section.

When you have done this, plan your answer to the question. Some candidates prefer to do this mentally, others to jot down the plan on paper. Either way the plan should consist of the major subject headings and notes on the content of each in the form of phrases or "key words."

Make sure that your answer covers everything on the plan. While answering any particular question, think only of that question; allowing your thoughts to wander on to a later question may result in disaster.

3. Problems.

Many examinations on Physical Chemistry contain problem questions. For some reason many candidates avoid these like the plague. This is quite illogical. Problem questions have definite answers, and if answered successfully will gain the full marks allotted to the question in the mark scheme. It is highly unlikely that any but the most competent candidate will approach this desirable state on a more general type of question. In fact questions of the type "Compare and contrast . . ." or "Describe the chemistry of . . ." are far more difficult to answer, and should be avoided by the mediocre candidate.

4. Timing.

The "instructions to candidates" section will tell you how many questions to answer. Answer this number; certainly no more, and preferably no fewer. Remember that a graph of *marks awarded* versus *time spent* on a question can be thought of as approximately exponential; *i.e.* you can obtain the first few marks with the expenditure of comparatively little time, but to obtain the last one or two marks will require a disproportionate amount of time. It is, therefore, unwise to run over your allotted time on any question in an attempt to pick up the last few marks; this time would be better spent on the next question, assuming that you can answer it. In general, the same amount of time should be allotted to each question and any time left after attempting all questions should be used in checking your answers and in extra work on those questions where it can be used to best advantage. If, through some miscalculation, you find yourself in a situation where, for example, you have ten minutes instead of forty minutes for the last question, write your answer in note form. Do not skimp on the planning, and try to write more legibly than you may have done previously; remember that the examiner will not now have the normal verbiage to give some contextual clue as to the meaning of the odd hieroglyphic.

5. Developing examination technique. The only way to become proficient is by constant practice. Taken literally this could lead to some practical difficulties; therefore the following method is recommended. When you have studied a topic, find a series of questions on it at the same (or similar) standard to that of the relevant examination. For each question jot down a plan for an answer consisting of subject headings and key words as in 2 above and then check it against the relevant text. For one or two questions on each topic actually try writing down the answer in the time allowed, simulating examination conditions as closely as possible.

EXAMINATION QUESTIONS

The following specimen questions have been taken from Part I
of the University of London B.Sc. General Degree (L.U.) and
from Part I of the Royal Institute of Chemistry graduate member-
ship examination (R.I.C.). Permission to use these questions is
gratefully acknowledged. In all instances units have been converted into SI units.

1. List the principal assumptions made in the derivation of the
expression

$$pV = mN\overline{c^2}/3$$

for N molecules of an ideal gas in volume V (a proof is not required). Discuss the distribution of molecular velocities which
gives rise to the expression $\overline{c^2} = 3kT/m$ for the mean square
velocity of molecules of mass m in the gas at temperature T.
Show how the above expressions lead to the gas laws of Boyle,
Charles, Avogadro, Dalton and Graham.

(*L.U.*, 1969)

2. Define the molar heat capacities of a gas. At 25°C the molar
heat capacities at constant volume of certain gases are: neon,
12·5, nitrogen 20·8, bromine 27·7, carbon dioxide 28·8, sulphur
dioxide 31·5 J K^{-1} mol^{-1}. Discuss these values. The gas constant
$R = 8·31$ J K^{-1} mol^{-1}.

(*L.U.*, 1965)

3. Explain what is meant by an equation of state and outline
the principles upon which the van der Waals equation is based.
Show how this equation may be written as a reduced equation of
state and discuss the advantages of this form of equation.

(*R.I.C.*, 1965)

4. What do you understand by the "critical point of a gas"?
Draw a diagram showing the pressure–volume relationships of a
gas near its critical point, explaining the significance of different
regions of the diagram.

(*L.U.*, 1963)

5. Explain what is meant by a thermodynamically reversible process. The volume of one mole of an ideal monatomic gas, initially at a volume of $11 \cdot 8$ dm^3 and a pressure of $202 \cdot 6$ kN m^{-2} at 15°C, can be doubled either by an isothermal reversible expansion or by an adiabatic reversible expansion.

(a) Calculate the final pressure in each case and (b) sketch each path on a $P - V$ diagram. (c) Calculate the work done by the system, the heat absorbed by the system and the increase in the internal energy of the system for each process. (Assume $R = 8 \cdot 31$ J K^{-1} mol^{-1}; $C_V = 12 \cdot 5$ J K^{-1} mol^{-1}.)

(L.U., 1970)

6. Show that Hess's law of constant heat summation is a consequence of the law of conservation of energy (first law of thermodynamics). Apply Hess's law to determine the enthalpy change in the reaction

$$CO_2(g) + H_2(g) = CO(g) + H_2O(g) \qquad \cdot \cdot \cdot (1)$$

given that

$$2CO(g) + O_2(g) = 2CO_2(g)(\Delta H = -565 \cdot 5 \text{ kJ}) \cdot \cdot \cdot (2)$$
$$2H_2(g) + O_2(g) = 2H_2O(l)(\Delta H = -571 \cdot 3 \text{ kJ}) \cdot \cdot \cdot (3)$$
$$H_2O(g) = H_2O(l)(\Delta H = -43 \cdot 9 \text{ kJ}) \qquad \cdot \cdot \cdot (4)$$

Use your result to predict the effect of changes of temperature on the position of equilibrium in reaction (1). How are the positions of equilibrium in reactions (1) and (2) affected by changes in the total pressure of the system?

(L.U., 1966)

7. Derive the Kirchoff equation describing the variation of an enthalpy change with temperature.

At 25°C the latent heat of sublimation of iodine is $62 \cdot 3$ J mol^{-1} and the standard enthalpy of formation of HI(g) is $24 \cdot 1$ J mol^{-1}. Calculate the enthalpy change which occurs when HI(g) is formed from the gaseous elements at 225°C.

Mean molar heat capacities over the temperature range 25–225°C:

$$H_2(g): C_p = 29 \cdot 1 \text{ J K}^{-1} \text{ mol}^{-1}$$
$$I_2(g): C_p = 33 \cdot 5 \text{ J K}^{-1} \text{ mol}^{-1}$$
$$HI(g): C_p = 29 \cdot 9 \text{ J K}^{-1} \text{ mol}^{-1}$$

(R.I.C., 1968)

8. Under equilibrium conditions at 227°C and $101 \cdot 3$ kN m^{-2} total pressure, the gas nitrosyl chloride is 27% dissociated, according to the reaction $2NOCl = 2NO + Cl_2$. Calculate the standard free energy change at this temperature. Show that for

very slight dissociation the percentage dissociation would be inversely proportional to the cube root of the equilibrium pressure. Explain briefly what additional information would be necessary to calculate the extent of dissociation at some other temperature.

(R.I.C., 1969)

9. Define the symbols used in the thermodynamic equations:

$$G = U + pV - TS \quad \text{and} \quad H = U + pV$$

Show how using these relations and the condition that, for an infinitesimal reversible process $dU = TdS - pdV$, it is possible to obtain a relationship between ΔH, ΔG, and $d\Delta G/dT$ for a chemical reaction. Discuss the application of this relationship.

(R.I.C., 1963)

10. State the phase rule, explaining clearly the terms used. Draw and comment upon the phase diagram for water at temperatures near the triple point. Explain why the triple point for water is at a different temperature from that of the normal melting point of ice.

(L.U., 1969)

11. Discuss the equilibrium between a pure liquid and its vapour from a kinetic viewpoint.

Describe the effects on the position of this equilibrium of (a) increase in temperature up to the critical temperature, and (b) the addition of volatile and of non-volatile solutes.

(L.U., 1969)

12. State Henry's law and Raoult's law, showing the relationship between them. Discuss the bearing of deviations from these laws on the separation of liquid mixtures by distillation.

(R.I.C., 1964)

13. Illustrate by means of examples how the Le Chatelier principle may be applied to chemical equilibria. Show that the equation

$$\frac{d \ln K}{dT} = \frac{\Delta H}{RT^2}$$

for the variation with temperature of the equilibrium constant of a reversible reaction is in accordance with this principle.

(L.U., 1969)

14. When A mol of acetic acid and B mol of alcohol were allowed to react until equilibrium was reached, C mol of ester and C mol of water were formed.

(a) Express the equilibrium constant in terms of A, B and C.

(b) Discuss the effect of the following changes of condition on the yield of ester: (i) removal of water from the reaction mixture (ii) addition of an inert solvent, (iii) addition of a small amount of mineral acid.

(L.U., 1968)

15. Define the terms (a) conductivity, (b) molar conductivity and (c) ionic mobility. State their units and show how they are related to one another.

Explain how conductance measurements can be made the basis of the determination of the solubility of a sparingly soluble electrolyte.

(R.I.C., 1967)

16. Give a concise account of the principles of one method for the determination of transport numbers in an electrolyte solution. Describe the apparatus and experimental procedure involved in the measurement.

(L.U., 1963)

17. What is meant by the term "weak electrolyte"? Derive an expression for the dissociation constant of a weak acid and hence for the pH of a solution of the acid. Calculate the pH of the following aqueous solutions at 25°C:

(a) a 0·01 M solution of acetic acid;
(b) a solution 0·01 M in acetic acid and 0·02 M in potassium acetate.

(pK_a for acetic acid at 25°C = 4·76.)

(L.U., 1969)

18. Write short notes on three of the following:

(a) the hydration of ions;
(b) the concept of pH;
(c) the hydrolysis of salts in aqueous solution;
(d) acid–base indicators;
(e) the hydrogen electrode.

(L.U., 1968)

19. What is meant by the term "electrode potential"? Give the electrode and cell reactions for the following reversible cell, assuming that the silver–silver chloride electrode is positive with respect to the hydrogen electrode

$$Pt, H_2 \mid HCl(0·1\ M) \mid AgCl, Ag$$

Describe how the cell e.m.f. would change upon the slow addition of 0·1 M aqueous solutions of (a) sodium hydroxide; (b) ammonia; (c) silver nitrate.

(L.U., 1969)

20. Given that the standard potentials of the Fe(III)/Fe(II) and Ce(IV)/Ce(III) systems are respectively 0·75V and 1·60V on the hydrogen scale, calculate and illustrate graphically the titration curve for the oxidation of iron(II) by cerium(IV) in aqueous solution at 25°C. $(2 \cdot 3\ RT/F = 0 \cdot 059V.)$

(R.I.C., 1968)

21. State (a) the Pauli exclusion principle, and (b) Hund's rule. Show how these concepts are used in building up the electronic configurations of atoms, using the elements from hydrogen (atomic number 1) to neon (atomic number 10) to illustrate your answer. What feature of electronic configuration characterises (c) a rare gas, and (d) a transition element ?

(L.U., 1966)

22. Explain the principles of the LCAO approximation for constructing molecular orbitals. What are meant by bonding and anti-bonding orbitals ? Illustrate your answer by reference to a diatomic molecule.

The bond strength in the nitrogen molecule is much greater than that in the fluorine molecule; how may this fact be explained in terms of MO theory ?

(R.I.C., 1966)

23. Describe the shapes of the following molecules and give reasons for the molecular configurations which they adopt: BF_3, PF_3, NH_3, CO_2, SO_2, ClO_2.

(R.I.C., 1964)

24. Show how ideas concerning the packing of spheres have been used, together with the ionic radii, to classify the structures of ionic solids. Illustrate your answer by reference to the halides of the alkali metals.

(L.U., 1967)

25. Write notes on three of the following:

(a) radius ratio rule;
(b) ionisation potential;
(c) lattice energy;
(d) electronegativity;
(e) types of crystal lattice.

(L.U., 1968)

26. Derive equations by means of which rate (velocity) constants may be calculated from experimental data for the following reactions:

(a) First-order, A $\longrightarrow$ Products;
(b) Second-order, A + B $\longrightarrow$ Products, in which the initial concentrations of reactants are unequal.

Give a specific example of a reaction of one of the above types, and describe briefly how you would obtain the necessary experimental data for the determination of the rate constant of this reaction.

(*L.U.*, 1965)

27. The following data refer to the decomposition of benzene diazonium chloride

$$C_6H_5N_2Cl = C_6H_5Cl + N_2$$

at a starting concentration of 10 g dm^{-3} in solution at 50°C.

Time (min.)	6	9	12	14	18	22	24	26	30
N$_2$ evolved (cm³)	19·3	26·0	32·6	36·0	41·3	45·0	46·5	48·4	50.4 58·3

Find the order, the rate constant and the half-life of the reaction.
(*R.I.C.*, 1967)

28. The neutralisation reaction of nitroethane in aqueous alkaline solution proceeds according to the rate equation:

$$-d[OH^-]/dt = -d[C_2H_5NO_2]/dt = k[C_2H_5NO_2][OH^-]$$

Experiments at 0°C, with initial reactant concentrations $[C_2H_5NO_2] = 0·01$ mol dm^{-3} and $[NaOH] = 0·01$ mol dm^{-3} give a value of 150 s for the reaction half-life. Deduce the dependence of the reaction half-life on initial concentration for a second-order reaction and calculate the corresponding rate constant at 0°C.

Experiments at 25°C gave a value of 354 dm³ mol^{-1} min^{-1} for the reaction rate constant; calculate the activation energy of the reaction.

(*R.I.C.*, 1968)

29. Show how the parameters A and $E^{\ddagger}$ in the Arrhenius expression for the rate constant of a bimolecular chemical reaction, $k = A \exp(-E^{\ddagger}/RT)$, may be interpreted on the basis of the collision theory.

The rate constant for a certain reaction at 450°C is $5·5 \times 10^{-4}$ s^{-1} and the activation energy for the reaction is 259 kJ mol^{-1}. What is the value of the rate constant at 500°C?

(*L.U.*, 1966)

30. Write an essay on adsorption at either a gas–solid or at a gas–liquid interface.

(*L.U.*, 1969)

31. Discuss the essential principles of catalysis. Illustrate your answer with examples of both heterogeneous and homogeneous catalysis.

(*R.I.C.*, 1967)

32. Write notes on three of the following topics:

    (*a*) the stabilities of lyophobic colloids;
    (*b*) electrokinetic phenomena;
    (*c*) colloidal electrolytes;
    (*d*) semi-permeable membranes.

(*L.U.*, 1969)

33. Write informative notes on *two* of the following:

    (*a*) the critical state;
    (*b*) the specific heats of gases;
    (*c*) osmotic pressure;
    (*d*) the Avogadro constant.

(*R.I.C.*, 1963)

34. Explain three of the following statements:

    (*a*) the solubility of sodium sulphate in water increases as the temperature rises to $32 \cdot 4°C$, but decreases with further rise of temperature;
    (*b*) steam distillation is often effective in purifying high boiling organic liquids which are immiscible with water;
    (*c*) ideal gases cannot be liquefied;
    (*d*) at constant temperature, the ionic product for water remains constant on the addition of (i) acid and (ii) alkali;
    (*e*) silver chloride is insoluble in water but soluble in aqueous ammonia;
    (*f*) the work obtained during the isothermal expansion of a gas is at a maximum when the process is carried out under thermodynamically reversible conditions.

(*L.U.*, 1969)

35. Write short accounts of two of the following:

    (*a*) buffer solutions;
    (*b*) the transition state;
    (*c*) quantisation of energy;
    (*d*) solubility product.

(*L.U.*, 1967)

SUGGESTED ANSWERS TO PROGRESS TESTS

Progress Test 1

1 (a) The dimensions of p and $n^2 a / V^2$ must be identical. Therefore the dimensions of a are those of pV^2/n^2

i.e. $$[ml^{-1}t^{-2}][l^3]^2/[n]^2 = [ml^5 t^{-2} n^{-2}]$$

On the SI system, units of a are $[\text{N m}^{-2}][\text{m}^3]^2/[\text{mol}]^{-2} = [\text{N m}^4 \text{ mol}^{-2}]$.

(b) A is obviously dimensionless, since $\ln p$ is dimensionless.

2(a) From Table 5, the units of e are [C] and of ϵ_0, $[\text{J}^{-1}\text{ C}^2\text{ m}^{-1}]$. Hence the units of E_p are $[\text{C}]^2/[\text{J}^{-1}\text{ C}^2\text{ m}^{-1}][\text{m}]$

i.e. [J]

(b) The units of E are $[\text{kg}][\text{m s}^{-1}]^2$, i.e. $[\text{kg m}^2\text{ s}^{-2}]$, i.e. [J].

3. The dimensions of $(\overline{c^2})^{\frac{1}{2}}$ are $[lt^{-1}]$
The dimensions of $(3RT/M)^{\frac{1}{2}}$ are $\{[ml^2 t^{-2}\, n^{-1}\, T^{-1}][T]/[m\, n^{-1}]\}^{\frac{1}{2}} = [l^2 t^{-2}]^{\frac{1}{2}} = [lt^{-1}]$
Hence the equation is dimensionally correct.

Progress Test 2

1. $$0.82 = 0.78 + 0.059 \lg \frac{[\text{Fe}^{3+}]}{40.015}$$

$$\frac{0.04}{0.059} = 0.6780 = \lg[\text{Fe}^{3+}] - \lg 0.015$$

$$\lg 0.015 = \bar{2}.1761 = -1.8239$$
$$\therefore \lg[\text{Fe}^{3+}] = 0.6780 - 1.8239 = -1.1459 = \bar{2}.8541$$
$$[\text{Fe}^{3+}] = \text{antilg } \bar{2}.8541 = 0.0715 \text{ M}$$

2. $$\text{pH} = -\lg[\text{H}_3\text{O}^+] = 4.31$$
$$\lg[\text{H}_3\text{O}^+] = -4.31 = \bar{5}.69$$
$$[\text{H}_3\text{O}^+] = \text{antilg } \bar{5}.69 = 4.9 \times 10^{-5} \text{ M}$$

3. Taking natural logarithms

$$\ln c = \ln c_0 + \ln \exp(-kt)$$
$$= \ln c_0 - kt$$
$$k = \frac{1}{t} \ln \frac{c_0}{c} = \frac{2 \cdot 303}{t} \lg \frac{c_0}{c}$$

4. Turn the equation upside down

$$\frac{1}{\theta} = \frac{1}{bp} + 1$$

The graph of $\frac{1}{\theta}$ against $\frac{1}{p}$ will be linear.

5. $$dq/dT = C = a + bT/2 + cT^2/3$$

6. Separating the variables

$$dN/N = -\lambda dt$$

Integrating $$\int_{N_0}^{N} dN/N = -\lambda \int_{0}^{t} dt$$
$$[\ln N]_{N_0}^{N} = -\lambda [t]_0^t$$
$$\ln(N/N_0) = -\lambda t$$

7. $$w = RT \int_{1}^{10} dV/V = RT[\ln V]_1^{10} = RT \ln 10$$

$$= 8 \cdot 314 \times 300 \times 2 \cdot 303 \times 1 = 5740 \text{ J}$$

Progress Test 3

1. From source (a), percentage of silver in silver oxide $= 0 \cdot 560/0 \cdot 602 = 93\%$.

From source (b), percentage of silver in silver oxide $= 1 \cdot 06/1 \cdot 14 = 93\%$, $i.e.$ the composition of silver oxide is independent of its source as demanded by the law of definite proportions.

2. In the first compound $32 \cdot 2$ g of A combine with $67 \cdot 8$ g of B. In the second compound $49 \cdot 0$ g of A combine with $51 \cdot 0$ g of B. $\therefore 32 \cdot 2$ g of A combine with $(32 \cdot 3/49 \cdot 0) \times 51 \cdot 0 = 33 \cdot 5$ g of B.

$$\frac{\text{mass of B in first compound}}{\text{mass of B in second compound}} = \frac{67 \cdot 8}{33 \cdot 5} = \frac{2 \cdot 0}{1 \cdot 0}$$

$i.e.$ the masses of B in the two compounds are in the ratio of small integers as demanded by the law of multiple proportions. Possible formulae for the two compounds are AB and A_2B, A_2B and A_4B, etc.

3. From Dulong and Petit's law, rough relative atomic mass
$= 27/0 \cdot 130 = 208$.

Valency $= 208/103 \cdot 5 = 2$ (to the nearest integer)
$\therefore$ Accurate relative atomic mass $= 103 \cdot 5 \times 2 = 207$

4. Vapour density of mixture $= 11 \cdot 3$ kg m^{-3} $= 11 \cdot 3$ g dm^{-3}.
$\therefore$ Average relative molecular mass of mixture $= 11 \cdot 3 \times 2 = 22 \cdot 6$.
The reactions involved are:

$$CH_4 + O_2 = CO_2 + 2H_2O$$
$$C_2H_4 + 3O_2 = 2CO_2 + 2H_2O$$
$$C_2H_2 + \tfrac{5}{2}O_2 = 2CO_2 + H_2O$$

All the water formed condenses, and the carbon dioxide is absorbed by the caustic potash. The $5 \cdot 5$ cm^3 remaining after sparking must consist solely of unused oxygen; this is confirmed by its absorption by alkaline pyrogallol. Thus 10 cm^3 of the mixture react with $24 \cdot 5$ cm^3 of oxygen, or, invoking Avogadro's hypothesis, 10 molecules of mixture react with $24 \cdot 5$ molecules of oxygen. This could be achieved by mixtures containing (i) 4 molecules of CH_4:3 of C_2H_4:3 of C_2H_2, (ii) 2 molecules of CH_4:1 of C_2H_4:7 of C_2H_2. The average relative molecular masses of these two mixtures can be calculated as follows, using the relative molecular masses of CH_4, C_2H_4 and C_2H_2 as 16, 28 and 26 respectively:

Average relative molecular mass mixture (i)
$$= (4 \times 16 + 3 \times 28 + 3 \times 26)/10 = 22 \cdot 5$$
Average relative molecular mass mixture (ii)
$$= (2 \times 16 + 1 \times 28 + 7 \times 26)/10 = 24 \cdot 2$$

Since the average relative molecular mass of the mixture has been found experimentally to be $22 \cdot 5$, mixture (i) must be correct, i.e. four volumes of CH_4:three of C_2H_4:three of C_2H_2.

5. From (III, 3), the mole fraction of sodium chloride, x'_2, is given by

$$x_2' = \Delta p / p_1^{\bullet}$$

Since x_2' is given by a ratio of pressures, the units of pressure are unimportant

$$\therefore x_2' = 1/30 = 0 \cdot 033$$

Now sodium chloride is completely dissociated into Na^+ and Cl^- (i.e. into two ions) in solution, and since the colligative properties depend on the *number* of solute particles, the actual mole fraction $x_2 = 0 \cdot 0165$.

The amount of water in 1 kg of water, $n_1 = 1/0 \cdot 018 = 55 \cdot 55$ mol
The amount of sodium chloride, n_2, is then given by

$$x_2 = n_2/(n_1 + n_2) \approx n_2/n_1 \text{ for a dilute solution}$$
$$\therefore 0 \cdot 0165 = n_2/55 \cdot 55$$
Hence $\qquad n_2 = 0 \cdot 92$ mol

The osmotic pressure of this solution is found from (III, 5)

$$\pi = RTx_2/V$$

where $R = 8 \cdot 31$ J K^{-1} mol^{-1} and $V = 18 \times 10^{-6}$ m^3 mol^{-1}

$$\therefore \pi = (8 \cdot 31 \times 298 \times 0 \cdot 0165)/18 \times 10^{-6}$$
$$= 2270 \text{ kN m}^{-2}$$

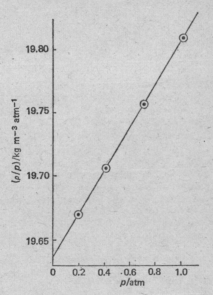

Fig. 123.—*Plot of (ρ/p) against p for carbon dioxide*

6. The graph of (ρ/p) against p is shown in Fig. 123. Extrapolation to zero pressure gives $(\rho/p)_0 = 1 \cdot 9636$ kg m^{-3} atm^{-1}. Note that the graph is plotted using atmospheres as pressure units. The reason for this will soon become apparent.

For CO_2: $\qquad M'_1 = (\rho_1/p_1)_0 RT$

Since the pressure is in atmospheres and not N m^{-2}, M_1' is not the molar mass but is nevertheless proportional to the molar mass

For O_2: $$M_2' = (\rho_2/p_2)_0 RT$$

$$\therefore \frac{M_1'}{M_2'} = \frac{M_1}{M_2} = \frac{(M_r)_1}{(M_r)_2} = \frac{(\rho_1/p_1)_0}{(\rho_2/p_2)_0}$$

$$\therefore (M_r)_1 = 31 \cdot 998 \times (1 \cdot 9636/1 \cdot 428) = 44 \cdot 00$$

Progress Test 4

1. If the data are presented as a plot of p against V the hyperbola shown in Fig. 124 (a) is obtained. Since all the data are recorded at a single temperature, the curve is called an *isotherm*.

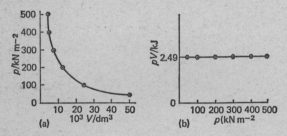

FIG. 124.—*Pressure–volume relationship for a gas: (a) p against v; (b) pV against p*

The isotherms for other temperatures will be a family of "parallel" hyperbolas. The general equation of a hyperbola is of the form $xy = c$, or in this case $pV = k$. However, it is difficult to tell whether the curve shown in Fig. 124 (a) deviates from the perfect hyperbola demanded by Boyle's law, and a better method of presenting the data would be as a plot of p against $1/V$, giving a straight line passing through the origin, or as a plot of pV against p giving a straight line of zero slope, as shown in Fig. 124 (b).

NOTE: The units of pV are kN m or kJ.

2. Using the ideal gas equation in the form

$$pV/RT = n = m/M$$

For the gases A and B:

$$\frac{m_A/M_A}{m_B/M_B} = 1$$

$$\therefore M_A/M_B = (M_r)_A/(M_r)_B = m_A/m_B = \tfrac{1}{2}$$

3. The solution to this problem is best illustrated by reference to Fig. 125. Using the ideal gas equation in the form $pV/RT = n$:

(*i*)   For the whole apparatus as in Fig. 125 (*a*)
   $50 \cdot 67 \times 2V/300R = n_1 + n_2$ (0·5 atm = 50·67 kN m^{-2})

(*ii*) For the bulb at 27°C (*see* Fig. 125 (*b*))

$$pV/300R = n_1$$

(*iii*) For the bulb at 127°C

$$pV/400R = n_2$$

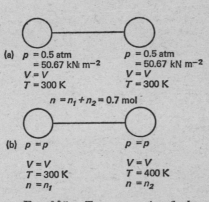

(a) p = 0.5 atm p = 0.5 atm
         = 50.67 kN m^{-2} = 50.67 kN m^{-2}
      $V = V$ $V = V$
      T = 300 K T = 300 K

         $n = n_1 + n_2$ = 0.7 mol

(b) $p = p$ $p = p$

      $V = V$ $V = V$
      T = 300 K T = 400 K
      $n = n_1$ $n = n_2$

FIG. 125.—*Two connecting flasks*

Thus, since (*ii*) + (*iii*) = (*i*)

$$(pV/300R) + (pV/400R) = 50 \cdot 67 \times 2V/300R$$
$$\therefore (p/300) + (p/400) = 101 \cdot 325/300$$

From which $p = 57 \cdot 9$ kN m^{-2}

To find n_1 and n_2 we make use of the fact that the pressures and volumes of both bulbs in Fig. 125 (*b*) are the same

$$\therefore p_1 V_1 = p_2 V_2$$
or $n_1 R T_1 = n_2 R T_2$
$$\therefore n_1 \times 300 = n_2 \times 400$$
or $n_1/n_2 = 4/3$

But $n_1 + n_2 = 0 \cdot 7$ mol

$$\therefore n_1 = 0 \cdot 4 \text{ mol}, n_2 = 0 \cdot 3 \text{ mol}$$

4. $PCl_5(g) = PCl_3(g) + Cl_2(g)$

From the stoichiometry of the equation, the partial pressures of PCl_3 and Cl_2 must be the same

$$\therefore p_{PCl_3} = p_{Cl_2} = 20 \cdot 26 \text{ kN m}^{-2}$$

Assuming the gases to behave ideally, this pressure may be converted into an amount of gas

$$n = pV/RT = (20 \cdot 62 \times 10^{-3})/(8 \cdot 314 \times 10^{-3} \times 500)$$
$$= 4 \cdot 88 \times 10^{-3} \text{ mol}$$

If none of the PCl_5 had dissociated, the amount of gas present would have been

$$n = m/M = 2 \cdot 08 \times 10^{-3}/208 \times 10^{-3} = 10^{-2} \text{ mol}$$

Since $4 \cdot 88 \times 10^{-3}$ mol of PCl_3 and Cl_2 are formed, $4 \cdot 88 \times 10^3$ mol of PCl_5 must have dissociated. Hence the amount of PCl_5 left undissociated must be $10^{-2} - 0 \cdot 488 \times 10^{-2} = 5 \cdot 12 \times 10^{-3}$ mol and the pressure exerted

$$p_{PCl_5} = nRT/V = (5 \cdot 12 \times 10^{-3} \times 8 \cdot 314 \times 10^{-3} \times 500)/10^{-3}$$
$$= 21 \cdot 24 \text{ kN m}^{-2}$$

The total pressure is

$$p_{PCl_5} + p_{PCl_3} + p_{Cl_2} = 21 \cdot 24 + 20 \cdot 26 + 20 \cdot 26$$
$$= 61 \cdot 76 \text{ kN m}^{-2}$$

5.
$$M\overline{c^2}/2 = 3RT/2$$
$$\therefore (\overline{c^2})^{\frac{1}{2}} = (3RT/M)^{\frac{1}{2}}$$

Note the difference between the root mean square speed and the average and most probable speeds, which are given by:

$$\bar{c} = (8RT/\pi M)^{\frac{1}{2}} \quad \text{and} \quad \hat{c} = (2RT/M)^{\frac{1}{2}}$$

For the electron

$$(\overline{c^2})^{\frac{1}{2}} = \left(\frac{3 \times 8 \cdot 314 \times 298}{5 \times 10^{-4} \times 10^{-3}} \right)^{\frac{1}{2}} = 1 \cdot 2 \times 10^5 \text{ m s}^{-1}$$

For the nitrogen molecule

$$(\overline{c^2})^{\frac{1}{2}} = \left(\frac{3 \times 8 \cdot 314 \times 298}{28 \times 10^{-3}} \right)^{\frac{1}{2}} = 5 \cdot 2 \times 10^2 \text{ m s}^{-1}$$

6(a)
$$E_t = 3kT/2$$
$$2 \times 10^{-21} = 3 \times 1 \cdot 38 \times 10^{-23} \times T/2$$
$$\therefore T = 96 \cdot 5 \text{ K}$$

(b)
$$\frac{dn}{n} = 4\pi \left(\frac{m}{2\pi kT} \right)^{\frac{3}{2}} \exp(-mc^2/2kT)c^2 dc \qquad \text{. . . (1)}$$

Putting
$$E_t = mc^2/2 \qquad \qquad \cdots \; (2)$$
$$c^2 = (2/m)E_t \qquad \qquad \cdots \; (3)$$
$$\therefore c = (2/m)^{\frac{1}{2}}E_t^{\frac{1}{2}}$$
$$dc = (2/m)^{\frac{1}{2}}(\tfrac{1}{2})E_t^{-\frac{1}{2}}dE_t$$
$$= (\tfrac{1}{2}m)^{\frac{1}{2}}E_t^{-\frac{1}{2}}dE_t \qquad \qquad \cdots \; (4)$$

Substituting (2), (3) and (4) in (1)

$$\frac{dn}{n} = 4\pi \left(\frac{m}{2\pi kT}\right)^{\frac{3}{2}} \exp(-E_t/kT)\left(\frac{2}{m}\right)E_t\left(\frac{1}{2m}\right)^{\frac{1}{2}}E_t^{-\frac{1}{2}}dE_t$$

$$= 2\pi \left(\frac{1}{\pi kT}\right)^{\frac{3}{2}} \exp(-E_t/kT)E_t^{\frac{1}{2}}dE_t$$

Substitution of the given values $E_t = 1 \cdot 98 \times 10^{-21}$ J, $dE_t = 0 \cdot 04 \times 10^{-21}$ J and $T = 96 \cdot 5$ K gives

$$dn/n = 0 \cdot 0515$$

Note that the Maxwell equation can be used in the above (differential) form only when the values of dE_t and dn are very small. For calculations involving larger values, the expression must first be integrated.

7.   At 100 K, $kT = 1 \cdot 38 \times 10^{-21}$ J
            At 300 K, $kT = 4 \cdot 14 \times 10^{-21}$ J

In order to use the equation

$$E_r = \frac{h^2}{8\pi^2 I} \times J(J + 1)$$

we must first calculate the moments of inertia, I. The general formula for the moment of inertia of a molecule is

$$I = \sum_i m_i r_t^2$$

where r_t is the distance of an atom of mass m_i from the centre of mass. For a homonuclear diatomic molecule this reduces simply to $I = md^2/2$, where d is the internuclear separation and the mass m to be used is the mass of a single atom.

$$m(H_2) = 10^{-3}/6 \times 10^{23} = 1 \cdot 67 \times 10^{-27}\,\text{kg}$$
and $$m(N_2) = 14 \times 10^{-3}/6 \times 10^{23} = 23 \cdot 3 \times 10^{-27}\,\text{kg}$$

$$\therefore I(H_2) = 1 \cdot 67 \times 10^{-27} \times (0 \cdot 075 \times 10^{-9})^2 = 0 \cdot 46 \times 10^{-47}\,\text{kg m}^2$$

and

$$I(N_2) = 23\cdot3 \times 10^{-27} \times (0\cdot109 \times 10^{-9})^2 = 14 \times 10^{-47} \, kg \, m^2$$

$$E_r \, (H_2) = \frac{h^2}{8\pi^2 I(H_2)} \times J \, (J + 1)$$

$$= \frac{(6\cdot5 \times 10^{-34})^2}{8\pi^2 \times 0\cdot46 \times 10^{-47}} \times J(J + 1)$$

$$= 1\cdot16 \times 10^{-21} J(J + 1) \, J$$

and

$$E_r \, (N_2) = \frac{h^2}{8\pi^2 I(N_2)} \times J(J + 1) = \frac{(6\cdot5 \times 10^{-34})^2}{8\pi^2 \times 14 \times 10^{-47}} \times J(J + 1)$$

$$= 0\cdot038 \times 10^{-21} J(J + 1) \, J$$

For hydrogen, the energies of the first two levels ($J = 1$ and 2) are $2\cdot32 \times 10^{-21}$ and $6\cdot96 \times 10^{-21}$ J respectively, so that only the first level lies below kT at 300 K. For nitrogen, the first ten levels have energies (in J $\times 10^{-21}$): $0\cdot076$, $0\cdot23$, $0\cdot46$, $0\cdot76$, $1\cdot14$, $1\cdot60$, $2\cdot12$, $2\cdot74$, $3\cdot42$, $4\cdot18$. Five levels lie below kT at 100 K and nine at 300 K. The spacings of the levels are shown graphically in Fig. 126 and demonstrate that at room temperature the equipartition principle provides a reasonable approximation to the rotational energy of nitrogen, but not to that of hydrogen.

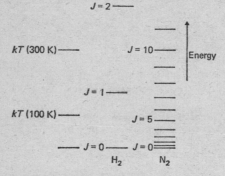

Fig. 126.—*Relative rotational energy levels in hydrogen and nitrogen*

Progress Test 5

1(a) Density of $CO_2 = 1\cdot981$ kg m^{-3}

$\therefore$ Number of molecules per m³,

$$N = 1\cdot981 \times 6\cdot023 \times 10^{23}/44 \times 10^{-3}$$
$$= 0\cdot271 \times 10^{26}$$

∴ Space inhabited by each molecule

$$= 1/0.271 \times 10^{26}$$
$$= 3.70 \times 10^{-26}\,\text{m}^3$$

If we consider this space to be a sphere of radius r with the molecule at its centre, then

$$4\pi r^3/3 = 3.70 \times 10^{-26}$$
$$\therefore r = 20.7 \times 10^{-10}\,\text{m}$$

The average distance between the centres of two molecules will be $2r$, *i.e.* $41.4 \times 10^{-10}\,\text{m}$.

(b) From V, 2, $b = 2\pi d^3 N/3$

Since b is quoted in $\text{m}^3\,\text{mol}^{-1}$, the value of N to be used in this relation is the Avogadro constant, *not* the value calculated in (a) above

$$\therefore d^3 = (4.27 \times 10^{-5} \times 3)/(2 \times 6.023 \times 10^{23} \times \pi)$$
Hence $d = 3.24 \times 10^{-10}\,\text{m}$.

Note that the average distance between molecules is rather greater than ten times the molecular diameter, indicating that carbon dioxide should behave in a near-ideal manner under these conditions.

(c) $Z = 2N^2 d^2(\pi RT/M)^{\frac{1}{2}}$

Substituting $N = 0.271 \times 10^{26}$ (from (a)), $d = 3.24 \times 10^{-10}\,\text{m}$ (from (b)), $M = 44 \times 10^{-3}\,\text{kg mol}^{-1}$ and $R = 8.314\,\text{J K}^{-1}\,\text{mol}^{-1}$

Find $Z = 6.2 \times 10^{34}\,\text{m}^{-3}\,\text{s}^{-1}$
(d) $\lambda = (\tfrac{1}{2})^{\frac{1}{2}}\pi N d^2$

Substituting for N and d as above

$$\lambda = 7.95 \times 10^{-8}\,\text{m}$$

2(a) $p_c = a/27b^2$ $T_c = 8a/27Rb$

$$\therefore T_c/p_c = 8b/R$$
i.e. $b \propto T_c/p_c$

∴ Gas A has the smaller value of b.

(b) $a = 27RbT_c/8$
i.e. $a \propto bT_c$
i.e. $a \propto (T_c/p_c)T_c$

∴ Gas A has the smaller value of a.

(c) $V_c = 3b$

∴ Gas A has the smaller critical volume.

(d) Gas A more closely approximates ideality, since B is below its critical temperature.

Progress Test 6

1. The volume of one mole of water vapour at 373·2 K and 101·3 kN m^{-2} will be $0·0224 \times 373·2/273·2 = 0·030\ 61$ m^3. The volume of one mole, $i.e.$ 0·018 kg, of liquid water under the same conditions will be $0·018/9·583 \times 10^2 = 1·88 \times 10^{-5}$ m^3

Change in volume on vaporisation

$$= 0·030\ 61 - 0·000\ 02 = 0·030\ 59 \text{ m}^3$$

The expansion work, $p\Delta V = 101·3 \times 0·030\ 59 = 3·10$ kJ.

2. From the first law $\Delta U = q - p\Delta V$
$$= (40·64 - 3·10) = 37·54 \text{ kJ}$$

3. The reaction is

$$CH_3OH(l) + \tfrac{3}{2} O_2(g) = CO_2(g) + 2H_2O(l)$$

Amount of gaseous reactants $= \tfrac{3}{2}$ mol
Amount of gaseous products $= 1$ mol

$$\Delta n = (1 - \tfrac{3}{2}) = -\tfrac{1}{2} \text{ mol}$$

Now $\Delta H = \Delta U + \Delta nRT$
$$= -724·9 + (-\tfrac{1}{2})(8·314 \times 298·2 \times 10^{-3})$$
$$= -724·9 - 1·2 = -726·1 \text{ kJ mol}^{-1}$$

When $\Delta n = 0$, ΔU and ΔH are identical (for ideal gases).

4. Since the gas is ideal, the amount of gas is given by

$$n = pV/RT = 99·8 \times 10^3 \times 10^{-3}/8·314 \times 298·2$$
$$= 0·4033 \text{ mol}$$

Work done, $w = nRT \ln(V_2/V_1)$
$$= 0·4033 \times 8·314 \times 298·2 \times 2·303 \lg(20/10)$$
$$= 693 \text{ J}$$

5. Substitution of $V = RT/p$ into (VI, 15) gives

$$C_V \ln(T_2/T_1) = R \ln\left(\frac{RT_1/p_1}{RT_2/p_2}\right)$$
$$= R \ln(T_1/T_2) + R \ln(p_2/p_1)$$
$$\therefore (C_V + R)\ln(T_2/T_1) = C_p \ln(T_2/T_1) = R \ln(p_2/p_1)$$

For an monatomic ideal gas $C_p = 5R/2$, and converting to lgs

$$(5R/2) \lg(T_2/300) = R \lg(1/10)$$
$$(5/2)(\lg T_2 - \lg 300) = \lg 1 - \lg 10$$
$$\therefore \lg T_2 = 2/5(-1 + 2·4771)$$

Hence $$T_2 = 114 \text{ K}$$

For the irreversible adiabatic expansion

$$\Delta U = q - w = -w = -p_2(V_2 - V_1)$$

where p_2 is the constant opposing pressure.
From (VI, 11), for a finite change

$$\Delta U = C_V \Delta T = C_V(T_2 - T_1)$$

Equating the two relationships for ΔU

$$C_V(T_2 - T_1) = -p_2(V_2 - V_1)$$
$$= -p_2\left(\frac{RT_2}{p_2} - \frac{RT_1}{p_1}\right) = -R\left(T_2 - \frac{T_1 p_2}{p_1}\right)$$

For a monatomic ideal gas, $C_V = 3R/2$
$$(3R/2)(T_2 - 300) = -R\{T_2 - (300/10)\}$$
From which $\qquad T_2 = 192$ K.

Thus the maximum cooling is obtained when the expansion is carried out reversibly.

Progress Test 7

1. From (VII, 10)

$$\Delta H^{\ominus}_{298} = 2\Delta H^{\ominus}_f(H_2O, l) + 3\Delta H^{\ominus}_f(S, \text{rhombic})$$
$$-2\Delta H^{\ominus}_f(H_2S, g) - \Delta H^{\ominus}_f(SO_2, g)$$
$$-234 \cdot 5 = -571 \cdot 8 + 0 - 2\Delta H^{\ominus}_f(H_2S, g) + 296 \cdot 9$$
Hence $\qquad \Delta H^{\ominus}_f(H_2S, g) = -20 \cdot 2$ kJ mol^{-1}

For reactions involving ideal gases, (VI, 6) is applicable

$$\therefore \ \Delta H^{\ominus}_f = \Delta U^{\ominus}_f + \Delta n RT \quad \text{where } \Delta n = -3$$
$$-234 \cdot 5 = \Delta U^{\ominus}_f - (3 \times 8 \cdot 314 \times 298 \cdot 2 \times 10^{-3})$$
$$= \Delta U^{\ominus}_f - 7 \cdot 4$$
Hence $\qquad \Delta U^{\ominus}_f = -227 \cdot 1$ kJ mol^{-1}

2. From (VII, 10)

$$\Delta H^{\ominus}_{298} = \Delta H^{\ominus}_f(Na^+, aq) + \Delta H^{\ominus}_f(Cl^-, aq) - \Delta H^{\ominus}_f(Na, s)$$
$$- \tfrac{1}{2}\Delta H^{\ominus}_f(Cl_2, g)$$
$$= -240 \cdot 6 - 166 \cdot 9 - 0 = -407 \cdot 5 \text{ kJ mol}^{-1}$$

3. This problem is similar to Example 3. The reaction for the formation of ethane from its atoms in the gas phase is

$$2C(g) + 6H(g) = C_2H_6(g) \qquad \Delta H_{298}$$
where $\qquad \Delta H_{298} = -(\Delta H_{\bar{D}}(C - C) + 6\Delta H_{\bar{D}}(C - H))$
$$= -(\Delta H_{\bar{D}}(C - C) + 6 \times 413) \qquad \ldots \ (5)$$

ΔH_{298} may also be found from heats of formation (see VII, **4**)
$$\Delta H_{298} = \Delta H_f(C_2H_6, g) - 2\Delta H_f(C, g) - 6\Delta H_f(H, g) \ \ldots \ (6)$$

The heats of formation of the atomic species are given and $\Delta H_f^\ominus(C_2H_6, g)$ may be found from the combustion reaction

$$C_2H_6(g) + \tfrac{7}{2}O_2(g) = 2CO_2(g) + 3H_2O(l) \quad \Delta H_{298}^\ominus = -1560 \text{ kJ}$$
$$\Delta H_{298}^\ominus = 2\Delta H_f^\ominus(CO_2, g) + 3\Delta H_f^\ominus(H_2O, l) - \Delta H_f^\ominus(C_2H_6, g) - \tfrac{7}{2}\Delta H_f^\ominus(O_2, g)$$

$$-1560 = (2 \times -393 \cdot 5) + (3 \times -285 \cdot 9) - \Delta H_f^\ominus(C_2H_6, g) - 0$$

Hence $\quad \Delta H_f^\ominus(C_2H_6, g) = -84 \cdot 7 \text{ kJ mol}^{-1}$

From (6)

$$\Delta H_{298} = -84 \cdot 7 - (2 \times 716) - (6 \times 218) = -2824 \cdot 7 \text{ kJ mol}^{-1}$$

Substituting this value of ΔH_{298} into (5) gives

$$-2824 \cdot 7 = -(\Delta H_{\overline{D}}(C - C) + 6 \times 413)$$

Hence $\quad \Delta H_{\overline{D}}(C - C) = 346 \cdot 7 \text{ kJ mol}^{-1}$.

4. $\quad \Delta C_p = C_p(NH_3) - \tfrac{1}{2}C_p(N_2) - \tfrac{3}{2}C_p(H_2)$
$\qquad\qquad = -20 \cdot 8 - 3 \cdot 6 \times 10^{-3}T + 21 \cdot 3 \times 10^{-6}T^2$

From (VII, 20)

$$\Delta H_{1000}^\ominus - \Delta H_{298}^\ominus = \int_{298}^{1000} \Delta C_p \, dT$$

$$\Delta H_{1000}^\ominus + 46\,190 = \int_{298}^{1000} (-20 \cdot 8 - 3 \cdot 6 \times 10^{-3}T + 21 \cdot 3 \times 10^{-6}T^2) dT$$

Integrating this equation

$$\Delta H_{1000}^\ominus + 46\,190 = [-20 \cdot 8T - 1 \cdot 8 \times 10^{-3}T^2 + 7 \cdot 1 \times 10^{-6}T^3]_{298}^{1000}$$
$$= -20 \cdot 8(1000 - 298) - 1 \cdot 8 \times 10^{-3}(1000^2 - 298^2) + 7 \cdot 1 \times 10^{-6}(1000^3 - 298^3)$$
$$= -14\,600 - 1640 + 6910 = -9380$$
$$\therefore \Delta H_{1000}^\ominus = -9330 - 46\,190 = -55\,520 \text{ J mol}^{-1}$$

The value is in reasonable agreement with that found in Example 4, where mean molar heat capacities were used to calculate $\Delta H_{1000}^\ominus$.

Progress Test 8

1. The sequence of changes may be shown diagrammatically:

$$H_2O(g) \ 373 \text{ K} \xrightarrow{\Delta S} H_2O(s) \ 273 \text{ K}$$

$$\downarrow \Delta S_1 \qquad\qquad\qquad \uparrow \Delta S_3$$

$$H_2O(l) \ 373 \text{ K} \xrightarrow{\Delta S_2} H_2O(l) \ 273 \text{ K}$$

$$\Delta S = \Delta S_1 + \Delta S_2 + \Delta S_3$$

$\Delta S_1 = -\Delta H_v/T = -2257 \cdot 5 \times 18/373$ (negative sign since condensation)

$\Delta S_2 = C_p \ln(T_2/T_1) = -C_p \ln(T_1/T_2)$
$$= -4 \cdot 18 \times 18 \times 2 \cdot 303 \times \lg(373/273)$$

$\Delta S_3 = -\Delta H_f/T = -333 \cdot 5 \times 18/273$ (negative sign since solidification)

$$\therefore \Delta S = 18(-6 \cdot 05 - 1 \cdot 31 - 1 \cdot 22)$$
$$= 18(-8 \cdot 58) = -154 \cdot 4 \text{ J K}^{-1}\text{g}^{-1}$$

2. From (VIII, 6)

$\Delta S = C_p \ln(T_2/T_1) + R \ln(p_1/p_2) = C_p \ln(T_2/T_1) - R \ln(p_2/p_1)$
$= (5 \times 8 \cdot 314/2)2 \cdot 303 \lg(1000/273) - 8 \cdot 314$
$$\times 2 \cdot 303 \lg(1013/101 \cdot 3)$$
$= 47 \cdot 87 \lg(3 \cdot 663) - 19 \cdot 15 \lg(10) = 7 \cdot 84 \text{ J K}^{-1}\text{g}^{-1}$

3. Since each N_2O molecule may be orientated in one of two ways there will be 2^L ways of building up a lattice consisting of L molecules, *i.e.* one mole of N_2O.

Hence $\qquad S = k \ln 2^L = Lk \ln 2 = R \ln 2$
$$= 5 \cdot 76 \text{ J K}^{-1} \text{mol}^{-1}$$

4. The combustion reactions are:

$C_2H_4(g) + 3O_2(g) = 2CO_2(g) + 2H_2O(l) \quad \Delta H^{\ominus}_{298} = -1410 \text{ kJ}$
$$\cdots (7)$$
$H_2(g) + \frac{1}{2}O_2(g) = H_2O(l) \qquad\qquad \Delta H^{\ominus}_{298} = -286 \text{ kJ}$
$$\cdots (8)$$
$C_2H_6(g) + \frac{7}{2}O_2(g) = 2CO_2(g) + 3H_2O(l) \quad \Delta H^{\ominus}_{298} = -1560 \text{ kJ}$
$$\cdots (9)$$

The combination $(7) + (8) - (9)$ gives

$C_2H_4(g) + H_2(g) = C_2H_6(g) \qquad\qquad \Delta H^{\ominus}_{298} = -136 \text{ kJ}$
$$\cdots (10)$$

The standard entropy change for (10) may be found using

$\Delta S^{\ominus}_{298} = S^{\ominus}(C_2H_6, g) - S^{\ominus}(C_2H_4, g) - S^{\ominus}(H_2, g)$
$$= 229 \cdot 5 - 219 \cdot 4 - 130 \cdot 6 = -120 \cdot 5 \text{ J K}^{-1} \text{mol}^{-1}$$

The standard free energy change is given by

$\Delta G^{\ominus}_{298} = \Delta H^{\ominus}_{298} - T \Delta S^{\ominus}_{298}$
$$= -136 - 298(-120 \cdot 5 \times 10^{-3}) = -100 \text{ kJ mol}^{-1}$$

Since $\Delta G^{\ominus}_{298}$ is negative, the reaction is thermodynamically feasible.

5. The dissociation reaction is

$$H_2O(g) \rightleftharpoons H_2(g) + \tfrac{1}{2}O_2(g)$$

Equilibrium
percentages: 100 2×10^{-3} 10^{-3}

From (VIII, 36)

$$K_p = \left\{ \frac{(p_{H_2}/p_{H_2}^{\ominus}) \times (p_{O_2}/p_{O_2}^{\ominus})^{\frac{1}{2}}}{(p_{H_2O}/p_{H_2O}^{\ominus})} \right\}_{\text{equil}}$$

$(p_{H_2}/p_{H_2}^{\ominus}) = 101 \cdot 3 \times 2 \times 10^{-5}/101 \cdot 3 = 2 \times 10^{-5}$
$(p_{O_2}/p_{O_2}^{\ominus}) = 101 \cdot 3 \times 2 \times 10^{-5}/2 \times 101 \cdot 3 = 10^{-5}$
$(p_{H_2O}/p_{H_2O}^{\ominus}) = 101 \cdot 3 \times 1 \cdot 000/101 \cdot 3 = 1 \cdot 000$

$$\therefore K_p = \frac{(2 \times 10^{-5}) \times (10^{-5})^{\frac{1}{2}}}{1 \cdot 000} = 6 \cdot 32 \times 10^{-8}$$

$\Delta G_{1500}^{\ominus} = -RT \ln K_p = RT \ln(1/K_p)$
     $= 2 \cdot 303 \times 8 \cdot 314 \times 10^{-3} \times 1500 \lg(1 \cdot 582 \times 10^{7})$
     $= 206 \cdot 7 \text{ kJ mol}^{-1}$

6. The synthetic reaction is

$$\tfrac{1}{2}N_2(g) + \tfrac{3}{2}H_2(g) \rightleftharpoons NH_3(g)$$

Equilibrium
percentages:

$93 \cdot 3/4$	$3 \times 93 \cdot 3/4$	$6 \cdot 7$	at 823 K
$98 \cdot 5/4$	$3 \times 98 \cdot 5/4$	$1 \cdot 5$	at 1023 K

$$K_p = \left\{ \frac{(p_{NH_3}/p_{NH_3}^{\ominus})}{(p_{N_2}/p_{N_2}^{\ominus})^{\frac{1}{2}} \times (p_{H_2}/p_{H_2}^{\ominus})^{\frac{3}{2}}} \right\}_{\text{equil}}$$

At 833 K $(p_{NH_3}/p_{NH_3}^{\ominus}) = 10130 \times 6 \cdot 7 \times 10^{-2}/101 \cdot 3 = 6 \cdot 7$
$(p_{N_2}/p_{N_2}^{\ominus}) = 10130 \times 93 \cdot 3 \times 10^{-2}/4 \times 101 \cdot 3$
         $= 23 \cdot 3$
$(p_{H_2}/p_{H_2}^{\ominus}) = 10130 \times 3 \times 93 \cdot 3 \times 10^{-2}/4 \times 101 \cdot 3$
         $= 70 \cdot 0$

$$\therefore K_p(825 \text{ K}) = \frac{6 \cdot 7}{(23 \cdot 3)^{\frac{1}{2}} \times (70 \cdot 0)^{\frac{3}{2}}} = 2 \cdot 32 \times 10^{-3}$$

At 1023 K $(p_{NH_3}/p_{NH_3}^{\ominus}) = 1 \cdot 5$
$(p_{N_2}/p_{H_2}^{\ominus}) = 24 \cdot 6$
$(p_{H_2}/p_{N_2}^{\ominus}) = 73 \cdot 9$

$$\therefore K_p(1023 \text{ K}) = \frac{1 \cdot 5}{(24 \cdot 6)^{\frac{1}{2}} \times (73 \cdot 9)^{\frac{3}{2}}} = 4 \cdot 76 \times 10^{-4}$$

From (VIII, 40)

$$\lg \frac{2 \cdot 32 \times 10^{-3}}{4 \cdot 76 \times 10^{-4}} = -\frac{\Delta H^{\ominus}}{2 \cdot 303 \times 8 \cdot 314} \left(\frac{1}{823} - \frac{1}{1023} \right)$$

$\lg 4 \cdot 87 = 0 \cdot 688 = -\Delta H^{\ominus} \times 0 \cdot 2317 \times 10^{-3}/19 \cdot 15$
From which $\Delta H^{\ominus} = -55 \cdot 6 \text{ kJ mol}^{-1}$

7. Integration of the Clausius–Clapeyron equation (VIII, 46) gives

$$\ln p = -\frac{\Delta H_t}{RT} + \text{constant}$$

or

$$\lg p = -\frac{\Delta H_t}{2 \cdot 303\,RT} + \text{constant}$$

Thus a plot of $\lg p$ against $1/T$ should yield a straight line with a slope equal to $-\Delta H_t/2 \cdot 303R$. The units of pressure given are mmHg. These may be converted to kN m^{-2} using the fact that 760 mmHg are equivalent to 101·3 kN m^{-2}. However, this is unnecessary, since $\lg p$ is to be plotted and the slope obtained will be independent of the units of p, i.e. multiplying the pressures by 101·3/760 will not affect the slope. The data are tabulated below.

p/mmHg	$\lg(p/\text{mmHg})$	T/K	$10^3\text{K}/T$
4·6	0·663	273	3·66
17·5	1·243	293	3·41
55·3	1·743	313	3·20
149·4	2·174	333	3·00
335·1	2·525	353	2·83
760·0	2·881	373	2·68

The plot of $\lg p$ against $1/T$ is shown in Fig. 127, and has a slope of $-2 \cdot 22/0 \cdot 98 \times 10^{-2} = -2 \cdot 26 \times 10^3$

$$\therefore \quad -\Delta H_t/2 \cdot 303 \times 8 \cdot 314 = -2 \cdot 26 \times 10^3$$

from which

$$\Delta H_t = 43 \cdot 3 \text{ kJ mol}^{-1}.$$

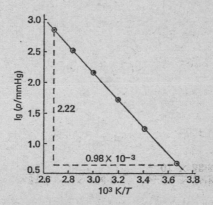

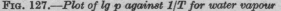

FIG. 127.—Plot of $\lg p$ against $1/T$ for water vapour

Progress Test 9

1. $$N_2(g) + 3H_2(g) \rightleftharpoons 2NH_3(g)$$

(a) Three distinct chemical species are present but the equation imposes a restriction on the system so that only two concentrations can be independently varied, the third concentration being determined by the other two. Thus there are two components.

(b) Again, there are three chemical species and one restriction, as in (a). In this case, however, there is also a further restriction, since all the hydrogen and nitrogen must have been formed from the ammonia originally introduced, i.e. $[N_2] = 3[H_2]$. Thus there is only one component.

2. This system is completely analogous to the sulphur system considered in IX, 22.

3. T_1 represents the temperature at which the first solid phase (either pure bismuth or pure cadmium) crystallises from the melt, and T_2 represents the temperature at which the second solid phase begins to crystallise. The first and last data points represent the freezing points of pure bismuth and pure cadmium respectively, and the eutectic composition is 60% bismuth, 40% cadmium. The phase diagram is shown in Fig. 128.

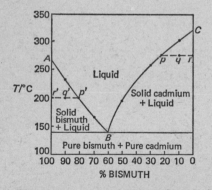

FIG. 128.—*Phase diagram for the bismuth–cadmium system*

4(a) From the tie line rqp in Fig. 128, the liquid contains 23% bismuth and the solid is pure cadmium. Using the lever rule

$$\frac{\text{Amount of liquid}}{\text{Amount of solid}} = \frac{qr}{pq} = \frac{10}{13}$$

(b) From the tie line $r'p'q'$ in Fig. 128, the liquid contains 81% bismuth and the solid is pure bismuth. The alternative possibility that the solid is pure cadmium and the liquid contains 48·5% bismuth, obtained by producing $r'p'q'$ to meet BC, can obviously be discounted. Using the lever rule

$$\frac{\text{Amount of liquid}}{\text{Amount of solid}} = \frac{q'r'}{p'q'} = \frac{10}{9}$$

(c) This is the eutectic composition and here the liquid and solid have identical compositions, i.e. 60% bismuth, 40% cadmium. The relative amounts of liquid and solid are infinitely variable.

Progress Test 10

1. At equilibrium the two solutions have the same vapour pressure and therefore contain equal mole fractions of the two solutes.

Amount of NaCl in 1% solution $= 0·001/0·0585 = 0·0171$ mol

Amount of H_2O in 1% solution $= 0·099/0·018 = 5·50$ mol

Since sodium chloride is completely dissociated, the actual amount of Na^+ and Cl^- ions present will be 0·0342.

$\therefore$ Effective mole fraction $= 0·0342/(0·0342 + 5·50) = 0·00619$

This must be the mole fraction of a 10·5% sucrose solution, i.e. if the amount of sucrose is n_2, then

$$0·00619 = \frac{n_2}{n_2 + (0·0895/0·018)}$$

$$\therefore n_2 = 0·03097 \text{ mol}$$

But the molar mass $M = m/n_2 = 0·0105/0·03097 = 0·339$ kg mol^{-1}. Thus the relative molecular mass of sucrose in water is 339 (the formula mass is 342·0).

2. Using (X, 8)

$$m = \Delta T/K_F = 0·8/5·12 = 0·1563 \text{ mol kg}^{-1}$$

Since 2 g were dissolved in 100 g of benzene, the mass per kg of benzene is 20 g. Using $m = m/M$,

$$M = m/m = 0·020/0·1563 = 0·128 \text{ kg mol}^{-1}$$

Thus the relative molecular mass is 128.

The compound contains $0·9376/0·012 = 78·0$ mol carbon to $0·0624/0·001 = 62·4$ mol hydrogen, i.e. for each carbon there are $62·4/78·0 = 0·8$ hydrogens.

The simplest formula is $C_1H_{0·8}$. Since the molar mass is 0·128 kg mol^{-1}, the empirical formula must be $C_{10}H_8$.

The compound is in fact naphthalene.

3. Using (X, 5)

$$x_2 = \frac{\Delta H_v}{R} \times \frac{\Delta T}{T_0^2} = \frac{29 \cdot 5 \times 10^3 \times 5 \cdot 0}{8 \cdot 134 \times (437)^2} = 0 \cdot 093$$

But $\quad x_2 = n_2/(n_1 + n_2)$

$$\therefore 0 \cdot 093 = \frac{n_2}{0 \cdot 1/0 \cdot 12) + n_2}$$

$$n_2 = 0 \cdot 085 \text{ mol}$$

Also $M = m/n_2 = 0 \cdot 002/0 \cdot 085 = 0 \cdot 023 \cdot 5 \text{ kg mol}^{-1}$
i.e. relative molecular mass of unknown compound $= 23 \cdot 5$.
To find ΔH_f for mesitylene we use (X, 7) in the form

$$\Delta H_f = x_2 R T_0^2/\Delta T \qquad \ldots \text{ (X, 7)}$$
$$= 0 \cdot 093 \times 8 \cdot 314 \times (221)^2/3 \cdot 6$$
$$= 10 \cdot 45 \text{ kJ mol}^{-1}$$

4. Using the van't Hoff equation (X, 11)

$$\pi V = n_2 R T$$

with $R = 8 \cdot 314 \text{ J K}^{-1} \text{ mol}^{-1}$, $V = 10^{-4} \text{ m}^3$, $T = 300 \text{ K}$,
$n_2 = 0 \cdot 001/M$

$$\therefore 4 \cdot 29 \times 10^3 \times 10^{-4} = 10^{-3} \times 8 \cdot 314 \times 300/M$$

Hence $M \approx 6 \cdot 0 \text{ kg mol}^{-1}$ and the relative molecular mass is
6000. The relative molecular mass is quoted to only two significant
figures, as certain corrections to the van't Hoff equation are
necessary for materials of high relative molecular mass.

5. Using the subscripts $_1$ and $_2$ to refer to benzene and toluene
respectively, and assuming Raoult's law (X, 1) is obeyed

$$p_1 = p_1^{\bullet} x_1 = 10 \cdot 00 \times 0 \cdot 5 = 5 \cdot 00 \text{ kN m}^{-2}$$
$$p_2 = p_2^{\bullet} x_2 = 2 \cdot 92 \times 0 \cdot 5 = 1 \cdot 46 \text{ kN m}^{-2}$$
$$\therefore \text{ Total pressure } p = p_1 + p_2 = 6 \cdot 46 \text{ kN m}^{-2}$$

The composition of the vapour is given by Dalton's law:

$$x_1 = p_1/p = 5 \cdot 00/6 \cdot 46 = 0 \cdot 772$$
$$x_2 = p_2/p = 1 \cdot 46/6 \cdot 46 = 0 \cdot 228$$

i.e. the vapour contains $0 \cdot 772$ mol fraction benzene and $0 \cdot 228$
mol fraction toluene.

The composition of the vapour obtained by condensation and
re-evaporation of this mixture is obtained by the same method as:

$$p_1 = p_1^{\bullet} x_1 = 10 \cdot 00 \times 0 \cdot 772 = 7 \cdot 72 \text{ kN m}^{-2}$$
$$p_2 = p_2^{\bullet} x_2 = 2 \cdot 92 \times 0 \cdot 228 = 0 \cdot 67 \text{ kN m}^{-2}$$
$$\therefore p = p_1 + p_2 = 8 \cdot 39 \text{ kN m}^{-2}$$

Composition of vapour:

$$x_1 = p_1/p = 7 \cdot 72/8 \cdot 39 = 0 \cdot 92$$
$$x_2 = p_2/p = 0 \cdot 67/8 \cdot 39 = 0 \cdot 08$$

i.e. the vapour contains $0 \cdot 92$ mol fraction benzene and $0 \cdot 08$ mol fraction toluene.

Progress Test 11

1. $K_b = \dfrac{(p_{CO}/p_{CO}^{\ominus})(p_{O_2}/p_{O_2}^{\ominus})^{\frac{1}{2}}}{(p_{CO_2}/p_{CO_2}^{\ominus})}$ $K_c = \dfrac{(p_{H_2O}/p_{H_2O}^{\ominus})}{(p_{H_2}/p_{H_2}^{\ominus})(p_{O_2}/p_{O_2}^{\ominus})^{\frac{1}{2}}}$

$$\therefore K_a = K_b K_c = \frac{(p_{CO}/p_{CO}^{\ominus})(p_{H_2O}/p_{H_2O}^{\ominus})}{(p_{CO_2}/p_{CO_2}^{\ominus})(p_{H_2}/p_{H_2}^{\ominus})}$$

$$= 1 \cdot 14 \times 10^{-5} \times 2 \cdot 5 \times 10^5 = 2 \cdot 85$$
$$\Delta G^{\ominus} = -RT \ln K = -8 \cdot 314 \times 1473 \times 2 \cdot 303 \lg 2 \cdot 85$$
$$= -12 \cdot 82 \text{ kJ mol}^{-1}$$

2. $$H_2(g) + I_2(g) \rightleftharpoons 2HI(g)$$

$$K_p = \frac{(p_{HI}/p_{HI}^{\ominus})^2}{(p_{H_2}/p_{H_2}^{\ominus})(p_{I_2}/p_{I_2}^{\ominus})} = 55 \cdot 3 \qquad \ldots \text{ (11)}$$

Here $\dfrac{(p_{HI} \, p_{HI}^{\ominus})^2}{(p_{H_2}/p_{H_2}^{\ominus})(p_{I_2}/p_{I_2}^{\ominus})} = \dfrac{(114 \cdot 6/101 \cdot 3)^2}{(12 \cdot 2/101 \cdot 3)(20 \cdot 3/101 \cdot 3)} = 53 \cdot 0$

Therefore the amount of HI must be increased to establish the equality (11). If an additional pressure, $2x$ kN m^{-2}, of HI is formed:

$$p_{H_2} = 12 \cdot 2 - x, \quad p_{I_2} = 20 \cdot 3 - x, \quad p_{HI} = 114 \cdot 6 + 2x$$
$$\therefore 55 \cdot 3 = \frac{(114 \cdot 6 + 2x/101 \cdot 3)^2}{(12 \cdot 2 - x/101 \cdot 3)(20 \cdot 3 - x/101 \cdot 3)}$$

From which $x = 0 \cdot 25$ kN m^{-2}

i.e. $0 \cdot 50$ kN m^{-2} of HI are formed.

3. $$CO(g) + Cl_2(g) \rightleftharpoons COCl_2(g)$$

If x kN m^{-2} of $COCl_2$ are formed:

$$p_{Cl_2} = 24 \cdot 32 - x, \quad p_{CO} = 36 \cdot 48 - x, \quad p_{COCl_2} = x$$

But total pressure $p = p_{Cl_2} + p_{CO} + p_{COCl_2}$
$$\therefore 41 \cdot 54 = 24 \cdot 32 - x + 36 \cdot 48 - x + x$$
$$x = 19 \cdot 26 \text{ kN m}^{-2}$$

$$\therefore p_{Cl_2} = 5 \cdot 06 \text{ kN m}^{-2}, \quad p_{CO} = 17 \cdot 22 \text{ kN m}^{-2},$$
$$p_{COCl_2} = 19 \cdot 26 \text{ kN m}^{-2}$$

$$K_p = \frac{(p_{COCl_2}/p_{COCl_2}^{\ominus})}{(p_{CO}/p_{CO}^{\ominus})(p_{Cl_2}/p_{Cl_2}^{\ominus})} = \frac{(19 \cdot 26/101 \cdot 3)}{(17 \cdot 22/101 \cdot 3)(5 \cdot 06/101 \cdot 3)} = 22 \cdot 4$$

4. This question is similar to Example 3. If the initial amount of $COCl_2$ is a, the degree of dissociation α, and the total equilibrium pressure p:

$$COCl_2(g) \rightleftharpoons CO(g) + Cl_2(g)$$

Initially: a mol   0   0   Total a mol

At equilibrium: $a(1 - \alpha)$ mol $a\alpha$ mol $a\alpha$ mol   Total $a(1 + \alpha)$ mol

Using Dalton's law:

$$p_{COCl_2} = p(1 - \alpha)/(1 + \alpha), \quad p_{CO} = p_{Cl_2} = p\alpha/(1 + \alpha)$$

$$K_p' = \frac{(p_{CO}/p_{CO}^\ominus)(p_{Cl_2}/p_{Cl_2}^\ominus)}{(p_{COCl_2}/p_{COCl_2}^\ominus)}$$

and K_p' for this equilibrium is simply the reciprocal of K_p found in the preceding question.

$$\therefore \frac{1}{22 \cdot 4} = \frac{[\alpha/(1 + \alpha)]^2[p/p^\ominus]^2}{[(1 - \alpha)/(1 + \alpha)][p/p^\ominus]} = \frac{\alpha^2}{(1 - \alpha^2)} \times \frac{p}{p^\ominus}$$

where $p_{CO}^\ominus = p_{Cl_2}^\ominus = p_{COCl_2}^\ominus = p^\ominus = 101 \cdot 3$ kN m^{-2}

(a) Substituting $p = 101 \cdot 3$ kN m^{-2}: $\alpha = 0 \cdot 207$.
(b) Substituting $p = 101 \cdot 3 - 76 \cdot 0 = 25 \cdot 3$ kN m^{-2}: $\alpha = 0 \cdot 39$.
(c) As would be qualitatively predicted by Le Chatelier's principle, decreasing the pressure (by dilution) has increased the degree of dissociation of $COCl_2$ from $20 \cdot 7\%$ to 39%.

5. If the initial amount of $SbCl_5$ is a, the degree of dissociation is α, the volume V, and the temperature T:

$$SbCl_5(g) \rightleftharpoons SbCl_3(g) + Cl_2(g)$$

Initially: a mol   0   0   Total a mol
At equilibrium: $a(1 - \alpha)$ mol $a\alpha$ mol $a\alpha$ mol   Total $a(1 + \alpha)$ mol

The initial mass of $SbCl_5$ is given by $m = aM$ where M is the molar mass of $SbCl_5$
Since the total mass cannot change during a chemical reaction m is also the mass of the mixture at equilibrium

i.e. $m = a(1 + \alpha)M'$

where M' is the apparent molar mass of the mixture

$$\therefore aM = a(1 + \alpha)M'$$
$$\therefore \alpha = (M/M') - 1$$
$$= (0 \cdot 299/0 \cdot 276) - 1 = 0 \cdot 083$$

6. From (8.40)

$$\lg(K_p)_1 - \lg(K_p)_2 = \frac{-\Delta H^{\ominus}}{2 \cdot 303\ R}\left(\frac{1}{T_1} - \frac{1}{T_2}\right)$$

$$\lg 0\cdot0266 - \lg(K_p)_{773} = \frac{51\cdot44 \times 10^3}{2\cdot303 \times 8\cdot314}\left(\frac{1}{623} - \frac{1}{773}\right)$$

Solving $\qquad K_p(773\ \text{K}) = 3\cdot64 \times 10^{-3}$

Note that K_p decreases markedly as the temperature is raised. It would therefore appear to be advantageous commercially to carry out the reaction at as low a temperature as possible, thus increasing the yield of ammonia. In practice, however, the reaction cannot be made to reach equilibrium in a reasonable time around room temperature, and a compromise has to be reached between the needs for high yield and the rapid attainment of equilibrium.

7. $\qquad$ $CH_3COOH + C_2H_5OH \rightleftharpoons CH_3COOC_2H_5 + H_2O$

Initially:	1 mol	0·33 mol	0	0
At equilibrium:	(1 − 0·293) mol	(0·33 − 0·293) mol	0·293 mol	0·293 mol

$$\therefore K_x = \frac{[CH_3COOC_2H_5][H_2O]}{[CH_3COOH][C_2H_5OH]} = \frac{(0\cdot293)^2}{0\cdot707 \times 0\cdot037} = 3\cdot3$$

Notice that it is not necessary to divide by the total amount of substance present (*i.e.* the total number of moles) to obtain K_x, because this factor cancels out.

If $[CH_3COOC_2H_5] = 0\cdot296$
$$K_x = (0\cdot296)^2/(0\cdot703)(0\cdot033) = 4\cdot1$$

i.e. a 1% error in measuring the ester concentration has led to a 25% error in the equilibrium constant. This emphasises the need for great care and accuracy in such measurements.

8. $\qquad$ $NH_4HS(s) \rightleftharpoons NH_3(g) + H_2S(g)$

$$K_p = (p_{NH_3}/p^{\ominus}_{NH_3})(p_{H_2S}/p^{\ominus}_{H_2S})$$

p_{NH_3} from dissociation $= p_{H_2S}$ from dissociation.

$\therefore$ Total pressure of NH_3, $p_{NH_3} = (101\cdot3 + p_{H_2S})$

$$\therefore K_p = [(101\cdot3 + p_{H_2S})/p^{\ominus}_{NH_3}][(p_{H_2S}/p^{\ominus}_{H_2S})]$$
$$p^{\ominus}_{NH_3} = p^{\ominus}_{H_2S} = p^{\ominus} = 101\cdot3\ \text{kN m}^{-2}$$
$$\therefore 0\cdot11 \times 101\cdot3^2 = 101\cdot3\ p_{H_2S} + p^2_{H_2S}$$

From which $\qquad p_{H_2S} = 10\cdot1\ \text{kN m}^{-2}$
Hence $\qquad p_{NH_3} = 111\cdot4\ \text{kN m}^{-2}$

9. 100 cm³ of water contain 0·5 g of $HgBr_2$ and

$$K_D = [HgBr_2]_{H_2O}/[HgBr_2]_{C_6H_6}$$

Thus if x g is extracted by the benzene:

(a) $$0·9 = (0·5 - x)/x$$
                               $$\therefore x = 0·263 \text{ g}$$

Notice that this problem can be worked directly in mass rather than in amount of substance. The molar mass is the same in both solvents and therefore cancels out.

(b) For the first 50 cm³ portion the calculation is the same, except that half the quantity of benzene is used.

$$0·9 = (0·5 - x)/2x$$
$$\therefore x = 0·179 \text{ g}$$

For the second 50 cm³ portion, the water now contains only $(0·5 - 0·179) = 0·321$ g of $HgBr_2$

$$\therefore 0·9 = (0·321 - x)/2x$$
$$x = 0·115$$

The total quantity extracted is therefore $0·179 - 0·115 = 0·294$ g. Thus an improvement of approximately 11% is obtained by using the same quantity of extracting solvent in two portions. This has important practical implications.

Progress Test 12

1. The electrode reaction is

$$\tfrac{1}{2}H_2O = \tfrac{1}{2}O + H^+ + e^-$$
or $$2H_2O = O_2 + 4H^+ + 4e^-$$

The amount of oxygen produced is given by (XII, 5)

$$n = 1 \times 2 \times 60 \times 60/4 \times 9·649 \times 10^4 = 1·86 \times 10^{-2} \text{ mol}$$

If it is assumed that oxygen behaves as an ideal gas, then at 25°C (298 K) and 101·3 kN m^{-2}, $1·86 \times 10^{-2}$ mol will occupy a volume of $1·86 \times 10^{-2} \times 298 \times 22·4/273 = 0·456$ dm³

2. Let HB and NaB represent benzoic acid and sodium benzoate respectively. Then

$$\Lambda^\infty(HB) = \Lambda^\infty(NaB) + \Lambda^\infty(HCl) - \Lambda^\infty(NaCl)$$
$$= 82·5 + 426·2 - 126·5 = 382·2 \ \Omega^{-1} \text{ cm}^2 \text{ mol}^{-1}.$$

From (XII, 10),

$$\Lambda = \kappa/c = 1.14 \times 10^{-4}/1.5 \times 10^{-6} = 76.0\ \Omega^{-1}\ \text{cm}^2\ \text{mol}^{-1}$$
$$\therefore\ \alpha = \Lambda/\Lambda^\infty = 76.0/382.2 = 0.199.$$

K_c, the dissociation constant, is given by (XII, 14)

$$K_c = \alpha^2 c/(1 - \alpha) = (0.199)^2 \times 1.5 \times 10^{-3}/(1 - 0.199)$$
$$= 7.41 \times 10^{-5}$$

3. It is first required to find the change in the amount of silver in the anode compartment. This amount must be associated with a fixed amount of water.

After electrolysis 23.312 g of the anode solution contain 0.0371 g of silver $= 0.0371 \times 169.9/107.9 = 0.0584$ g of silver nitrate.

$\therefore$ 0.0584 g of silver nitrate is associated with $23.312 - 0.0584 = 23.254$ g of water.

Initially 0.02 mol $= 3.329$ g of silver nitrate were associated with 1000 g of water.

$\therefore$ 23.254 g of water contained $3.398 \times 23.254/1000 = 0.790$ g of silver nitrate.

$\therefore$ Gain in electrolyte in the anode compartment $= 0.0790 - 0.0584 = 0.0206$ g $= 1.21 \times 10^{-4}$ mol of silver nitrate.

The passage of 9.649×10^4 C results in the anode compartment gaining 1 mol of Ag^+ by electrode reaction and losing t_+ mol of Ag^+ by migration, i.e. there is a net gain of $1 - t_+ = t_-$ mol of Ag^+. Hence the passage of $0.02 \times 20 \times 60 = 24$ C results in a net gain of $24t_-/9.649 \times 10^4$ mol of Ag^+.

But the observed gain is 1.21×10^{-4} mol of Ag^+

$$\therefore\ 24t_-/9.649 \times 10^4 = 1.21 \times 10^{-4}$$
Hence $$t_- = 0.486 \quad \text{and} \quad t_+ = 0.514$$

4. From XII, 27

$$t_+ = Vc \times 9.649 \times 10^4/Q$$

Now $V = \pi \times (0.5)^2 \times 14.7 = 11.5\ \text{cm}^2 = 11.5 \times 10^{-3}\ \text{dm}^3$
$\therefore\ t_+ = 11.5 \times 10^{-3} \times 0.01 \times 9.649 \times 10^4/5 \times 10^{-3} \times 60$
$\qquad \times 60$
$\qquad = 0.490$
$\quad t_- = 0.510$

5. $\Lambda^\infty(BaCl_2) = 2\lambda^\infty(Cl^-) + \lambda^\infty(Ba^{2+})$
$\therefore\ \lambda^\infty(Ba^{2+}) = 280.0 - 2 \times 76.4 = 127.2\ \Omega^{-1}\ \text{cm}^2\ \text{mol}^{-1}$

From (XII, 23)

$u^\infty(Ba^{2+}) = 127.2/2 \times 9.649 \times 10^4 = 6.59 \times 10^{-4}\ \text{cm}^2\ \text{s}^{-1}\ \text{V}^{-1}$
$u^\infty(Cl^-) = 76.4/9.649 \times 10^{-4} = 7.91 \times 10^{-4}\ \text{cm}^2\ \text{s}^{-1}\ \text{V}^{-1}$

From (XII, 24) and (XII, 25)

$$t\infty(\text{Ba}^{2+}) = 6·59 \times 10^{-4}/14·5 \times 10^{-4} = 0·454$$
$$t\infty(\text{Cl}^-) = 7·91 \times 10^{-4}/14·5 \times 10^{-4} = 0·546$$

6. $\kappa(\text{AgCl}) = \kappa(\text{AgCl saturated solution}) - \kappa(\text{water})$
$$= (3·24 - 1·44) \times 10^{-6} = 1·80 \times 10^{-6}\,\Omega^{-1}\,\text{cm}^{-1}$$

$$\Lambda\infty(\text{AgCl}) = \lambda\infty(\text{Ag}^+) + \lambda\infty(\text{Cl}^-) = 61·9 + 76·3$$
$$= 138·2\,\Omega^{-1}\,\text{cm}^2\,\text{mol}^{-1}$$

Since the solution is so dilute, $\Lambda = \Lambda\infty = \kappa/c$

$$c = \kappa/\Lambda = 1·80 \times 10^{-6}/138·2 = 1·30 \times 10^{-8}\,\text{mol cm}^{-3}$$
$$= 1·30 \times 10^{-5}\,\text{mol dm}^{-3}$$

7(a) The conductance curve is shown in Fig. 129 (a). The initial conductance is low, owing to the small amount of ionisation

$$\text{HAc} + \text{H}_2\text{O} \rightleftharpoons \text{H}_3\text{O}^+ + \text{Ac}^-$$

On addition of NaOH the conductance initially falls as H_3O^+ ions are replaced by Na^+ ions. As the titration proceeds the conductance increases as Na^+ ions are added and Ac^- ions are produced from the reaction

$$\text{HAc} + \text{OH}^- \rightleftharpoons \text{Ac}^- + \text{H}_2\text{O}$$

After the end-point Na^+ and OH^- ions are being added and, since OH^- is more highly conducting than Ac^-, the conductance increases rapidly.

(b) Initially the hydrochloric acid alone is neutralised and the conductance falls (see XII, 32). Second, the acetic acid is neutralised and, as in (a) above, the conductance increases. After the end-point the conductances increases rapidly, for the reasons given in (a) above (see Fig. 129 (b)).

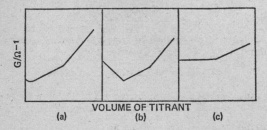

Fig. 129.—*Conductimetric titrations.* (a) *Acetic acid with sodium hydroxide.* (b) *Hydrochloric acid + acetic acid with sodium hydroxide.* (c) *Sodium chloride with silver nitrate*

(c) The reaction is

$$NaCl + AgNO_3 = AgCl(s) + NaNO_3$$

Initially the solution contains only Na^+ and Cl^- ions and the conductance is high. On addition of Ag^+ and NO_3^- ions, Cl^- is effectively replaced by NO_3^- (since AgCl is precipitated). These two ions have similar conductivities and hence the conductance of the solution is virtually unchanged. After the end-point, Ag^+ and NO_3^- ions are being added, and the conductance increases (see Fig. 129 (c)).

Progress Test 13

1. The equilibrium occurring is

$$CH_2ClCO_2H + H_2O \rightleftharpoons H_3O^+ + CH_2ClCO_2^-$$

If α is the degree of ionisation, and the concentration of H_3O^+ ions formed by the self-ionisation of water can be neglected, we may use (XIII, 16)

$$\alpha = \frac{-K_a + (K_a{}^2 + 4cK_a)^{\frac{1}{2}}}{2c}$$

$$= \frac{-1 \cdot 5 \times 10^{-3} + \{(1 \cdot 5 \times 10^{-3})^2 + 4 \times 0 \cdot 1 \times 1 \cdot 5 \times 10^{-3}\}^{\frac{1}{2}}}{2 \times 0 \cdot 1}$$

$$= 0 \cdot 115$$

Also
$$[H_3O^+] = \alpha c = 1 \cdot 15 \times 10^{-2} \, M$$
$$\therefore pH = 2 - \lg 1 \cdot 15 = 1 \cdot 94$$

2. The equilibrium occurring is

$$NH_4^+ + OH^- \rightleftharpoons NH_3 + H_2O$$
for which
$$K = 1/K_b = 1/1 \cdot 8 \times 10^{-5}$$

Since K is large the equilibrium is well over to the right, and we may assume to a good approximation that

$$[NH_3] = 0 \cdot 1 - [NH_4^+] \approx 0 \cdot 1 \, M$$

Also, we may neglect the concentration of OH^- ions formed by the self-ionisation of water, since this is bound to be less than $10^{-7} \, M$.

$$\therefore [OH^-] = 0 \cdot 2 - 0 \cdot 1 = 0 \cdot 1 \, M$$
$$K_b = [NH_4^+][OH^-]/[NH_3]$$
$$\therefore [NH_4^+] = K_b[NH_3]/[OH^-] = (1 \cdot 8 \times 10^{-5} \times 0 \cdot 1)/0 \cdot 1$$
$$= 1 \cdot 8 \times 10^{-5} \, M$$

3. From (XIII, 23),

$$K_h = K_w \, K_b = 10^{-14}/1 \cdot 8 \times 10^{-5} = 5 \cdot 6 \times 10^{-10}$$

If the degree of hydrolysis is α and the concentration of H_3O^+ ions formed by self-ionisation of water is neglected, the equilibrium occurring is

$$NH_4^+ \;\; + H_2O \rightleftharpoons H_3O^+ + NH_3$$

Equilibrium
concentrations: $0 \cdot 1(1 - \alpha)$M $\qquad\qquad 0 \cdot 1\alpha$ M $\quad 0 \cdot 1\alpha$ M

$$\therefore K_h = 0 \cdot 1\alpha^2/(1 - \alpha)$$

and using the value of K_h calculated above gives $\alpha = 7 \cdot 5 \times 10^{-5}$

$$[H_3O^+] = 0 \cdot 1\alpha = 7 \cdot 5 \times 10^{-6} \text{ M}$$
$$\therefore \text{pH} = 6 - \lg 7 \cdot 5 = 5 \cdot 1$$

The pH could also have been calculated without knowing α from (XIII, 26)

$$\text{pH} = \tfrac{1}{2}\text{p}K_w - \tfrac{1}{2}\text{p}K_b - \tfrac{1}{2}\lg c$$

4. Notice that for the second ionisation

$$H_2PO_4^- + H_2O \rightleftharpoons H_3O^+ + HPO_4^{2-}$$

the pK_2 is approximately equal to the pH of the buffer. Thus (XIII, 25) gives

$$7 \cdot 5 = 7 \cdot 2 + \lg[HPO_4^{2-}]/[H_2PO_4^-]$$

and solving this equation $[HPO_4^{2-}] = 2[H_2PO_4^-]$
Since 100 cm³ of $0 \cdot 1$ M buffer are required

$$[H_2PO_4^-] + [HPO_4^{2-}] = 0 \cdot 01 \text{ M}$$

Therefore if x is the concentration of $H_2PO_4^-$ required in 100 cm³

$$x + 2x = 0 \cdot 01$$
$$\therefore x = 0 \cdot 0033 \text{ M}$$

$\therefore$ mass of $Na_2HPO_4 \cdot 2H_2O$ required $= 2 \times 0 \cdot 0033 \times 178 \times 10^{-3}$
$$= 1 \cdot 18 \times 10^{-3} \text{ kg}$$
and mass of $NaH_2PO_4 \cdot H_2O$ required $= 0 \cdot 0033 \times 138 \times 10^{-3}$
$$= 4 \cdot 6 \times 10^{-4} \text{ kg}$$

5. The equilibria occurring are:

$H_2S + H_2O \rightleftharpoons H_3O^+ + HS^- \qquad K_1 = 1 \cdot 0 \times 10^{-7} \text{ mol dm}^{-3}$
$HS^- + H_2O \rightleftharpoons H_3O^+ + S^{2-} \qquad K_2 = 1 \cdot 3 \times 10^{-13} \text{ mol dm}^{-3}$

If the concentration of H_3O^+ ions by self-ionisation of water is neglected

$$[H_3O^+] = 0 \cdot 1 \text{ M}$$

Also, since K_1 is small, very nearly all the H_2S is unionised, and we may make the further approximation that

$$[H_2S] = 0.1 - [HS^-] \approx 0.1\ \text{M}$$

$$K_1 = [H_3O^+][HS^-]/[H_2S] \qquad K_2 = [H_3O^+][S^{2-}]/[HS^-]$$

$$\therefore K_1K_2 = [H_3O^+]^2[S^{2-}]/[H_2S]$$

$$[S^{2-}] = K_1K_2[H_2S]/[H_3O^+] = (1.0 \times 10^{-7} \times 1.3 \times 10^{-13}$$
$$\times\ 0.1)/0.1^2$$
$$= 1.3 \times 10^{-19}\ \text{M}$$

Calculations of this type are important in selective precipitation, as illustrated in question 6 below.

6. The solubility product is defined here by the equation.

$$[\text{cation}][S^{2-}] = K_s$$

and if the product of the concentration of the ions exceeds K_s precipitation will occur. The concentration of sulphide ions obtained from question 5 is 1.3×10^{-19} M and the cation concentration is fixed at 10^{-2} M. Thus if the ratio $K_s/[S^{2-}]$ (= [cation] remaining in solution) for a particular cation is less than 10^{-2}, precipitation must have occurred, *i.e.*

(a) For Mn^{2+} : $2.5 \times 10^{-10}/1.3 \times 10^{-19} > 10^{-2}$ no precipitation.

(b) For Fe^{2+} : $1.0 \times 10^{-19}/1.3 \times 10^{-19} > 10^{-2}$ no precipitation.

(c) For Pb^{2+} : $1 \times 10^{-29}/1.3 \times 10^{-19} < 10^{-2}$ precipitation.

(d) For Zn^{2+} : $4.5 \times 10^{-24}/1.3 \times 10^{-19} < 10^{-2}$ precipitation.

Progress Test 14

1. The cell diagram is

$$\text{Pt,}\ \overset{-}{H_2}\ \big|\ H^+\ \big\|\ Cl^-\ \big|\ \overset{+}{Hg_2Cl_2}(s),\ Hg$$

Right hand electrode: $Hg_2Cl_2(s) + 2e^- = 2Hg(l) + 2Cl^-$
Left hand electrode: $\qquad\qquad H_2(g) = 2H^+ + 2e^-$

It should be apparent that the cell e.m.f. is simply the potential of the right hand calomel electrode.

(a) $\Delta G^{\ominus} = -nFE^{\ominus}$
$\qquad = -2 \times 9.649 \times 10^4 \times 0.2676 = -51.64\ \text{kJ mol}^{-1}$

(b) $\Delta S^{\ominus} = nF(\partial E^{\ominus}/\partial T)_p$
$\qquad = -2 \times 9.649 \times 10^4 \times 3.2 \times 10^{-4}$
$\qquad = -61.7\ \text{J K}^{-1}\ \text{mol}^{-1}$

(c) $\Delta H^{\ominus} = \Delta G^{\ominus} + T\Delta S^{\ominus}$
$\qquad = -51.64 - (298 \times 61.7 \times 10^{-3}) = -70.02\ \text{kJ mol}^{-1}$

2. The cell reaction is

$$Pb(s) + Cu^{2+} = Pb^{2+} + Cu(s)$$
$$\therefore E^{\ominus} = V_R - V_L = 0.337 + 0.126 = 0.463 \text{ V}$$

Now
$$E = E^{\ominus} + \frac{2.303\,RT}{2F}\,\lg\frac{[Cu^{2+}]}{[Pb^{2+}]}$$

$$E = 0.463 + \frac{0.059}{2}\,\lg\frac{10^{-2}}{10^{-3}}$$

$$= 0.463 + \frac{0.059 \times 1}{2} = 0.493 \text{ V}$$

3. Right hand electrode: $AgCl(s) + e^- = Ag(s) + Cl^-$
$$V^{\ominus}(AgCl/Ag, Cl^-) = 0.222 \text{ V}$$

Left hand electrode: $Ag^+ + e^- = Ag(s)$
$$V^{\ominus}(Ag^+/Ag) = -0.799 \text{ V}$$

The cell reaction: $AgCl(s) = Ag^+ + Cl^-$
$$E^{\ominus} = -0.577 \text{ V}$$

The equilibrium constant for the cell reaction is simply the solubility product of AgCl.

$$\lg K_s = nFE^{\ominus}/2.303\,RT = -5.77/0.059$$
$$\therefore \lg(1/K_s) = 5.77/0.059 = 9.78$$
$$1/K_s = 6.03 \times 10^9 \text{ and } K_s = 1.66 \times 10^{-10}$$

4. The cell diagram is

$$\overset{-}{Pt} \,\bigg|\, Fe^{2+}, Fe^{3+} \,\bigg|\bigg|\, Mn^{2+}, MnO_4^- \,\bigg|\, \overset{+}{Pt}$$

The cell e.m.f. is given by

$$E^{\ominus} = \frac{2.303\,RT}{5F}\,\lg K = \frac{0.059}{5} \times 63 = 0.743 \text{ V}$$

Also $E^{\ominus} = V_R^{\ominus} - V_L^{\ominus} = V^{\ominus}(MnO_4^-/Mn^{2+}) - V^{\ominus}(Fe^{3+}/Fe^{2+})$
$$\therefore V^{\ominus}(MnO_4^-/Mn^{2+}) = 0.743 + 0.771 = 1.514 \text{ V}$$

5(a) The cell diagram is

$$\overset{-}{Pt} \,\bigg|\, Fe^{2+}, Fe^{3+} \,\bigg|\bigg|\, Ce^{4+}, Ce^{3+} \,\bigg|\, \overset{+}{Pt}$$

$$E^{\ominus} = V^{\ominus}(Ce^{4+}/Ce^{3+}) - V^{\ominus}(Fe^{3+}/Fe^{2+})$$
$$= 1.61 - 0.771 = 0.839 \text{ V}$$
$$\lg K = nFE^{\ominus}/2.303\,RT = 0.839/0.059 = 14.22$$
$$\therefore K = 1.5 \times 10^{14}$$

(b) At the equivalence point, exactly equivalent amounts of Fe^{2+} and Ce^{4+} have reacted

$$\therefore [Fe^{2+}] = [Ce^{4+}] \quad \text{and} \quad [Fe^{3+}] = [Ce^{3+}] \quad . \quad . \quad . \quad (12)$$

From XIV, 27 it is apparent that at the equivalence point the potential of the right hand electrode can be written in two ways.

$$V = V(Fe^{3+}/Fe^{2+}) = V^{\ominus}(Fe^{3+}/Fe^{2+}) + \frac{2 \cdot 303\, RT}{F} \lg \frac{[Fe^{3+}]}{[Fe^{2+}]}$$

$$V = V(Ce^{4+}/Ce^{3+}) = V^{\ominus}(Ce^{4+}/Ce^{3+}) + \frac{2 \cdot 303\, RT}{F} \lg \frac{[Ce^{4+}]}{[Ce^{3+}]}$$

Adding, $2V = V^{\ominus}(Fe^{3+}/Fe^{2+}) + V^{\ominus}(Ce^{4+}/Ce^{3+})$

$$+ \frac{2 \cdot 303\, RT}{F} \lg \frac{[Fe^{3+}][Ce^{4+}]}{[Fe^{2+}][Ce^{3+}]}$$

From (12), $[Fe^{3+}][Ce^{4+}]/[Fe^{2+}][Ce^{3+}] = 1$ at the equivalence point and the above equation becomes

$$V(\text{equivalence point}) = \frac{V^{\ominus}(Fe^{3+}/Fe^{2+}) + V^{\ominus}(Ce^{4+}/Ce^{3+})}{2}$$

$$= \frac{0 \cdot 771 + 1 \cdot 61}{2} = 1 \cdot 19\, V$$

6. The cell reaction is

$$H_2(g) + Hg_2Cl_2(s) = 2Hg(l) + 2Cl^- + 2H^+$$

and the cell potential is given by

$$E = V(\text{calomel}) - V(H_2/H^+)$$

Now $$V(H_2/H^+) = V^{\ominus}(H_2/H^+) + \frac{2 \cdot 303\, RT}{2F} \lg \frac{a_{H_2}}{a^2_{H^+}}$$

Since $$V^{\ominus}(H_2/H^+) = 0,$$

$$V(H_2/H^+) = \frac{2 \cdot 303\, RT}{2F} \lg a_{H_2} - \frac{2 \cdot 303\, RT}{2F} \lg a^2_{H^+}$$

The partial pressure of hydrogen is $101 \cdot 3$ kN m^{-2} and hence $a_{H_2} = 1$

$$\therefore V(H_2/H^+) = \frac{2 \cdot 303\, RT}{F} \text{pH}$$

$$\therefore E = V(\text{calomel}) - 0 \cdot 059\, \text{pH}$$

Rearranging $$\text{pH} = \frac{V(\text{calomel}) - E}{0 \cdot 059} = \frac{0 \cdot 244 + 0 \cdot 073}{0 \cdot 059} = 5 \cdot 37$$

Progress Test 15

1. The following points should be borne in mind when answering this question:

(i) The maximum number of electrons each type of orbital can accommodate (Table 20).

(ii) The order in which the orbitals are filled (XV, 36).

Bearing in mind these two points the electronic structures may be readily determined.

Atomic number	Electronic structure
3	$1s^2$ $2s^1$
15	$1s^2$ $2s^2$ $2p^6$ $3s^2$ $3p^3$
26	$1s^2$ $2s^2$ $2p^6$ $3s^2$ $3p^6$ $3d^6$ $4s^2$
36	$1s^2$ $2s^2$ $2p^6$ $3s^2$ $3p^6$ $3d^{10}$ $4s^2$ $4p$
51	$1s^2$ $2s^2$ $2p^6$ $3s^2$ $3p^6$ $3d^{10}$ $4s^2$ $4p^6$ $4d^{10}$ $5s^2$ $5p^3$
67	$1s^2$ $2s^2$ $2p^6$ $3s^2$ $3p^6$ $3d^{10}$ $4s^2$ $4p^6$ $4d^{10}$ $4f^{11}$ $5s^2$ $5p^6$ $6s^2$
92	$1s^2$ $2s^2$ $2p^6$ $3s^2$ $3p^6$ $3d^{10}$ $4s^2$ $4p^6$ $4d^{10}$ $4f^{14}$ $5s^2$ $5p^6$ $5d^{10}$ $5f^3$ $6s^2$ $6p^6$ $6d^1$ $7s^2$

All the structures are quite straightforward except for the element with an atomic number of 92, where the configuration would have been expected to include . . . $5f^4$ . . . instead of . . . $5f^3$ . . . $6d^1$. However, the $5f$ and $6d$ orbitals are very similar in energy, so the presence of the single $6d$ electron is not altogether unexpected.

2. The ionisation of the ground-state hydrogen atom corresponds to the transition of an electron from $n_1 = 1$ to $n_2 = \infty$.

$$1/\lambda = R_H(1/1^2 - 1/\infty^2) = R_H = 1 \cdot 097\ 373\ 1 \times 10^7\ m^{-1}$$

The energy required for this transition is the ionisation energy and is given by

$$\begin{aligned} \Delta H_I &= h\nu = hc/\lambda \\ &= 6 \cdot 625 \times 10^{-34} \times 2 \cdot 998 \times 10^8 \times 1 \cdot 097 \times 10^7 \\ &= 2 \cdot 178 \times 10^{-18}\ J\ atom^{-1} \end{aligned}$$

The number of atoms in 1 mol is $6 \cdot 023 \times 10^{23}$, hence

$$\Delta H_I = 2 \cdot 178 \times 10^{-18} \times 6 \cdot 023 \times 10^{23} = 1323\ kJ\ mol^{-1}$$

Also $\Delta H_I = (2 \cdot 178 \times 10^{-18})/(1 \cdot 602 \times 10^{-19}) = 13 \cdot 6\ eV\ atom^{-1}$

The ionisation potential is $13 \cdot 6$ V.

Progress Test 16

1. From (XVI, 10)

$$\begin{aligned} \Delta H_f(\text{CaF}) &= 153 \cdot 5 + 590 + 151 \cdot 3/2 - 349 - 803 \\ &= -331 \cdot 9\ kJ\ mol^{-1} \end{aligned}$$

This suggests that CaF(s) would be stable at room temperature. However, this is not so, and the reason is that CaF(s) disproportionates into the stable difluoride and the metal

$$2CaF(s) = 2CaF_2(s) + Ca(s)$$

The energy of disproportionation is -658 kJ mol^{-1}, indicating that CaF(s) is unstable with respect to CaF$_2$(s) and Ca(s). Thus although ΔH_f may be negative, other considerations must be taken into account before it can be concluded with certainty that any given compound is unstable. However, if ΔH_f were positive, then the compound would be unstable with respect to decomposition into its constituent elements.

Since CaF does not exist as an ionic solid, the assignment of a lattice energy requires some explanation. The lattice energy of CaF(s) was taken to be the same of that of the stable compound formed by the adjacent element in group I of the periodic table, i.e. KF.

Progress Test 17

1. Using the bond line notation (see XV, 1)

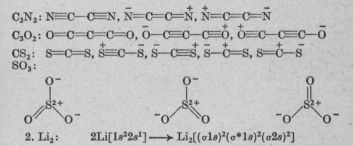

C_2N_2: $N{\equiv}C-C{\equiv}N$, $\overset{-}{N}{=}C{=}C{=}\overset{+}{N}$, $\overset{+}{N}{=}C{=}C{=}\overset{-}{N}$

C_3O_2: $O{=}C{=}C{=}C{=}O$, $\overset{-}{O}-C{\equiv}C-C{\equiv}\overset{+}{O}$, $\overset{+}{O}{\equiv}C-C{\equiv}C-\overset{-}{O}$

CS_2: $S{=}C{=}S$, $\overset{+}{S}{\equiv}C-\overset{-}{S}$, $\overset{-}{S}-C{\equiv}\overset{+}{S}$, $\overset{+}{S}-\overset{-}{C}{=}S$, $S{=}\overset{+}{C}-\overset{-}{S}$

SO_3:

2. Li$_2$: $\qquad$ $2Li[1s^2 2s^1] \longrightarrow Li_2[(\sigma 1s)^2(\sigma^* 1s)^2(\sigma 2s)^2]$

There are two net bonding electrons (four bonding, two antibonding) to form a σ bond. Li$_2$ is stable.

Li$_2^+$ will have the configuration $[(\sigma 1s)^2(\sigma^* 1s)^2(\sigma 2s)^1]$. There is one net bonding electron, giving rise to a very weak bond. Li$_2^+$ has only been observed spectroscopically. Li$_2^{2+}$ will have the configuration $[(\sigma 1s)^2(\sigma^* 1s)^2]$, i.e. the same as He$_2$, and for the same reason will not exist under normal conditions.

Be$_2$: $\qquad$ $2Be[1s^2 2s^2] \longrightarrow Be_2[(\sigma 1s)^2(\sigma^* 1s)^2(\sigma 2s)^2(\sigma^* 2s)^2]$.

There are no net bonding electrons and Be$_2$ will be unstable.

B$_2$: $2B[1s^2 2s^2 2p^1] \longrightarrow B_2[(\sigma 1s)^2(\sigma^* 1s)^2(\sigma 2s)^2(\sigma^* 2s)^2(\sigma 2p_x)^2]$

This configuration would be expected, on the basis of the order of stability of the molecular orbitals given in XVIII, 15. However, (a) the $\sigma 2p_x$ orbital in B_2 is of higher energy than the $\pi 2p_y$ and $\pi 2p_z$ orbitals, and (b) the two electrons which go into the orbitals must have their spins unpaired (Hund's rules). Hence B_2 has the configuration $[(\sigma 1s)^2(\sigma^* 1s)^2(\sigma 2s)^2(\sigma^* 2s)^2(\pi 2p_y)^1(\pi 2p_z)^1]$. There are two net bonding electrons, one in each $\pi 2_p$ orbital, and B_2 is stable.

$$C_2 \quad 2C[1s^2 2s^2 2p^2] \longrightarrow C_2[(\sigma 1s)^2(\sigma^* 1s)^2(\sigma 2s)^2(\sigma^* 2s)^2$$
$$(\sigma 2p_x)^2(\pi 2p_y)^1(\pi 2p_z)^1]$$

Again, the $\sigma 2p_x$ orbital is of higher energy than the $\pi 2p$ orbitals and the actual configuration of C_2 is $[(\sigma 1s)^2(\sigma^* 1s)^2(\sigma 2s)^2(\sigma 2s)^2 (\pi 2p_y)^2(\pi 2p_z)^2]$.

There are four net bonding electrons which constitute two π bonds, and C_2 is stable.

3. The electronic structure of the selenium atom is $[(Ar\ core)\ 3d^{10}\ 4s^2\ 4p_x^2\ 4p_y^1\ 4p_z^1]$. If only the half-filled $4p_y$ and $4p_z$ orbitals are available for σ bond formation with the $1s$ orbitals of the hydrogen atom, the bond angle will be 90°.

The oxygen atom with the electronic structure $[(He\ core)\ 2s^2\ 2p_x^2\ 2p_y^1\ 2p_z^1]$ might also be expected to utilise p orbitals in bonding, resulting in a bond angle of 90° in water. However, this is not the case and sp^3 hybridisation takes place (see XVII, 21).

Progress Test 18

1. The calculations are analogous to that performed for the HCl molecule in XVIII, 4. The answers are FCl, 11·3%; BrCl, 5·5%; ICl, 4·8%.

2. The dipole moment vector diagram for the H_2S molecule is shown in the accompanying diagram.

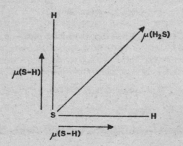

From this diagram it is obvious that the molecular dipole moment $\mu(H_2S)$ is the hypotenuse of a right angled triangle, the other sides of which represent the bond dipole moments $\mu(S–H)$. Thus

$$[\mu(S–H)]^2 + [\mu(S–H)]^2 = [\mu(H_2S)]^2 = (3\cdot10 \times 10^{-30})^2$$
$$= 9\cdot61 \times 10^{-60}$$

Hence
$$\mu(S–H) = 2\cdot19 \times 10^{-30}\,C\,m$$

Progress Test 19

1. Since the intercepts are at 3/4, 3/2, 1 multiples of the unit distances

$$a/m:b/n:c/p = h:k:l = 4/3:2/3:1 = 4:2:3$$

and the Miller index of the plane is (4, 2, 3).

2(a) In the ccp structure the sphere at the centre of a face diagonal just touches the two spheres at the corners of the face (Fig. 130(a)). Thus if the radius of the spheres is R, the length of the face diagonal must be $4R$. But the length of a face diagonal of a cube of side a is $(2)^{\frac{1}{2}}a$

$$\therefore (2)^{\frac{1}{2}}a = 4R$$
$$R = (2)^{\frac{1}{2}}a/4 = 0\cdot354a$$

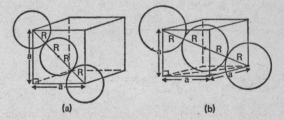

(a)   (b)

FIG. 130.—*Relationships between positions of spheres and the diagonals of a cube in (a) cubic close-packed structure, (b) body-centred structure*

The unit cell contains eight spheres at the corners (each of which is shared by eight unit cubes) and six spheres at the centres of faces (each of which is shared by two unit cubes). There are therefore $8(\frac{1}{8}) + 6(\frac{1}{2}) = 4$ spheres per unit cell in the lattice.

$$\therefore \text{Volume of spheres} = 4[4\pi(0\cdot354a)^3/3] = 0\cdot7406a^3$$

But volume of unit cube, side $a = a^3$

$$\therefore \text{There is } 25\cdot94\% \text{ unoccupied volume.}$$

(b) In the body-centred arrangement, spheres are in contact along the body diagonals (Fig. 130 (b))

$$\therefore (3)^{\frac{1}{2}}a = 4R$$
$$R = (3)^{\frac{1}{2}}a/4 = 0.0596a$$

Number of spheres $= 8(\frac{1}{8}) + 1 = 2$

$$\therefore \text{Volume of spheres} = 2[4\pi(0.0596a)^3/3] = 0.6803a^3$$

There is 31.97% unoccupied volume.

3. The spheres are in contact along the face diagonals. Therefore, if the radius of the spheres is R, the length of the face diagonal is $2R$.

$$\therefore \text{Side of cube} = (2)^{\frac{1}{2}}R$$
$$\therefore \text{Length of body diagonal} = (6)^{\frac{1}{2}}R$$

The radius of the tetrahedral hole is obtained by assuming a sphere of radius r to be placed at the centre of the body diagonal. Since this sphere touches the large spheres at both ends of the body diagonal:

$$2r + 2R = \text{length body diagonal}$$
$$\therefore r = (6)^{\frac{1}{2}}R/2 - R$$
$$= 0.2247\,R$$

4. Using (XIX, 2) and comparing ratios of $\sin\theta$ values for the $(1, 0, 0)$, $(1, 1, 0)$ and $(1, 1, 1)$ planes:

	$Sin\ \theta$	$1/sin\ \theta$	$Ratio$
$7°10'$	0.1248	8.014	1
$5°4'$	0.0884	11.31	1.414
$10°12'$	0.2164	4.621	0.5766

i.e. $d_{100}:d_{110}:d_{111} = 1:1.414:0.577$

The crystal has a body-centred cubic lattice.

5. This problem is solved by calculating the volume of the unit cell from the density and from the X-ray data and equating the two answers. Thus

density $=$ molar mass/molar volume

i.e.
$$\rho = M/V_m$$
$$\therefore V_m = M/\rho = 0.05844/2.17 \times 10^3 = 2.70 \times 10^{-5}\,\text{m}^3\,\text{mol}^{-1}$$

The number of molecules which occupy this volume is given by Avogadro's constant, L.

Hence the volume per molecule $= 2.70 \times 10^{-5}/L\ \text{m}^3$

But the unit cell of sodium chloride contains four Na^+ and $4Cl^-$ ions, i.e. effectively four "molecules" of sodium chloride per unit cell.

∴ Volume of unit cell $= 4 \times 2 \cdot 70 \times 10^{-5}/L = 1 \cdot 08 \times 10^{-4}/L$ m³
From the X-ray data using (XIX, 1)

$$d_{100} = \lambda/2 \sin \theta = (0 \cdot 1537 \times 10^{-9})/(2 \times 0 \cdot 2723)$$
$$= 2 \cdot 82 \times 10^{-10} \text{ m}$$

For a face-centred cube of side a, $d_{100} = a/2$

$$\therefore a = 5 \cdot 64 \times 10^{-10} \text{ m}$$
∴ Volume of unit cube $= (5 \cdot 64 \times 10^{-10})^3$ m³
$$= 1 \cdot 795 \times 10^{-28} \text{ m}^3$$

Equating the two values:

$$1 \cdot 08 \times 10^{-4}/L = 1 \cdot 795 \times 10^{-28}$$
$$\therefore L = 6 \cdot 02 \times 10^{23} \text{ mol}^{-1}$$

Progress Test 21

1. From XXI, 13, a plot of $\lg\{(R_0 - R_\infty)/(R_t - R_\infty)\}$ against t should give a straight line of slope $k/2 \cdot 303$.

$$R_0 - R_\infty = 99 \cdot 3°$$

Time/hr	0	4	12	24	48
$(R_t - R_\infty)$/deg	99·3	86·25	66·75	43·6	19·190
$\lg\{(R_0 - R_\infty)/(R_t - R_\infty)\}$	0	0·061	0·179	0·360	0·714

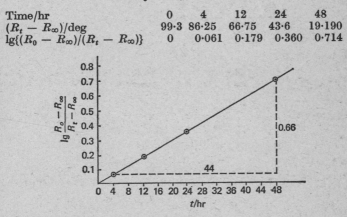

FIG. 131.—*Inversion of sucrose*

The graph is shown in Fig. 131 and has a slope given by

$$k/2 \cdot 303 = 0 \cdot 66/44$$

Hence $k = 0 \cdot 66 \times 2 \cdot 303/44 = 3 \cdot 4 \times 10^{-2}$ hr$^{-1} = 9 \cdot 5 \times 10^{-6}$ s^{-1}

2(a) Let p_0 be the initial pressure of A and y the decrease in pressure after time t. The pressure of A after time $t = p_A = p_0 - y$

Since two moles of A are consumed per mole of A_2 formed, the total pressure at any instant is given by

$$p = p_A + p_{A_2} = (p_0 - y) + y/2 = p_0 - y/2$$

Hence
$$y = 2(p_0 - p) \quad \text{and} \quad p_A = 2p - p_0$$

Since the reaction involves only one reacting species and is second-order, (XXI, 15) is applicable. Now a_0 is proportional to p_0, x to y, *i.e.* to $2(p_0 - p)$, and $a_0 - x$ to p_A, *i.e.* to $2p - p_0$. Hence (XXI, 15) becomes

$$2(p_0 - p)/(2p - p_0) = p_0 k_2 t$$

k_2 may be found from a plot of $2(p_0 - p)/(2p - p_0)$ against t,

$$p_0 = 600 \text{ mmHg}$$

Time/s	0	100	200	300	400	500
$2(p_0 - p)/(2p - p_0)$	0	0·5	1·0	1·5	2·0	2·5

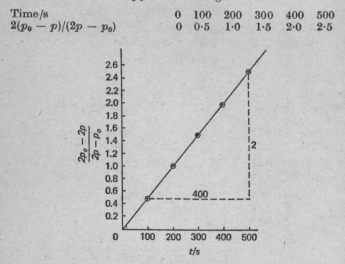

FIG. 132.—*Dimerisation* $2A \longrightarrow A_2$

The plot is shown in Fig. 132 and has a slope given by

$p_0 k_2 = 2/400 = 5 \times 10^{-3} \text{ s}^{-1}$
$\therefore k_2 = 5 \times 10^{-3}/600 = 8\cdot3 \times 10^{-6} \text{ mmHg}^{-1}\text{ s}^{-1}$ . . . . (13)

(*b*) The rate constant may be expressed in $\text{dm}^3 \text{ mol}^{-1} \text{ s}^{-1}$ if it is assumed that A behaves as an ideal gas. Since

$$pV = nRT, \quad n/V = c = p/RT$$

Now $p_0 = 600$ mmHg $= 101\cdot325 \times 10^3 \times 600/760$ N m^{-2}
$n/V = (101\cdot325 \times 10^3 \times 600)/(760 \times 8\cdot314 \times 673)$mol m^{-3}
$\qquad = 1\cdot43$ mol m$^{-3} = 1\cdot43 \times 10^{-3}$ mol dm^{-3}

Since this is the initial concentration of A, (13) becomes

$$k_2 = 5 \times 10^{-3}/1\cdot43 \times 10^{-3} = 3\cdot50 \text{ dm}^3 \text{ mol}^{-1} \text{s}^{-1}$$

3. The time taken is given by (XXI, 18)

$$t = \frac{2\cdot303}{k_2(a_0 - b_0)} \lg \frac{b_0(a_0 - x)}{a_0(b_0 - x)}$$

where
$\qquad a_0 = [\text{CH}_3\text{CH}_2\text{NO}_2] = 4 \times 10^{-3}$ mol dm^{-3}
$\qquad b_0 = [\text{NaOH}] = 5 \times 10^{-3}$ mol dm^{-3}
$\qquad x = 4 \times 10^{-3} \times 80/100 = 3\cdot2 \times 10^{-3}$ mol dm^{-3}

$$\therefore t = \frac{2\cdot303}{6\cdot5(4 - 5) \times 10^{-3}} \lg \frac{5(4 - 3\cdot2)}{4(5 - 3\cdot2)}$$

$$= \frac{-2\cdot303}{6\cdot5 \times 10^{-3}} \lg 0\cdot55$$

$$= \frac{-2\cdot303}{6\cdot5 \times 10^{-3}} \times (-0\cdot26) = 92\cdot1 \text{ s}$$

4. The rates of the two reactions are

$$R_1 = k_1 a \quad \text{and} \quad R_2 = k_2 a^2$$

(a) If a_0 is the initial concentration, then the initial rates are

$$R_1 = k_1 a_0 \quad \text{and} \quad R_2 = k_2 a_0^2$$
$$\therefore R_1/R_2 = k_1/k_2 a_0$$

Since the half-lives are the same

$$t_{\frac{1}{2}} = 0\cdot693/k_1 = 1/k_2 a_0$$
$$\therefore k_1/k_2 = 0\cdot693 \, a_0$$
Hence
$$R_1/R_2 = 0\cdot693 \, a_0/a_0 = 0\cdot693$$

(b) At the time of one half-life

$$R_1 = k_1 a_0/2 \quad \text{and} \quad R_2 = k_2 a_0^2/4$$
$$\therefore R_1/R_2 = 2k_1/k_2 a_0 = 2 \times 0\cdot693 \, a_0/a_0 = 1\cdot386$$

For the first-order reaction the concentration after one half-life will be $a_0/2$, after two half-lives $a_0/4$, etc. For the second-order reaction the concentration after one half-life will again be $a_0/2$. However, the half-life for a second-order reaction is inversely proportional to the initial concentration. Thus if the initial concentration is taken as $a_0/2$, then the half-life for this concentration will be twice as great as the original half-life, *i.e.* $2t_{\frac{1}{2}}$. The two cases are illustrated in Fig. 133.

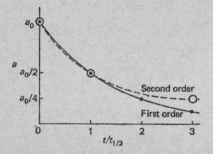

FIG. 133.—*Comparison of first- and second-order reactions*

5. Integrating the rate expression between the limits $x = 0$ at $t = 0$ and $x = x$ at $t = t$

$$\int_0^x \frac{\mathrm{d}x}{(a_0 - x)^n} = k_n \int_0^t \mathrm{d}t$$

$$= \left[-\frac{1}{(n - 1)(a_0 - x)^{n-1}} \right]_0^x = k_n[t]_0^t$$

$$= \frac{1}{n - 1} \left\{ \frac{1}{(a_0 - x)^{n-1}} - \frac{1}{a_0^{n-1}} \right\} = k_n t$$

Substituting the condition that $x = a_0/2$ when $t = t_{\frac{1}{2}}$ into the above expression gives

$$\frac{1}{n - 1} \left\{ \frac{1}{(a_0/2)^{n-1}} - \frac{1}{a_0^{n-1}} \right\} = k_n t_{\frac{1}{2}}$$

$$\frac{1}{n - 1} \left\{ \frac{2^{n-1}}{a_0^{n-1}} - \frac{1}{a_0^{n-1}} \right\} = k_n t_{\frac{1}{2}}$$

$$\therefore t_{\frac{1}{2}} = \frac{2^{n-1} - 1}{k_n a_0^{n-1}(n - 1)}$$

The above equation may be written

$$t_{\frac{1}{2}} = \frac{2^{n-1} - 1}{k_n(n - 1)} \times \frac{1}{a_0^{n-1}}$$

$$\therefore \lg t_{\frac{1}{2}} = \lg \left\{ \frac{2^{n-1} - 1}{k_n(n - 1)} \right\} - (n - 1) \lg a_0$$

Hence a plot of $\lg t_{\frac{1}{2}}$ against $\lg a_0$ will give a straight line of slope $-(n-1)$, from which n may be determined.

6. Using the expression for the half-life derived in question 5 above

(a) when $a_0 = 1 \text{ mol dm}^{-3}$

$$t_{\frac{1}{2}} = \frac{2^2 - 1}{1 \cdot 2 \times 10^{-3}(2)(1)^2} = 1 \cdot 25 \times 10^3 \text{ s}$$

(b) when $a_0 = 10^{-1} \text{ mol dm}^{-3}$

$$t_{\frac{1}{2}} = \frac{2^2 - 1}{1 \cdot 2 \times 10^{-3}(2)(10^{-1})^2} = 1 \cdot 25 \times 10^5 \text{ s}$$

7. From (XXI, 33)

$$K = k_1/k_{-1} = x_e/(a_0 - x_{-e}) = 60/(100 - 60) = 1 \cdot 5$$

From (XXI, 25) $\quad k_1 = \dfrac{2 \cdot 303 \, x_e}{a_0 t} \lg \dfrac{x_e}{x_e - x}$

After 100 s, $\quad x = 100 - 75 \cdot 6 = 24 \cdot 4\%$.

$$\therefore k_1 = \frac{2 \cdot 303 \times 60}{100 \times 100} \lg \left(\frac{60}{60 - 24 \cdot 4} \right)$$
$$= 1 \cdot 382 \times 10^{-2} \lg 1 \cdot 685 = 1 \cdot 382 \times 10^{-2} \times 0 \cdot 2268$$
$$= 3 \cdot 13 \times 10^{-3} \text{ s}^{-1}$$

$$k_{-1} = k_1/K = 3 \cdot 13 \times 10^{-3}/1 \cdot 5 = 2 \cdot 18 \times 10^{-3} \text{ s}^{-1}$$

8. Doubling the initial pressure, which is equivalent to doubling the concentration, decreases the half-life by a factor of two. The reaction must therefore be second-order.

The second-order rate constant is given by

$$k_2 = 1/t_{\frac{1}{2}} a_0$$

a_0, the initial concentration, is proportional to the initial pressure, hence

$$k_2 = 1/800 \times 200 = 6 \cdot 25 \times 10^{-6} \text{ mmHg s}^{-1}$$

The units may be converted to the more conventional $\text{dm}^3 \text{ mol}^{-1} \text{ s}^{-1}$ if it is assumed that acetaldehyde behaves as an ideal gas. The procedure for doing this is given in question 2 above. Thus

$$a_0 = n/V = p_0/RT = (101 \cdot 325 \times 10^3 \times 200)/(760 \times 8 \cdot 314 \times 773)$$
$$= 0 \cdot 415 \text{ mol m}^{-3} = 0 \cdot 415 \times 10^{-3} \text{ mol dm}^{-3}$$
$$\therefore k_2 = 1/(800 \times 0 \cdot 415 \times 10^{-3})$$
$$= 0 \cdot 276 \times 10^3 \text{ dm}^3 \text{ mol}^{-1} \text{ s}^{-1}.$$

9. Comparing experiments Nos. 1 and 2, we see that doubling [A] at constant [B] doubles the rate. The reaction is therefore first-order in A. From a similar comparison of experiments Nos. 1 and 3 we see that reducing [B] by a factor of 4 at constant [A] decreases the rate by an identical factor, *i.e.* the order with respect to B is also 1. The rate expression is

$$\text{Rate} = k[\text{A}][\text{B}]$$

Substituting the values from any of the experiments into the rate expression gives $k = 2 \cdot 3 \times 10^3 \, \text{dm}^3 \, \text{mol}^{-1} \, \text{s}^{-1}$

Progress Test 22

1(a) The logarithmic form of the Arrhenius equation is

$$\lg k = \lg A - E^{\ddagger}/2 \cdot 303 \, RT \qquad . \quad . \quad . \quad (14)$$

A plot of $\lg k$ against $1/T$ will have a slope of $-E^{\ddagger}/2 \cdot 303 \, R$ and an intercept of $\lg A$. The values of $\lg k$ and $1/T$ are shown below and plotted in Fig. 134 (a).

$\lg(k/\text{s}^{-1})$	$-4 \cdot 461$	$-3 \cdot 303$	$-2 \cdot 313$
T/K	298	318	338
$10^3 \text{K}/T$	3·36	3·14	2·96

Slope from Fig. 134 $= -2 \cdot 15/0 \cdot 4 \times 10^{-3} = E^{\ddagger}/2 \cdot 303 \times 8 \cdot 314$
Hence $\qquad\qquad E^{\ddagger} = 103 \, \text{kJ mol}^{-1}$

(b) Because of the long extrapolation involved, the position (M in Fig. 134 (b)) at which the line ZX cuts the $\lg k$ axis cannot

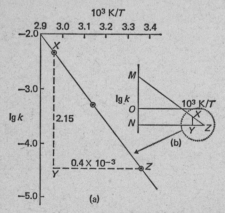

FIG. 134.—(a) *Plot of lg k against 1/T.* (b) *Expanded plot of lg k against 1/T*

be accurately determined. Note that OM must be a positive value, since A is a positive number.

From Fig. 134 (b), by the theory of similar triangles

$$NM/XY = NZ/XZ$$

i.e.

$$\frac{NM}{(4 \cdot 461 - 2 \cdot 312)} = \frac{3 \cdot 36 \times 10^{-3}}{(3 \cdot 36 - 2 \cdot 96) \times 10^{-3}}$$

Hence
$$NM = 18 \cdot 04$$
$$\therefore OM = \lg A = NM - NO = 18 \cdot 04 - 4 \cdot 46$$
$$= 13 \cdot 59$$
$$\therefore A = 3 \cdot 89 \times 10^{13} \text{ s}^{-1}$$

Alternatively, A may be found by substituting the data at one particular temperature into (14).

2(a) From the collision theory

$$\text{Rate} = Z \exp(-E^{\ddagger}/RT)$$
$$= 6 \times 10^{37} \exp(-184 \times 10^3/8 \cdot 314 \times 556)$$

Taking logarithms:

$$\lg \text{Rate} = \lg(6 \times 10^{37}) - (184 \times 10^3/8 \cdot 314 \times 556 \times 2 \cdot 303)$$
$$= 37 \cdot 778 - 17 \cdot 282 = 20 \cdot 496$$
$$\therefore \text{Rate} = 3 \cdot 13 \times 10^{20} \text{ molecules m}^{-3} \text{ s}^{-1}$$

(b) To convert to mol dm^{-3} s^{-1} it is necessary to divide by $6 \cdot 023 \times 10^{23} \times 10^3$

$$\text{Rate} = 3 \cdot 13 \times 10^{20}/6 \cdot 023 \times 10^{23} \times 10$$
$$= 5 \cdot 7 \times 10^{-7} \text{ mol dm}^{-3} \text{ s}^{-1}$$

3(a) From (XXII, 4)

$$\lg k_2/k_1 = \frac{E^{\ddagger}}{2 \cdot 303 \, R} \left(\frac{1}{T_1} - \frac{1}{T_2} \right) = \frac{E^{\ddagger}}{2 \cdot 303 \, R} \left(\frac{T_2 - T_1}{T_1 T_2} \right)$$

$$\lg \frac{2 \cdot 28 \times 10^{-3}}{1 \cdot 70 \times 10^{-5}} = \frac{E^{\ddagger}}{2 \cdot 303 \times 8 \cdot 314} \left(\frac{456 - 404}{404 \times 456} \right)$$

$$\therefore \lg 1 \cdot 34 \times 10^2 = \frac{E^{\ddagger}}{19 \cdot 15} \times \frac{52}{18 \cdot 42 \times 10^4}$$

From which $E^{\ddagger} = 145 \text{ kJ mol}^{-1}$

(b) $$k = A \exp(-E^{\ddagger}/RT)$$

Taking logarithms and rearranging

$$\lg A = \lg k + E^{\ddagger}/2 \cdot 303 \, RT$$

Substituting the values $k = 1 \cdot 70 \times 10^{-5}\,\text{s}^{-1}$ and $T = 404\,\text{K}$ into this equation gives

$$\lg A = \lg(1 \cdot 70 \times 10^{-5}) + (145 \times 10^3/8 \cdot 314 \times 404 \times 2 \cdot 303)$$
$$= -4 \cdot 77 + 18 \cdot 66 = 13 \cdot 89$$

Hence $A = 7 \cdot 76 \times 10^{13}\,\text{s}^{-1}$

(c) At 423 K, $k_{423} = A \exp(-E/R \times 423)$
$$\lg k_{423} = \lg A - (145 \times 10^3/8 \cdot 314 \times 423 \times 2 \cdot 303)$$
$$= 13 \cdot 89 - 17 \cdot 82 = -3 \cdot 93 = \bar{4} \cdot 07$$
$$\therefore k_{423} = 1 \cdot 18 \times 10^{-4}\,\text{s}^{-1}$$

For a first-order reaction,

$$t_{\frac{1}{2}} = 0 \cdot 693/k$$
$$= 0 \cdot 693/1 \cdot 18 \times 10^{-4} = 5 \cdot 5 \times 10^3\,\text{s}$$

(d) From the transition state theory,

$$A = \frac{kT}{h} \exp(\Delta S^{\ddagger}/R)$$

$$\therefore \lg A = \lg(kT/h) + \Delta S^{\ddagger}/2 \cdot 303\,R$$

At 423 K,

$$13 \cdot 89 = \lg(1 \cdot 38 \times 10^{-23} \times 423/6 \cdot 624 \times 10^{-34}) + \Delta S^{\ddagger}/2 \cdot 303$$
$$\times 8 \cdot 314$$

$$13 \cdot 89 = 12 \cdot 95 + \Delta S^{\ddagger}/19 \cdot 15$$
$$\therefore S^{\ddagger} = 19 \cdot 15(13 \cdot 89 - 12 \cdot 95)$$
$$= 19 \cdot 15(0 \cdot 94) = 17 \cdot 8\,\text{J K}^{-1}\,\text{mol}^{-1}$$

INDEX

absolute zero of temperature, 52–3
acid
  catalysis by, 397–8
  conjugate, 220, 223
  see also acid–base
acid–base
  Arrhenius classification, 219
  Brønstead–Lowry classification, 220
  equilibria, 221–33
  heat of neutralisation, 106–7
  indicators, 232–3
  Lewis classification, 221
  strength of, 222
  titrations, 230–2, 260
activated complex, 387–8, 392–4
activation energy, 386–7
active centres, 413
activity, 128–9, 221, 235, 245
  coefficient, 128–9, 221
adiabatic process, 88, 95
adsorption
  chemical (chemisorption), 407–8
  from solution, 410
  isotherm, 408–9
  of gases by solids, 407–14
  physical, 407–8
allotropy, 147
ampere, definition of, 4
angular
  momentum quantum number, 277
  probability distribution, 280–2
anion, 196
anode, 196
area, determination by integration, 25–6
Arrhenius
  equation, 386–7
  factor, 387
  theory of ionisation, 203–5
associated colloids, 422
atom, models of, 265
atomic
  energy states, 272
  mass, 266
  number, 266
  orbital, 276, 308
    d-, 277, 280–1, 286
    f-, 277, 286
    hybrid, 313–17
    linear combination of, 309

    overlap, 312–13
    p-, 277, 280–1, 282, 284–6
    s-, 277, 280, 284–6
  spectra, 270–1
  structure, 265, 282–4
  weight, *see* relative atomic mass
aufbau principles, 284
autoprotolysis, 223
Avogadro constant, 5
Avogadro's hypothesis (law), 35, 53
azeotrope, 170, 172

Balmer series, 270, 271, 273
band model of solids, 324–7
base
  catalysis by, 397–8
  conjugate, 220, 223
  see also acid–base
Beattie–Bridgman equation of state, 83
benzene, structure of, 308, 317
Berthelot equation of state, 83
black body radiation, 268–9
body-centred cubic structure, 338
Bohr theory of hydrogen atom, 271–4
boiling point, 142–3
  against composition diagram, 166–7, 169–70, 172
  elevation of, 157, 158–60
Boltzmann
  constant, 54
  factor, 60
bond
  covalent, 294, 302–18, 320
  dative (co-ordinate), 295
  energy, 109–11, 303
  hydrogen, 321, 322
  ionic, 294, 295–301
  metallic, 323–7
Born–Haber cycle, 296–7
Boyle
  law, 51–2
  temperature, 74
Bragg
  equation, 345–7
  X-ray spectrometer, 347
Brownian motion, 62
bubble cap column, 167
buffer solution, 228–9

caesium chloride (CsCl) structure, 342

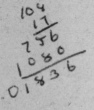